程序员硬核技术丛书

剑指大数据
Hadoop学习精要

尚硅谷教育 ◎ 编著

电子工业出版社

Publishing House of Electronics Industry

北京·BEIJING

内 容 简 介

Hadoop 是使用最广泛的大数据处理框架之一，在大数据领域有着极其重要的地位，掌握 Hadoop 可以让学习者对大数据的理解更进一步。本书是基于 Hadoop 3.1.3 编写的，从大数据的特点和处理难点入手，逐步讲解 Hadoop 的起源和发展。从搭建 Hadoop 的学习环境开始，依次对 Hadoop 的三大功能模块进行重点讲解，并且结合大量案例，细致地讲解 HDFS、MapReduce、YARN 的内核原理和调优方法，还会扩展讲解 Hadoop 的高可用实现、在生产环境中的调优方法及源码解读。

本书广泛适用于大数据的学习者与从业人员，是大数据学习的必备书籍。

未经许可，不得以任何方式复制或抄袭本书之部分或全部内容。
版权所有，侵权必究。

图书在版编目（CIP）数据

剑指大数据：Hadoop 学习精要 / 尚硅谷教育编著. —北京：电子工业出版社，2022.11

（程序员硬核技术丛书）

ISBN 978-7-121-44392-3

Ⅰ. ①剑… Ⅱ. ①尚… Ⅲ. ①数据处理软件 Ⅳ. ①TP274

中国版本图书馆 CIP 数据核字（2022）第 185876 号

责任编辑：李　冰　　　　　特约编辑：田学清
印　　刷：三河市鑫金马印装有限公司
装　　订：三河市鑫金马印装有限公司
出版发行：电子工业出版社
　　　　　北京市海淀区万寿路 173 信箱　　　邮编：100036
开　　本：850×1168　　1/16　　印张：19.25　　字数：610 千字
版　　次：2022 年 11 月第 1 版
印　　次：2022 年 11 月第 1 次印刷
定　　价：105.00 元

凡所购买电子工业出版社图书有缺损问题，请向购买书店调换。若书店售缺，请与本社发行部联系，联系及邮购电话：（010）88254888，88258888。

质量投诉请发邮件至 zlts@phei.com.cn，盗版侵权举报请发邮件至 dbqq@phei.com.cn。

本书咨询联系方式：libing@phei.com.cn。

前 言

Hadoop 的诞生，为后来的大数据处理框架提供了一个思考角度、一个切入点和一个垫脚石。有人可能会说，Hadoop 已经不再流行，我们要学习更新、更快的技术，但最终通常都会重新回到 Hadoop 这个起点。Hadoop 中的很多数据处理理念，如分而化之、shuffle、分布式存储等，在大数据处理领域都非常经典，值得反复琢磨。

很多读者在初学 Hadoop 时，面对众多的 Hadoop 图书都会有点困惑，到底哪一本才是最全面、最适合自己的？在编写本书时，编者站在一个大数据初学者的角度去思考：到底什么样的书，可以让初学者轻松入门，理解 Hadoop 的精妙之处，不被晦涩的源码和专业术语劝退，并且可以真正掌握这门技术，将其应用到实践中？为了解决这个问题，编者编写本书的最高准则就是深入而不晦涩，简洁而无疏漏，实践与理论并重。

本书首先会为读者演示如何搭建一个简单的 Hadoop 学习环境，以此为基础展开 Hadoop 的学习过程。在讲解具体内容时，力求针对每个知识点设计一个案例，通过讲解案例，让读者更容易掌握相关的理论知识：大数据精妙的设计理念是如何应用到实际的数据处理领域的，在实际工作中如何编写代码才可以获得更高的工作效率。Hadoop 自诞生以来，已经迭代更新了很多版本，本书选用编写期间最新的稳定版本，希望能给读者提供更多指导。

本书共分 9 章，具体内容如下。
- 第 1 章：大数据的特征、发展前景和生态体系，以及 Hadoop 的发展史和架构。
- 第 2 章：搭建 Hadoop 的 Linux 运行环境。
- 第 3 章：安装配置 Hadoop，并且运行基础的 Hadoop 案例程序。
- 第 4~6 章：详解 Hadoop 的三大模块——HDFS、MapReduce、YARN，这是 Hadoop 学习中的重要部分，穿插了大量的案例和命令行操作。
- 第 7 章：Hadoop 的高可用实现。
- 第 8 章：Hadoop 的生产环境调优。
- 第 9 章：进阶部分，学习如何逐步追踪源码，从源码中理解 Hadoop 的核心理念。

大数据和 Hadoop 的初学者可以按照章节顺序进行阅读，建议跟随书中案例实际操作，编写代码。对 Hadoop 有一定了解的读者可以选择重点章节进行阅读。

本书是尚硅谷集合多年的 Hadoop 教学经验，充分考虑初学者的认知规律，综合多年教案编写而成的。多年的教学经验，使编者更懂得 Hadoop 学习的重点和难点，希望本书能为读者带来良好的学习体验，让更多的人知道，Hadoop 其实并不难学。

读者可以关注尚硅谷教育微信公众号：atguigu，在聊天窗口发送关键字"hadoopbook"，免费获取本书配套的视频教程及资料包。如果不具备相关基础知识，则可以在尚硅谷教育公众号的聊天窗口发送关键字

"大数据",免费获取尚硅谷关于大数据的全套视频教程及学习路线图。

感谢电子工业出版社的李冰编辑,是您的精心指导让本书得以付梓面世。也感谢所有为本书内容编写提供技术支持的老师们所付出的努力。

关于我们

尚硅谷是一家专业的 IT 教育培训机构,现拥有位于北京、深圳、上海、武汉、西安的五处分校,开设有 Java EE、大数据、HTML5 前端开发、UI/UE 设计等多门课程,累计发布视频教程三千多小时,广受赞誉。通过面授课程、视频分享、在线学习、直播课堂、图书出版等多种方式,满足了全国编程爱好者对多样化学习场景的需求。

尚硅谷一直坚持"技术为王,课比天大"的发展理念,设有独立的研究院,与多家互联网大厂的研发团队保持技术交流,保障教学内容始终基于研发一线,坚持聘用名校、名企的技术专家,进行源码级技术讲解。

希望通过我们的努力,让天下没有难学的技术,为中国的 IT 人才培养尽一点绵薄之力。

尚硅谷教育

目 录

第1章 大数据概论 1
1.1 大数据的特征 1
1.2 大数据的发展前景 2
1.2.1 大数据的应用场景 2
1.2.2 大数据的未来发展 3
1.3 大数据生态体系与Hadoop 4
1.3.1 Hadoop的发展史 4
1.3.2 大数据生态体系 5
1.3.3 Hadoop架构 7
1.4 本章总结 10

第2章 环境准备 11
2.1 安装VMware 11
2.2 安装CentOS 11
2.3 安装远程终端 19
2.3.1 安装Xshell 19
2.3.2 安装SecureCRT 21
2.4 虚拟机配置 23
2.4.1 网络配置 23
2.4.2 网络IP地址配置 24
2.4.3 主机名配置 25
2.4.4 防火墙配置 25
2.4.5 一般用户配置 26
2.4.6 克隆虚拟机 26
2.5 本章总结 27

第3章 Hadoop快速上手 28
3.1 集群角色 28
3.1.1 Hadoop集群的主要角色 28
3.1.2 YARN的主要组成部分 28
3.2 本地模式 30
3.2.1 安装 30
3.2.2 运行官方示例程序 32
3.3 完全分布式模式 33

	3.3.1	SSH 免密登录	33
	3.3.2	shell 脚本准备	34
	3.3.3	集群配置	36
	3.3.4	NameNode 格式化问题	42
	3.3.5	配置历史服务器与日志聚集功能	44
	3.3.6	Hadoop 集群启停脚本	47
	3.3.7	集群时间同步	48
3.4	本章总结		50

第 4 章 分布式文件系统 HDFS ... 51

- 4.1 HDFS 概述 ... 51
  - 4.1.1 HDFS 背景及定义 ... 51
  - 4.1.2 HDFS 的基本架构 ... 52
- 4.2 HDFS 的 shell 操作 ... 53
  - 4.2.1 命令大全 ... 53
  - 4.2.2 命令行命令实操 ... 54
- 4.3 HDFS 的 API 操作 ... 58
  - 4.3.1 客户端环境准备 ... 58
  - 4.3.2 HDFS 文件上传案例 ... 61
  - 4.3.3 HDFS 文件下载案例 ... 62
  - 4.3.4 HDFS 文件重命名案例 ... 63
  - 4.3.5 HDFS 文件删除案例 ... 63
  - 4.3.6 HDFS 文件详情查看案例 ... 63
  - 4.3.7 HDFS 文件和文件夹判断案例 ... 64
- 4.4 HDFS 的读/写流程 ... 65
  - 4.4.1 HDFS 中的数据块大小 ... 65
  - 4.4.2 写数据流程 ... 65
  - 4.4.3 读数据流程 ... 68
- 4.5 HDFS 的工作机制 ... 69
  - 4.5.1 NameNode 和 SecondaryNameNode 的工作机制 ... 69
  - 4.5.2 EditLog 和 FsImage 文件解析 ... 70
  - 4.5.3 检查点时间设置 ... 76
  - 4.5.4 DataNode 的工作机制 ... 76
  - 4.5.5 数据完整性 ... 77
- 4.6 本章总结 ... 78

第 5 章 分布式计算 MapReduce ... 79

- 5.1 MapReduce 概述 ... 79
  - 5.1.1 MapReduce 定义 ... 79
  - 5.1.2 MapReduce 核心思想 ... 80
- 5.2 MapReduce 编程入门 ... 81
  - 5.2.1 官方示例程序 WordCount 源码 ... 81
  - 5.2.2 编程规范 ... 82
  - 5.2.3 WordCount 案例实操 ... 82

- 5.3 Hadoop 的序列化 89
  - 5.3.1 序列化概述 89
  - 5.3.2 Writable 接口 89
  - 5.3.3 序列化案例实操 90
- 5.4 MapReduce 框架原理之 InputFormat 数据输入 96
  - 5.4.1 切片与 MapTask 并行度决定机制 96
  - 5.4.2 Job 提交流程源码和 FileInputFormat 切片源码详解 98
  - 5.4.3 FileInputFormat 切片机制总结 101
  - 5.4.4 TextInputFormat 101
  - 5.4.5 CombineTextInputFormat 切片机制 102
  - 5.4.6 CombineTextInputFormat 案例实操 103
- 5.5 MapReduce 框架原理之 shuffle 机制 104
  - 5.5.1 shuffle 机制 104
  - 5.5.2 分区 105
  - 5.5.3 分区案例实操 106
  - 5.5.4 WritableComparable 排序 110
  - 5.5.5 WritableComparable 排序案例实操（全排序）...111
  - 5.5.6 WritableComparable 排序案例实操（区内排序）...116
  - 5.5.7 Combiner 合并 119
  - 5.5.8 Combiner 合并案例实操 120
- 5.6 MapReduce 框架原理之 OutputFormat 数据输出 121
  - 5.6.1 OutputFormat 接口的实现类 121
  - 5.6.2 自定义 OutputFormat 类的案例实操 122
- 5.7 MapReduce 工作流程 126
- 5.8 Join 127
  - 5.8.1 Reduce Join 127
  - 5.8.2 Reduce Join 案例实操 127
  - 5.8.3 Map Join 133
  - 5.8.4 Map Join 案例实操 134
- 5.9 数据清洗 137
- 5.10 Hadoop 中的数据压缩 139
  - 5.10.1 数据压缩概述 139
  - 5.10.2 压缩参数配置 140
  - 5.10.3 压缩案例实操 141
- 5.11 本章总结 145

第 6 章 资源调度器 YARN 146
- 6.1 YARN 概述 146
  - 6.1.1 基本架构 147
  - 6.1.2 工作机制 148
- 6.2 YARN 的资源调度器和调度算法 150
  - 6.2.1 FIFO 调度器 150
  - 6.2.2 容量调度器 150

VII

6.2.3 公平调度器 152
　6.3 YARN 实操 156
　　　6.3.1 常用的命令行命令 156
　　　6.3.2 核心参数 158
　　　6.3.3 核心参数配置案例 159
　　　6.3.4 容量调度器配置案例 163
　　　6.3.5 公平调度器配置案例 168
　　　6.3.6 Tool 接口案例 171
　6.4 本章总结 174

第 7 章 高可用 HA 175
　7.1 ZooKeeper 详解 175
　　　7.1.1 ZooKeeper 入门 175
　　　7.1.2 ZooKeeper 安装 178
　　　7.1.3 ZooKeeper 的内部原理 180
　　　7.1.4 ZooKeeper 的命令操作 181
　7.2 HA 概述 188
　　　7.2.1 什么是 HA 188
　　　7.2.2 HDFS HA 的工作机制 188
　7.3 Hadoop HA 集群的搭建 189
　　　7.3.1 HDFS HA 手动故障转移 189
　　　7.3.2 HDFS HA 自动故障转移 192
　　　7.3.3 YARN HA 195
　　　7.3.4 Hadoop HA 集群规划 199
　7.4 本章总结 199

第 8 章 生产调优手册 200
　8.1 HDFS 的核心参数 200
　　　8.1.1 NameNode 的内存生产配置 200
　　　8.1.2 NameNode 心跳并发配置 202
　　　8.1.3 启用回收站功能 203
　8.2 HDFS 集群压测 204
　　　8.2.1 测试 HDFS 的写性能 205
　　　8.2.2 测试 HDFS 的读性能 207
　8.3 HDFS 的多目录配置 208
　　　8.3.1 NameNode 的多目录配置 208
　　　8.3.2 DataNode 的多目录配置 208
　　　8.3.3 集群数据均衡之磁盘之间的数据均衡 209
　8.4 HDFS 集群的扩容及缩容 209
　　　8.4.1 添加白名单 209
　　　8.4.2 服役新服务器 212
　　　8.4.3 服务器之间的数据均衡 214
　　　8.4.4 黑名单退役服务器 214
　8.5 HDFS 的存储优化策略 216

8.5.1 纠删码	216
8.5.2 异构存储	218
8.6 HDFS 的故障排除	223
8.6.1 NameNode 故障处理	223
8.6.2 集群安全模式&磁盘数据损坏	224
8.6.3 慢磁盘监控	226
8.6.4 小文件存档	227
8.7 MapReduce 的生产经验	228
8.7.1 MapReduce 程序运行较慢的原因	228
8.7.2 MapReduce 的常用调优参数	229
8.7.3 MapReduce 的数据倾斜	231
8.8 Hadoop 的综合调优	232
8.8.1 Hadoop 的小文件优化方法	232
8.8.2 测试 MapReduce 的计算性能	233
8.8.3 企业开发场景案例	234
8.9 本章总结	239

第 9 章 源码解析 240

9.1 RPC 通信原理	240
9.2 NameNode 启动源码解析	243
9.2.1 查看源码的准备工作	243
9.2.2 启动 9870 端口服务	246
9.2.3 加载镜像文件和编辑日志文件	247
9.2.4 初始化 RPC 服务器端	248
9.2.5 检查资源	249
9.2.6 检测心跳信息并进行超时判断	252
9.2.7 退出安全模式	255
9.3 DataNode 启动源码解析	257
9.3.1 查看源码的准备工作	257
9.3.2 初始化 DataXceiverServer	259
9.3.3 初始化 HTTP 服务	260
9.3.4 初始化 RPC 服务器端	261
9.3.5 DataNode 向 NameNode 注册	262
9.3.6 DataNode 向 NameNode 发送心跳信息	266
9.4 HDFS 写数据流程的源码解析	268
9.4.1 查看源码的准备工作	269
9.4.2 Client 向 NameNode 发起写请求	269
9.4.3 NameNode 处理 Client 的写请求	270
9.4.4 DataStreamer 启动流程	271
9.4.5 向 DataStreamer 的队列中写数据	274
9.4.6 建立管道之机架感知	276
9.4.7 建立管道之 socket 发送	278
9.4.8 建立管道之 socket 接收	280

  9.4.9 客户端接收 DataNode 的写数据响应 283
9.5 YARN 源码解析 284
  9.5.1 查看源码的准备工作 284
  9.5.2 创建 YARN 客户端并提交任务 286
  9.5.3 启动 MRAppMaster 288
  9.5.4 调度器任务执行 291
9.6 Hadoop 的源码编译 294
  9.6.1 前期准备工作 295
  9.6.2 安装工具包 295
  9.6.3 编译源码 297
9.7 本章总结 298

第1章

大数据概论

随着互联网行业的迅猛发展，各类互联网应用渗透进了我们生活中的各个角落。我们体验到了越来越便捷的服务，也生成了海量的数据。大数据实际上已经在我们的生活中发挥了重要的作用。Hadoop 与大数据联系紧密，要学习 Hadoop，必须先了解大数据的特征。在充分了解大数据后，即可对 Hadoop 的底层架构设计有更透彻的理解。本章将围绕大数据的概念展开，介绍大数据的特征和应用场景，然后带领读者初步认识 Hadoop。

1.1 大数据的特征

在现实生活中，我们时时刻刻都在跟数据打交道。我们在浏览购物网站、为喜欢的内容点赞、打开感兴趣的商品页面、处理电子邮件时，都会产生一条条的数据。随着互联网用户群体规模的扩大，数据量呈爆炸式增长。最初，人们认为产生的这些数据是系统的负累，无用、繁杂且占用大量空间，但是慢慢地，人们开始意识到，这些数据中可能蕴藏着巨大的信息资源。

但是要如何分析海量的数据，挖掘其中的价值呢？传统的数据处理技术都具有局限性，要了解局限性在哪里，首先需要了解大数据的定义。国际顶级权威咨询机构麦肯锡提出：大数据涉及的数据集规模已经超过了传统数据库软件获取、存储、管理和分析数据的能力。根据这句话可知，大数据的定义是十分主观的，并没有固定标准。因为随着技术的不断发展，符合大数据标准的数据集容量也会增长。此外，在不同的行业中，大数据的标准不同，因为通常不同行业的数据集规模不同，使用的分析软件也不同。

根据以上大数据的定义可知，传统数据库软件在大数据的获取、存储、管理和分析方面都存在较强的局限性。随着时代的发展，符合大数据标准的数据集规模会越来越大，并且不同行业的大数据标准不尽相同。

我们将大数据的特征总结为 4 点，分别为数据量大（Volume）、高速（Velocity）、多样（Variety）、低价值密度（Value），也就是 4 个 V。

1. 数据量大

目前，典型的个人计算机的硬盘容量规模为 TB 级，而一些大企业的数据量规模已经接近 EB 级。面对越来越大的数据规模，企业的存储压力和计算压力也越来越大，急于寻求解决方案。

2. 高速

高速体现在多个方面，数据的生成速度越来越快，需要的数据处理速度也越来越快。根据 IDC 的数字宇宙研究报告，预计到 2025 年，全球数据使用量将达到 163ZB。在如此海量的数据面前，提高数据处理效率是企业的重要任务。提高数据处理效率，一方面需要及时获取生成的变动数据，另一方面需要提高数据计算引擎的处理速度。对数据的处理不能局限于数据获取后的静态处理，还应考虑流动数据的分析计算。

3．多样

在大数据时代，数据的种类不再局限于结构化数据。我们可以将数据初步划分为结构化数据、半结构化数据、准结构化数据和非结构化数据。与存储于传统关系型数据库中的结构化数据相比，半结构化数据、准结构化数据和非结构化数据越来越多，如用户浏览页面生成的日志信息、系统运行生成的日志文件、文本文档、图像、视频等数据。数据的多样化对数据存储能力和数据处理能力都提出了更高的要求。

4．低价值密度

数据的价值密度与数据量成反比。当我们说大数据中蕴藏着信息资源时，实际上并不能让人直观地感受到，需要使用数据挖掘工具挖掘这些数据中的重要信息。如何快速地对有价值的数据进行提纯，是目前大数据背景下亟待解决的难题。

大数据的四大特征都对数据处理工具提出了挑战，海量数据如何存储？如何对海量数据进行及时、高效的处理？如何挖掘数据中的隐藏价值？这些问题将在后面的内容解答。

1.2 大数据的发展前景

大数据分析最初应用于超市营销，人们走进超市通常会发现，在售卖纸尿裤的货架旁边摆放着啤酒。这是因为，超市通过统计发现，当家庭中的爸爸来超市购买纸尿裤时，通常会顺便给自己买一些啤酒。这是一个典型的通过分析用户行为模式更改营销策略的案例。

现在，大数据的应用已经不再局限于超市营销，它扩展到各行各业，并且悄无声息地改变着我们的生活。

1.2.1 大数据的应用场景

大数据现在在各行各业都有广泛的应用，电商、游戏、娱乐、物流等行业都在大数据的快速发展中受到了影响。

1．电商行业

前面，我们通过一个传统零售行业的案例说明了大数据的最初应用场景。在零售行业中，通过分析用户的行为模式，可以为用户提供更便捷的购物体验，从而提升商品销量。

现在，电商行业的飞速发展改变了很多用户的消费习惯，电商网站为了提升销售额、扩展用户群体，会根据大数据分析用户的行为模式，从而制订营销策略。用户在浏览电商网站时，会留下各种行为痕迹，如浏览了哪些商品、收藏了哪些商品、将哪些商品加入了购物车、在哪个商品页面停留的时间最长，这些都体现了用户的购物偏好。根据用户偏好推荐用户最想购买的商品，根据用户已经购买的商品预测用户将要购买的商品，可以大幅提升销售额。根据用户的行为，结合用户的注册信息，可以勾勒出多种用户画像。构建电商网站的用户画像体系，可以为电商网站的用户群体拓展提供决策建议。

2．游戏行业

基于大数据分析，越来越多的大型游戏脱颖而出、风靡全球。游戏玩家在游戏中体验着种种乐趣，同时生成了大量数据。游戏开发人员会针对这些数据进行分析和计算。通过对游戏玩家的行为进行分析，判断哪些是活跃玩家，哪些是不太活跃的玩家。针对活跃玩家，分析其偏爱的游戏道具、规则，在新推出游戏道具、游戏活动、游戏功能时，分析玩家反馈，是增加了活跃玩家，还是造成了玩家的流失。针对不太活跃的玩家，分析什么样的活动可以刺激玩家，通过赠送道具等措施增强玩家黏性。

根据游戏玩家的行为分析，对游戏玩家进行分类，然后分析不同玩家的行为模式，进一步根据数据分析的结果，调整游戏设计与运营的手段，增加活跃玩家，提高玩家黏性。

3．娱乐行业

短视频 App 的兴起同样离不开大数据。我们在沉浸式观看短视频时，会有源源不断的令自己感兴趣的视频涌现出来。这是因为，短视频 App 会根据用户行为模式，分析用户的偏好，进而针对不同用户推送其感兴趣的内容。

各种娱乐 App 根据用户的行为模式，分析其特点，判断其喜好，从而有针对性地推荐不同用户感兴趣的内容，使用户更长时间地停留在平台上。

4．物流行业

电商行业的蓬勃发展离不开物流的提速，物流企业会对物流数据进行跟踪和分析，根据分析结果为物流企业提出决策建议。大数据在物流行业的应用包括地域分析、物流供给与需求匹配、物流资源优化与配置、市场变化实时分析、配送路线优化、物流中心选址、仓储选址等。大数据在物流行业的应用可以降低物流成本、提高物流效率。

1.2.2　大数据的未来发展

虽然大数据的应用已经非常广泛了，但在未来，大数据将有更加广阔的发展空间。各行各业的开发人员从大数据的应用中获得了好处，越来越多的后来者们都希望能借助大数据得到更长远的发展。

大数据的未来发展方向是充满无限可能的，建设智慧城市、打造智能家居、开发人工智能，以及最近的热门话题元宇宙，都离不开大数据。

建设智慧城市是指以新一代信息技术和通信技术为支撑，通过透明、充分的信息获取，广泛、安全的信息传递，科学、有效的信息处理，提升城市管理水平，提高市民的生活质量，促进产业升级和经济增长。智慧城市的建设可以有效缓解现有的城市管理难题，是城市建设的重要方向。智慧城市建设与传统城市建设的不同之处，在于要充分认识大数据在城市建设中的重要性。在智慧城市建设中，大数据发挥着重要作用：分析历史数据，做出科学预测，为决策提供依据；分析实时数据，及时做出响应。

智能家居之前只是一个概念，现在已经开始走进千家万户。智能家居通俗来说，就是将家居中的各种设备通过网络连接在一起，实现自动控制，从而提升家居生活的便利性、舒适性和安全性。智能家居可以通过各种传感器的信号及用户的控制，根据环境指标、用户习惯等，调整环境温度、湿度，开启或关闭影音系统，远程控制家电运行情况。智能家居离不开物联网和大数据，大数据帮助硬件厂商挖掘用户行为模式，建立用户画像，从而针对不同的用户提供个性化的智能体验，指导硬件升级，制订营销策略。

在很久之前，人们已经提出了人工智能的概念，但是人工智能从设想到实现还有很长的路要走，可以说，最终使人工智能变为现实的是大数据的不断发展。互联网可以生成海量的数据，数据就是资源。通过对大量、多种类的数据进行分析，可以获得更加全面、生动的事实刻画，人工智能可以从中不断学习、优化。人工智能目前已经有了一定的发展，但是随着大数据技术的升级，人工智能将给人类社会带来革命性的改变。

元宇宙（Metaverse）是最近异军突起的新概念，成了一个新的风口，大量互联网公司纷纷涌入跑道。元宇宙一词最早诞生于科幻小说，小说中描绘了一个庞大的虚拟世界，即元宇宙，人们在元宇宙中以数字为化身，在其中体验近乎真实的世界。如果说真实世界是由物质元素构成的，那么元宇宙是由数据打造的。元宇宙可以说是数据应用的极致，在其中，每个节点都由数据构成，节点的行动又会产生海量的数据，通过对新产生数据进行学习，元宇宙可以自己发展变化。元宇宙是一个大胆的设想，随着各大互联网公司的加入，相信离真正实现也不远了。

大数据的未来发展充满无限可能，为未来世界的描绘提供了多样的色彩。

1.3 大数据生态体系与 Hadoop

前面已经介绍了大数据的 4 个特征，即数据量大、高速、多样、低价值密度。针对这 4 个特征，传统的数据存储和处理工具都具有极强的局限性。大数据的体量远远超过了单台计算机的存储量，传统数据库也无法做到对多种类的海量数据提供高速、准确的计算服务。Hadoop 的诞生，为我们提供了高效、可靠、可扩展的数据存储系统和分布式的计算引擎。可以说，Hadoop 就是为大数据而生的。

1.3.1 Hadoop 的发展史

2002 年，Apache Lucene（一个应用广泛的文本搜索系统库）的创始人 Doug Cutting 计划开发一个大型网络搜索引擎 Nutch。很快，Nutch 的开发遇到了瓶颈，要实现一个支持 10 亿网页的搜索引擎，其硬件成本和运维成本是惊人的。

2003 年和 2004 年，谷歌发表了两篇论文。一篇论文是"The Google File System"，描述的是谷歌的分布式文件系统，简称 GFS。GFS 是一个可扩展的分布式文件系统，应用于大型的、分布式的、可以对海量数据进行访问的应用程序，可以运行于廉价的机器上，有很高的容错性。另一篇论文是"MapReduce—Simplified Data Processing On Large Clusters"，描述的是谷歌的大数据分布式计算方式，其核心思想是将计算任务分解，然后在多台计算机上同时处理。这两篇论文的推出为 Doug Cutting 提供了灵感，根据这两篇论文，他用了两年的时间，在 Nutch 上实现了分布式文件系统 NDFS 和 MapReduce，使 Nutch 的性能飙升。开发人员认识到了 NDFS 和 MapReduce 的优秀之处，将其从 Nutch 中独立出来，并且将其命名为 Hadoop。Hadoop 是 Doug Cutting 的孩子为一只毛绒玩具小象起的名字，在这之后，Hadoop 生态体系的其他框架也沿用了这种命名习惯，名称来源于动物，并且通俗、易流通。2006 年，谷歌发表了另一篇论文"Bigtable—A Distributed Storage System for Structured Data"，这篇论文中的理论为很多 NoSQL 奠定了基础，包括 Hadoop 生态下的 HBase。

这 3 篇论文被称为谷歌的三驾马车。虽然谷歌并没有公布这 3 个系统的源码，但是其中的理论思想，为后续的大数据体系发展奠定了坚实的基础。

在 Hadoop 独立出来后，Doug Cutting 携 Hadoop 加入雅虎，Apache Hadoop 项目正式启动，支持 MapReduce 和 HDFS 独立发展。Hadoop 在雅虎获得了极大的重视，在 2008 年 2 月，雅虎宣布，雅虎搜索引擎使用的索引是在一个拥有 1 万个内核的 Hadoop 集群上构建的。2008 年 4 月，Hadoop 打破了世界纪录，成为最快的 TB 级数据排序系统，并且在之后不断创造着新的纪录。从那之后，Hadoop 正式走进人们的视线，跃升为企业主流的部署系统。

在 Hadoop 正式发行后，有一系列活跃版本面世，但是主流版本为 Hadoop 1.x 和 Hadoop 2.x，其中，Hadoop 1.x 是 Hadoop 0.20 发行版本系列的延续，Hadoop 2.x 是 Hadoop 0.23 发行版本系列的延续。此外，Hadoop 的发行版本比较特殊，高版本并不完全包含低版本的特性。

Hadoop 1.x 是一个较早的稳定版本，这个版本的 Hadoop 中包含一个分布式文件存储系统 HDFS 和一个分布式文件计算系统 MapReduce，其中，HDFS 由一个 NameNode 和多个 DataNode 构成，MapReduce 由一个 JobTracker 和多个 TaskTracker 构成。Hadoop 1.x 可以稳定运行，但是具有一定的局限性，MapReduce 程序由 JobTracker 和 TaskTracker 进行调度执行，集群资源调度也由这两个组件负责，集群资源不能为其他计算任务服务，扩展性、容错性和多框架支持方面存在不足。

与 Hadoop 1.x 相比，Hadoop 2.x 有非常大的改变，它解决了 Hadoop 1.x 中的多种局限性问题。Hadoop 2.x 提出了 HDFS Federation，用于解决 HDFS 的扩展性瓶颈问题。针对 Hadoop 1.x 中 MapReduce 系统的扩展性和多框架支持不足问题，Hadoop 2.x 将系统的资源调度功能独立出来，通用的资源管理框架 YARN 诞生了。YARN 的出现，使用户可以在 Hadoop 集群上运行各种各样的计算程序，而不再局限于只运行

MapReduce 程序。

在推出 Hadoop 2.x 后，Hadoop 得到了越来越多的重视，在后续的更新中，新版本不断推出，新功能不断完善。

在最新推出的 Hadoop 3.x 中，新增了很多功能，增加了对纠删码功能的支持、优化了 HDFS 的高可用配置、冷热数据分离存储、磁盘间数据均衡等功能，这些更新使 Hadoop 的功能更加完善，并且体现了 Hadoop 的发展活力与用户的广泛程度。

不只社区的 Apache Hadoop 会发布 Hadoop 的发行版本，一些大公司也会发布 Hadoop 的发行版本，如 Cloudera、Hortonworks 和 MapR。

2008 年成立的 Cloudera 是最早将 Hadoop 商用的公司，Hadoop 的创始人 Doug Cutting 也加盟了 Cloudera。Cloudera 可以为合作伙伴提供 Hadoop 的商用解决方案，主要包括技术支持、咨询服务和使用培训等。Cloudera 的主要产品为 CDH、Cloudera Manager、Cloudera Support。CDH 是 Cloudera 的 Hadoop 发行版本，完全开源，比 Apache Hadoop 在兼容性、安全性和稳定性上都有所增强。Cloudera Manager 是集群的软件分发及管理监控平台，可以在几个小时内部署好一个 Hadoop 集群，并且对集群中的节点和服务进行实时监控。

2011 年成立的 Hortonworks 是雅虎与硅谷风投公司 Benchmark Capital 合资组建的，在成立之初就吸纳了大约 25～30 名专门研究 Hadoop 的雅虎工程师，这些工程师均在 2005 年就开始协助雅虎开发 Hadoop，为 Hadoop 贡献了 80%的代码。Hortonworks 的主打产品是 Hortonworks Data Platform，简称 HDP，也是完全开源的产品。HDP 中包括 Ambari，一款开源的安装和管理系统。2018 年，Hortonworks 被 Cloudera 公司收购，然后推出了新的产品 CDP。

1.3.2 大数据生态体系

Hadoop 在创建之初只是一个包含 HDFS 和 MapReduce 的大数据处理系统，但是现在已经形成了一个完善的生态体系。围绕着 Hadoop，诞生了很多大数据组件，这些组件功能不同，但是都依赖于 Hadoop 平台，都应用大数据处理技术。大数据生态体系的大体构成如图 1-1 所示。

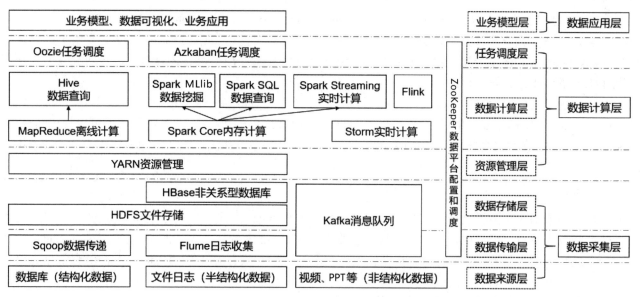

图 1-1 大数据生态体系的大体构成

图 1-1 中的大部分项目受 Apache 软件基金会支持，下面简单介绍一些重点项目。

1. Hive

Hive 是 Hadoop 生态体系的数据仓库工具。Hive 管理 HDFS 中存储的数据，并且提供标准的 SQL 功能。Hive 的解析引擎可以将标准的 SQL 语句解析成 MapReduce 程序并执行，对存储于 HDFS 中的数据集进行读、写、管理等操作。Hive 的优点在于学习成本低，不需要学习 MapReduce 程序的编写，只要会最基础的 SQL 查询语言，就可以进行大规模数据的统计与分析，大大降低了 Hadoop 的使用门槛。

2. HBase

HBase 是一个分布式的列式数据库，来源于谷歌的论文"BigTable—A Distributed Storage System for Structured Data"。HBase 可以使用本地文件系统运行，如果基于 HDFS 运行，则可以最大限度地发挥 HBase 对大型数据集的处理能力。HBase 的设计目标是使用廉价的硬件设施，构建能处理由成千上万的行和列构成的大型数据集。HBase 中存储的是松散型数据，这些数据是 key-value 的映射关系。HBase 中存储的数据可以由很多列构成，并且可以动态增加列。基于 HBase 的列式存储，HBase 可以提供强大的查询功能。基于 HDFS 的 MapReduce 计算引擎可以应对海量数据计算任务，但是无法满足大规模数据的实时处理需求，HBase 弥补了这个不足。

3. Spark

Apache Spark（简称 Spark）是一个开源的集群计算引擎。Spark 最初由加州大学伯克利分校的 AMP 实验室开发，后来被捐献给了 Apache 软件基金会，成了 Apache 的一个开源项目。Spark 的开发借鉴了 MapReduce 的理念，继承了其分布式计算的优点并弥补了 MapReduce 的缺陷。Spark 拥有 DAG 执行引擎，支持在内存中对数据进行迭代计算，官方数据表明，如果计算数据来源于磁盘，那么其计算速度可以是 Hadoop 的 MapReduce 计算速度的 10 倍以上；如果计算数据从内存中读取，那么其计算速度可以是 Hadoop 的 MapReduce 的计算速度的 100 倍以上。

Spark 程序可以使用 Scala 语言开发，这种语言灵活性很高，能够使用灵活的代码处理复杂的逻辑问题。Spark 还提出了一种数据结构——弹性分布式数据集（Resilient Distributed Dataset，RDD），通过操作 RDD，可以像操作本地数据集一样轻松操作分布式数据集。对于 RDD，Spark 也提供了多种多样的操作类型，调用形式简单、灵活，避免了编写复杂的 MapReduce 程序。

Spark 拥有很多组件，包括 Spark Core、Spark SQL、Spark Streaming、Spark MLlib 和 GraphX 等，Spark Core 主要用于进行内存离线计算，Spark SQL 主要用于进行即席查询，Spark Streaming 主要用于进行实时数据计算，Spark MLlib 主要用于进行机器学习，Spark GraphX 主要用于进行图计算。这些功能强大且全面的组件统一了离线计算与实时计算。

Spark 具有很强的适应性。Spark 运行需要一个资源管理机制和文件存储系统作为支撑。在资源管理机制方面，Spark 可以作为独立集群进行管理，也可以使用 Hadoop 的 YARN 进行管理，还可以使用 Apache Mesos 进行管理。在存储系统方面，Spark 可以与多种文件存储系统对接，包括 HDFS、HBase、Cassandra、Amazon S3 等。

4. Flink

Flink 是 Apache 软件基金会旗下的一个开源大数据处理框架。目前，Flink 已经成为大部分公司大数据实时处理的发力重点。在国内，大部分互联网企业都为 Flink 社区贡献了大量源码。Flink 对自己的定位是一个分布式数据处理引擎，主要用于对有界和无界数据流进行有状态的计算。Flink 可以在所有常见的集群环境中运行，具有优秀的计算速度，并且可以执行 PB 级数据规模的计算任务。Flink 具有很多优点，包括高吞吐、低延迟、有状态计算、精确的一致性保证等，这些特性使 Flink 在流处理领域脱颖而出。在近期的版本更新中，Flink 将批处理与流处理合二为一，不再明显区分，体现了 Flink 在大数据计算领域的深谋远虑。

5. Flume

Flume 是一个分布式、可靠且可用的服务，主要用于高效地收集、汇总和移动大量日志数据。在实际生产环境中，日志等数据的生成是源源不断的，并且可能生成在服务器集群中的各个节点上，通过 Flume 可以将数据不断地采集至 HDFS 集群中。Flume 并不局限采集的数据源，可以通过自定义数据源采集各种来源的数据。Flume 还可以通过自定义组件对采集到的源数据进行分流、清洗等操作，并且将数据写入各种数据接收方。

6. Sqoop

Sqoop（SQL to Hadoop）主要用于在结构化的数据存储设备与 Hadoop 之间传输数据。也就是说，Sqoop 可以将关系型数据库（MySQL、Oracle 等）中的数据导入 Hadoop 的 HDFS，也可以将 HDFS、Hive 中的数据导出至关系型数据中。Sqoop 的操作全部通过命令行与标准 SQL 完成，可以将 Sqoop 的操作转换成简单的 MapReduce 程序并执行，十分简洁、高效。

7. ZooKeeper

ZooKeeper 是一种分布式、高可用的协调服务，分布式应用程序可以基于 ZooKeeper 实现数据的发布订阅、负载均衡、分布式协调通知、集群管理、分布式锁等功能。

Hadoop 生态体系中还有很多组件，这些组件不断迭代更新，实现了大数据处理中的各种功能。

1.3.3 Hadoop 架构

目前，各大企业应用的 Hadoop 都以 Hadoop 2.x 版本为主，本书讲解的 Hadoop 架构也是自 Hadoop 2.x 以来的经典架构。Hadoop 2.x 主要由 4 个基本模块构成，如图 1-2 所示。

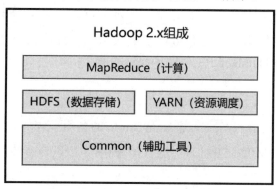

图 1-2　Hadoop 2.x 的主要组成部分

- Common：支持其他 Hadoop 模块的通用程序包。
- HDFS：分布式文件存储系统，主要用于实现海量数据的高可靠存储。
- YARN：作业调度和资源管理框架。
- MapReduce：Hadoop 提供的分布式数据计算模型。

其中，HDFS、YARN 和 MapReduce 是 Hadoop 的核心组成部分。

1. HDFS

HDFS（Hadoop Distributed Filesystem）是 Hadoop 的分布式文件存储系统，在 Hadoop 的生态体系中，是最基础的部分之一。Hadoop 的 MapReduce（计算模型）、Hive（数据库）、Spark（计算引擎）等都依赖于 HDFS 中存储的文件。HDFS 通过将大型数据集分布式地存储在多个存储节点上，解决了海量数据的存储难题。

HDFS 通过分布式存储技术使 Hadoop 集群可以存储 PB 级、EB 级甚至更大规模的数据集；通过多副本的模式，可以提供高可靠的数据存储性能。

HDFS 采取经典的主从架构，由一个主节点和多个数据存储节点构成，基本架构如图 1-3 所示。根据图 1-3 可知，HDFS 主要由组件 NameNode、DataNode、SecondaryNameNode 和 Client 组成。

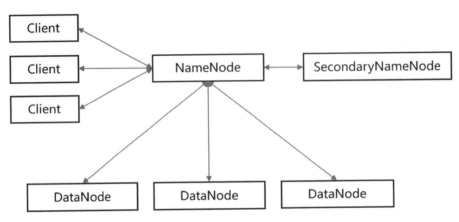

图 1-3　HDFS 的基本架构

1）NameNode。

NameNode 主要用于存储文件的元数据，如文件名、文件目录结构、文件副本数量、文件权限等信息，还包括每个文件的块列表、数据块所在的 DataNode 等数据信息。NameNode 是整个 HDFS 集群的管理者，管理着 HDFS 的目录结构和 DataNode 的健康状态。

2）DataNode。

DataNode 是 HDFS 中的数据存储节点，在本地文件系统中存储数据及数据块的校验和。DataNode 还负责定期向 NameNode 汇报数据块信息和节点健康状态。文件在被上传至 HDFS 集群中时，会被切分成多个数据块存储在不同的 DataNode 上，为了保证数据的可靠性，每个数据块中都会存储多个数据副本。

3）SecondaryNameNode。

SecondaryNameNode 是辅助节点，并不是 NameNode 的备用节点，主要用于辅助 NameNode 定期生成元数据检查点并将其传递给 NameNode。当 NameNode 发生故障时，可以通过 SecondaryNameNode 恢复数据。

4）Client。

用户通过 Client（客户端）与 NameNode 和 DataNode 交互，从而访问 HDFS 中的文件。

2. YARN

YARN 是 Hadoop 2.x 中新增加的资源管理系统。在 Hadoop 1.x 中，MapReduce 任务的资源调度由 JobTracker 组件和 TaskTracker 组件负责，这两个组件既要负责资源管理，又要负责作业执行，但在可扩展性、资源利用率和多框架支持方面存在不足。在 Hadoop 2.x 中，将资源管理功能独立出来，衍生出了一个新的资源统一管理平台 YARN。YARN 的出现，使 Hadoop 的集群计算资源可以开放给更多的计算框架。

YARN 采用的也是主从架构，主要由主节点 ResourceManager 和从节点 NodeManager 组成，其中，ResourceManager 负责全局的资源调度，NodeManager 负责本节点的资源管理，每个应用程序都单独有一个 ApplicationMaster，负责本任务的进度管理。YARN 的基本架构如图 1-4 所示。

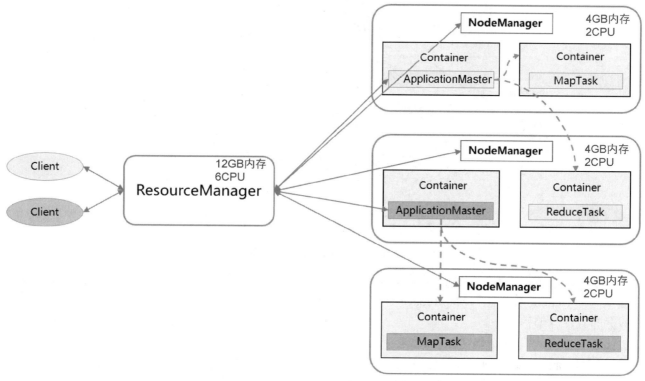

图 1-4 YARN 的基本架构

1）ResourceManager。

ResourceManager 主要负责整个系统的资源管理和分配工作，与客户端进行交互，处理来自客户端的请求。例如，查询应用的运行情况，响应客户端提交的任务请求，启动和管理各个应用的 ApplicationMaster，管理 NodeManager，响应 ApplicationMaster 的资源请求，分配调度资源。

2）NodeManager。

NodeManager 是每个节点上的资源和任务管理器，会定时向 ResourceManager 汇报本节点上的资源使用情况和各个 Container 的运行状态，并且接收和处理来自 ApplicationMaster 的启动或停止 Container 的请求。

3）Container。

Container 是 YARN 中的资源抽象单位，封装了节点中的内存、CPU、磁盘、网络等资源。当 ApplicationMaster 向 ResourceManager 申请资源时，ResourceManager 返回的资源就是 Container。Container 是一个容器，在容器上可以运行不同种类的任务。

4）ApplicationMaster。

用户每提交一个应用程序，都需要包含一个 ApplicationMaster。ApplicationMaster 主要负责本应用与 YARN 集群之间的交互。例如，向 ResourceManager 申请资源，分配调度应用内部的任务，与 NodeManager 进行交互以启动或停止 Container，监控任务运行状态。

3. MapReduce

MapReduce 是 Hadoop 提供的一个分布式计算模型，如图 1-5 所示。根据图 1-5 可知，MapReduce 将大数据计算分成 map 阶段和 reduce 阶段，在 map 阶段并行处理输入数据，在 reduce 阶段对 map 阶段的输出结果进行汇总等计算。

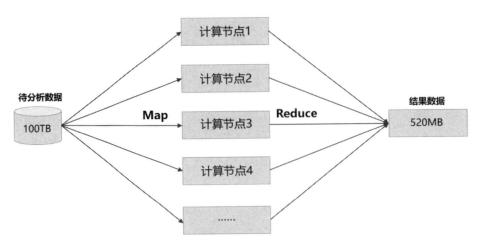

图 1-5　MapReduce 计算模型

MapReduce 将用户编写的业务逻辑代码和自带的默认组件整合成一个完整的分布式计算程序，并发地运行在一个 Hadoop 集群上。MapReduce 计算模型使大数据计算更加易于编程，通过实现一些接口，就能完成分布式计算编程。

Hadoop 具有以下 4 个优势。

- 高可靠性：Hadoop 底层通过维护多个数据副本，提高了数据的可靠性。当数据块中的其中一个副本发生损坏时，可以通过存储在其他节点上的数据副本恢复数据。
- 高扩展性：Hadoop 以集群的形式运行，可以方便地进行集群扩展，可扩展至数以千计的节点。通过扩展集群规模，可以大幅提高集群的文件存储性能和计算性能。
- 高效性：通过 MapReduce 计算模型向 YARN 集群提交任务，可以将海量数据的计算任务分配给众多集群节点并行执行任务，从而加快任务处理速度。
- 高容错性：在 MapReduce 任务的执行过程中，如果有任务失败，则可以将失败的任务重新分配。

1.4　本章总结

本章内容主要围绕大数据的概念展开，通过讲解大数据的特征及发展前景引出 Hadoop 的相关知识。可以说，Hadoop 的发展史就是大数据的发展史，Hadoop 的推出解决了大数据开发的诸多难题。以 Hadoop 为中心，发展出的 Hadoop 生态体系，使大数据的发展走上了高速路。通过对本章内容的学习，相信读者已经对 Hadoop 有了初步的认识，在后续的具体学习过程中，可以进一步体会 Hadoop 的优秀之处。

第 2 章 环境准备

在正式讲解 Hadoop 前，我们需要搭建 Hadoop 的学习环境。在企业的实际生产环境中，Hadoop 通常是构建在 Linux 操作系统中的。企业一般都拥有一定规模的服务器集群，为了可以最大限度还原生产环境，需要在个人计算机上搭建一个小型的服务器集群。本章带领读者进行基本的环境准备工作，在这个过程中熟悉 Linux 操作系统命令行的使用方法。

2.1 安装 VMware

本节介绍的虚拟机软件是 VMware，VMware 可以使用户在一台计算机上同时运行多个操作系统，还可以像 Windows 应用程序一样来回切换不同的操作系统。用户可以像操作真实安装的操作系统一样操作虚拟机操作系统，甚至可以在一台计算机上将几个虚拟机操作系统连接为一个局域网或连接到互联网。

在虚拟机系统中，每一台虚拟产生的计算机都被称为虚拟机，存储所有虚拟机的计算机被称为宿主机。使用 VMware 虚拟机软件安装虚拟机可以减少因安装新操作系统导致的数据丢失问题，还可以使用户方便地体验各种操作系统，用于进行学习和测试。

VMware 支持多种平台，可以安装在 Windows、Linux 等操作系统中，初学者通常使用 Windows 操作系统，可以下载 VMware Workstation for Windows 版本。VMware 的安装非常简单，与其他 Windows 软件的安装方法类似，本书不进行详细讲解。值得一提的是，在 VMware 的安装过程中，安装的类型包括典型安装和自定义安装，笔者建议初学者选择典型安装。

在 VMware 安装完成并启动后，即可进行 Linux 操作系统的安装部署。

推荐使用版本：VMware Workstation Pro、VMware Workstation Player 15.1 及更高版本。其中，VMware Workstation Player 版本供个人用户使用，非商业用途，是免费的，其他的 VMware 版本在此不进行过多介绍。

2.2 安装 CentOS

在安装 CentOS 前，用户需要检查本机 BIOS 是否支持虚拟化。在开机后进入 BIOS 界面（不同计算机进入 BIOS 界面的操作方法有所不同），然后进入 Security 下的 Virtualization，选择 Enable 即可。

启动 VMware，进入主界面，进行新虚拟机的设置，设置配置类型，此处选择"自定义（高级）"单选按钮，如图 2-1 所示。

单击"下一步"按钮，进入"选择虚拟机硬件兼容性"界面，选择本机使用的 VMware Workstation 版本，如图 2-2 所示。

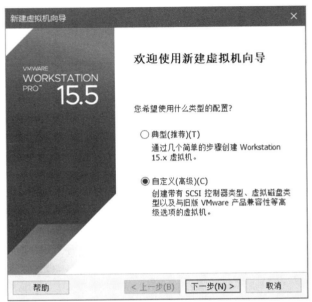

图 2-1　设置配置类型

图 2-2　"选择虚拟机硬件兼容性"界面

单击"下一步"按钮，进入"安装客户机操作系统"界面，选择"稍后安装操作系统"单选按钮，如图 2-3 所示。

单击"下一步"按钮，然后在"版本"下拉列表中选择要安装的 Linux 版本，此处选择"CentOS 7 64 位"选项，如图 2-4 所示。

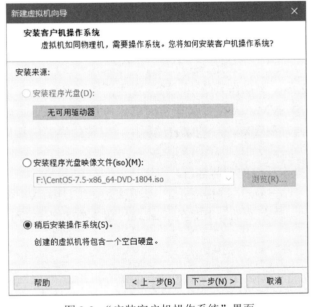

图 2-3　"安装客户机操作系统"界面

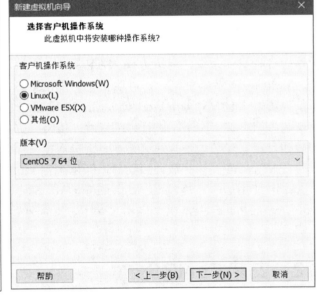

图 2-4　"选择客户机操作系统"界面

单击"下一步"按钮，进入"命名虚拟机"界面，给虚拟机起一个名字（在虚拟机创建完成后可以更改），然后单击"浏览"按钮，选择虚拟机操作系统安装文件的存储路径，如图 2-5 所示。

单击"下一步"按钮，进入"处理器配置"界面，为虚拟机配置处理器，处理器总核数不应多于本机处理器总核数，如图 2-6 所示。

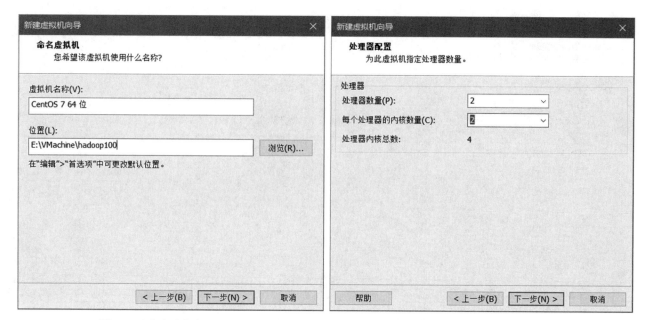

图 2-5 "命名虚拟机"界面　　　　　　　　　图 2-6 "处理器配置"界面

单击"下一步"按钮，进入"此虚拟机的内存"界面，配置为虚拟机分配的内存大小，最小 1GB，后续也可以根据需要进行修改，如图 2-7 所示。

单击"下一步"按钮，进入"网络类型"界面，选择虚拟机网络连接类型，选择"使用网络地址转换（NAT）"，如图 2-8 所示。

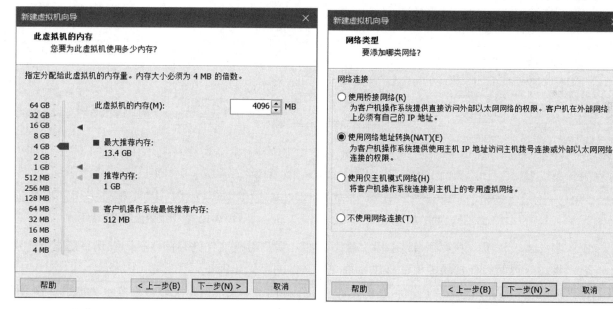

图 2-7 "此虚拟机的内存"界面　　　　　　　图 2-8 "网络类型"界面

单击"下一步"按钮，进入"选择 I/O 控制器类型"界面，采用默认参数设置，如图 2-9 所示。
单击"下一步"按钮，进入"选择磁盘类型"界面，采用默认参数设置，如图 2-10 所示。
单击"下一步"按钮，进入"选择磁盘"界面，采用默认参数设置，如图 2-11 所示。
单击"下一步"按钮，进入"指定磁盘容量"界面，此处建议将"最大磁盘大小"设置为 50.0GB，以便满足数据仓库项目对服务器的存储要求，如图 2-12 所示。

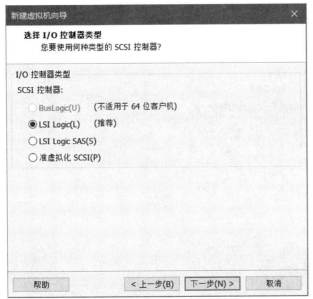

图 2-9 "选择 I/O 控制器类型"界面　　　　图 2-10 "选择磁盘类型"界面

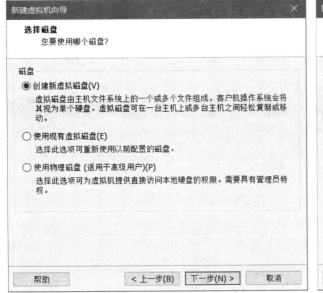
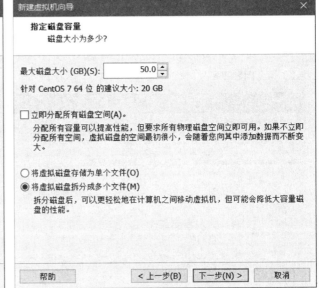

图 2-11 "选择磁盘"界面　　　　图 2-12 "指定磁盘容量"界面

单击"下一步"按钮,进入"指定磁盘文件"界面,指定磁盘文件的存储路径,默认将磁盘文件存储于"命名虚拟机"界面中指定的虚拟机操作系统安装文件的存储路径下,如图 2-13 所示。

单击"下一步"按钮,进入"已准备好创建虚拟机"界面,单击"自定义硬件"按钮,如图 2-14 所示,弹出"虚拟机设置"对话框,默认选择"硬件"选项卡,在左侧的列表框中选择 CD/DVD(IDE)选项,在右侧的"连接"选区中选择"使用 ISO 映像文件"单选按钮,单击"浏览"按钮,找到 ISO 映像文件的存储路径即可,如图 2-15 所示。

在配置完映像文件后,单击"确定"按钮,返回"已准备好创建虚拟机"界面,单击"完成"按钮,开始安装,如图 2-16 所示。

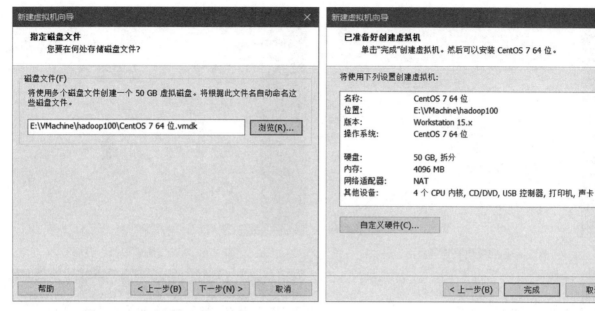

图 2-13 "指定磁盘文件"界面　　　　　　　图 2-14 "已准备好创建虚拟机"界面

图 2-15 "虚拟机设置"对话框

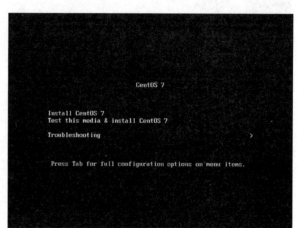

图 2-16 开始安装虚拟机

使用键盘中的上下方向键移动光标,使用回车键确定,这里选择 Install CentOS 7 选项,然后进入安装语言选择界面,选择简体中文作为安装语言,如图 2-17 所示。

单击"继续"按钮,进入"安装信息摘要"界面,将"日期和时间"设置为"亚洲/上海 时区",将"键盘"设置为"汉语",将"语言支持"设置为"简体中文(中国)",如图 2-18 所示。

在"安装信息摘要"界面选择"软件选择"选项,进入"软件选择"界面,进行软件安装配置,选择"GNOME 桌面"单选按钮,如图 2-19 所示。

在"安装信息摘要"界面选择"安装位置"选项,进入"手动分区"界面,进行手动分区配置,如图 2-20 所示。

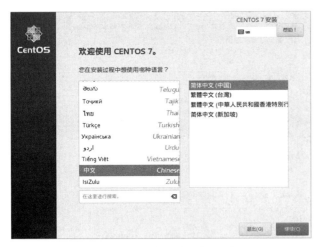
图 2-17　安装语言选择界面

图 2-18　"安装信息摘要"界面

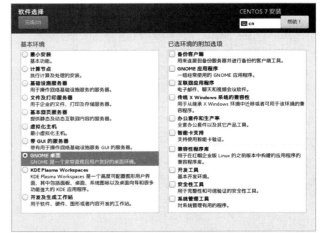

图 2-19　软件安装配置

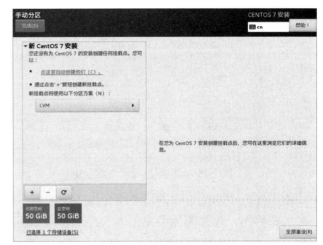

图 2-20　手动分区配置

在"手动分区"界面中单击"+"按钮，打开"添加新挂载点"面板，将"挂载点"设置为"/boot"，将"期望容量"设置为"1G"，然后单击"添加挂载点"按钮，即可配置/boot分区；返回"手动分区界面"，在右侧将"设备类型"设置为"标准分区"，将"文件系统"设置为"ext4"，如图 2-21 所示。

按照上述流程，手动配置 swap 分区和根目录，分别如图 2-22 和图 2-23 所示。

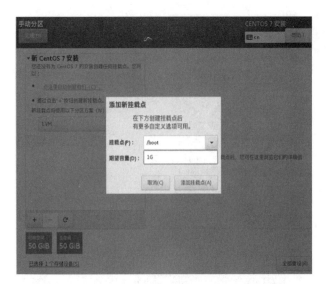 

图 2-21　手动配置/boot 分区

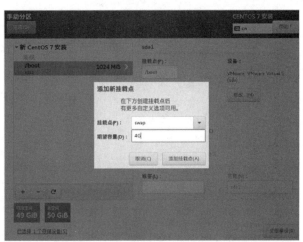 

图 2-22　手动配置 swap 分区

图 2-23　手动配置根目录

在手动配置完分区后，单击"手动分区"界面中的"完成"按钮，打开"更改摘要"面板，如图 2-24 所示，单击"接受更改"按钮，表示接受更改分区。

图 2-24 "更改摘要"面板

在"安装信息摘要"界面中选择 KDUMP 选项,进入 KDUMP 界面,进行 KDUMP 配置,如图 2-25 所示。

图 2-25 进行 KDUMP 配置

在"安装信息摘要"界面中选择"网络和主机名"选项,进入"网络和主机名"界面,进行网络和主机名配置,如图 2-26 所示。

图 2-26 进行网络和主机名配置

在网络和主机名配置完成后，在"安装信息摘要"界面中单击"开始安装"按钮，进入"配置"界面，如图 2-27 所示。

在"配置"界面中选择"ROOT 密码"选项，进入"ROOT 密码"界面，配置 root 用户的密码，如图 2-28 所示。

图 2-27 "配置"界面

图 2-28 "ROOT 密码"界面

在 root 用户的密码配置完成后，等待安装结束，整个安装过程大约需要 20 分钟。在安装完成后，单击"重启"按钮，如图 2-29 所示。

在重启后的界面中，根据界面提示，依次接收许可证、选择系统语言、关闭位置服务、选择时区、跳过关联账号、创建普通用户、设置普通用户密码，即可开始使用，这些步骤由读者自行配置，故不进行图片引导，配置完成后的界面如图 2-30 所示。

图 2-29 在安装完成后重启

图 2-30 配置完成后的界面

注意：在虚拟机和宿主机之间，鼠标是不能同时起作用的，如果从宿主机进入虚拟机，则需要将鼠标指针移入虚拟机；如果从虚拟机返回宿主机，则按 Ctrl+Alt 快捷键退出。

2.3 安装远程终端

大部分服务器都是通过远程管理工具进行日常管理操作的。远程管理方法包括 VNC 的图形远程管理、Webmin 的基于浏览器的远程管理、命令行操作等。下面介绍两个远程管理工具的安装方法。

2.3.1 安装 Xshell

Xshell 是一款功能非常强大的安全终端模拟软件，也是目前市场上比较主流、应用比较广泛的远程管理工具。Xshell 的功能非常丰富，支持 SSH1、SSH2 及 Microsoft Windows 平台的 TELNET 协议，给用户提供了很好的终端用户体验。Xshell 的安装非常简单，并且为普通用户提供了免费版本，可以放心使用。

用户可以在 Xshell 官方网站下载 Xshell 6.0 及更高版本的安装包，在本书提供的学习资料中，也附有 Xshell 的安装包，直接双击即可进行安装。在开始使用后，需要用户选择使用类型，选择"免费为家庭/学校"单选按钮，即可免费使用。

打开 Xshell，单击左上角的新建会话按钮，如图 2-31 所示。

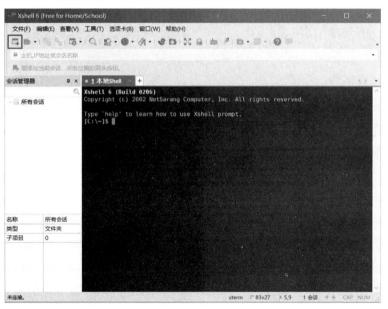

图 2-31　打开 Xshell 并单击左上角的新建会话按钮

弹出"新建会话属性"对话框，默认在左侧的"类别"列表框中选择"连接"选项，然后在右侧的"常规"选区中编辑会话名称，填写主机 IP 地址，如图 2-32 所示。在主机的 IP 地址栏中，可以直接填写服务器的 IP 地址，也可以在系统已经修改过主机映射文件的前提下直接填写主机映射名。

图 2-32　"新建会话属性"对话框（一）

在左侧的"类别"列表框中选择"用户身份验证"选项，然后在右侧填写连接的用户名和密码，如图 2-33 所示。在填写完成后，单击"确定"按钮，即可创建新的连接。

图 2-33 "新建会话属性"对话框（二）

在连接创建完成后，在连接列表中双击该连接，在第一次进行连接时，会弹出"SSH 安全警告"对话框，单击"接受并保存"按钮，接受主机密钥，如图 2-34 所示。

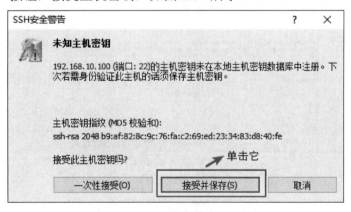

图 2-34 "SSH 安全警告"对话框

通过以上方式可以创建其他主机的连接，读者也可以自行探索 Xshell 的其他功能。

2.3.2 安装 SecureCRT

SecureCRT 将 SSH（Secure Shell）的安全登录、数据传送性能和 Windows 终端仿真提供的可靠性、可用性、易配置性结合在一起。如果需要管理多台服务器，那么使用 SecureCRT 可以很方便地记住多个地址，并且可以通过配置设置自动登录，方便远程管理，效率很高。缺点是 SecureCRT 需要安装，并且是一款共享软件，如果不付费注册，则不能使用。

在安装并启动 SecureCRT 后，单击"快速连接"按钮，弹出"快速连接"对话框，如图 2-35 所示，输入要连接的主机名和用户名，单击"连接"按钮，然后按照提示输入密码即可登录。

SecureCRT 默认不支持中文，中文会显示为乱码，解决方法如下。

在建立连接后，选择"选项"→"会话选项"命令，在弹出的对话框左侧的"分类"列表框中选择"终端"→"仿真"选项，在右侧的"终端"下拉列表中选择 Xterm 选项，勾选"ANSI 颜色"复选框，表示支持颜色显示，单击"确定"按钮，如图 2-36[①]所示。

图 2-35 "快速连接"对话框

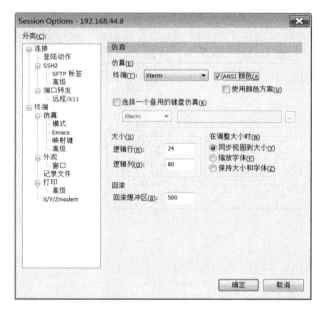

图 2-36 SecureCRT 的仿真设置

在左侧的"分类"列表框中选择"终端"→"外观"选项，在右侧的"当前颜色方案"下拉列表中选择 Traditional 选项，将"标准字体"和"精确字体"均设置为"新宋体 11pt"，将"字符编码"设置为"UTF-8"（CentOS 默认使用中文字符集 UTF-8），取消勾选"使用 Unicode 线条绘制字符"复选框，单击"确定"按钮，如图 2-37 所示。

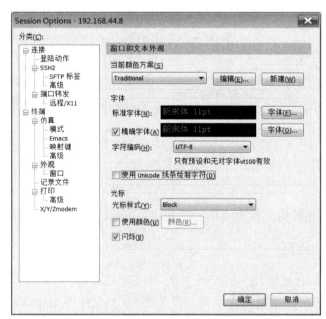

图 2-37 SecureCRT 的窗口和文本外观设置

至此，我们就初步搭建好了 Hadoop 的学习环境。

---

① 图 2-36 中"登陆动作"的正确写法为"登录动作"。

2.4 虚拟机配置

2.4.1 网络配置

对安装好的 VMware 进行网络配置，方便虚拟机连接网络，本次设置建议使用 NAT 模式，要求宿主机的 Windows 操作系统和虚拟机的 Linux 操作系统能够进行网络连接，同时虚拟机的 Linux 操作系统可以通过宿主机的 Windows 操作系统进入互联网。

选择"编辑"→"虚拟网络编辑器"命令，如图 2-38 所示，对虚拟机进行网络配置。在弹出的"虚拟网络编辑器"对话框中，选择"NAT 模式（与虚拟机共享主机的 IP 地址）"单选按钮，并且修改虚拟机的子网 IP 地址，如图 2-39 所示。

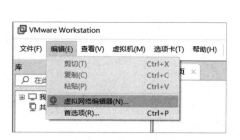

图 2-38 选择"虚拟网络编辑器"命令

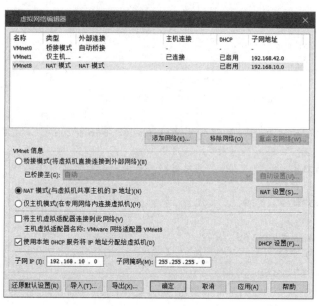

图 2-39 "虚拟网络编辑器"对话框

单击"NAT 设置"按钮，弹出"NAT 设置"对话框，用于查看网关设置，如图 2-40 所示。

在控制面板中选择"网络和 Internet"→"网络连接"命令，可以查看 Windows 操作系统中的 vmnet8 网络配置，如图 2-41 所示。

图 2-40 "NAT 设置"对话框

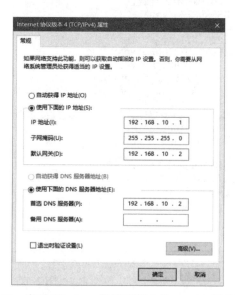

图 2-41 查看 Windows 操作系统中的 vmnet8 网络配置

2.4.2 网络 IP 地址配置

修改网络 IP 地址为静态 IP 地址，避免 IP 地址经常变化，从而方便节点服务器之间互相通信。

[root@hadoop100 桌面]#vim /etc/sysconfig/network-scripts/ifcfg-ens33

以下加粗的项必须修改，有这些项的按照下面的相关设置进行修改，没有这些项的需要添加这些项。

```
TYPE="Ethernet"        #网络类型（通常是Ethernet）
PROXY_METHOD="none"
BROWSER_ONLY="no"
BOOTPROTO="static"  #IP 的配置方法[none|static|bootp|dhcp]（引导时不使用协议|静态分配 IP|BOOTP 协议|DHCP 协议）
DEFROUTE="yes"
IPV4_FAILURE_FATAL="no"
IPV6INIT="yes"
IPV6_AUTOCONF="yes"
IPV6_DEFROUTE="yes"
IPV6_FAILURE_FATAL="no"
IPV6_ADDR_GEN_MODE="stable-privacy"
NAME="ens33"
UUID="e83804c1-3257-4584-81bb-660665ac22f6"     #随机id
DEVICE="ens33"  #接口名（设备,网卡）
ONBOOT="yes"    #系统启动时的网络接口是否有效（yes/no）
#IP 地址
IPADDR=192.168.10.100
#网关
GATEWAY=192.168.10.2
#域名解析器
DNS1=192.168.10.2
```

修改 IP 地址后的结果如图 2-42 所示，执行:wq 命令，保存文件并退出。

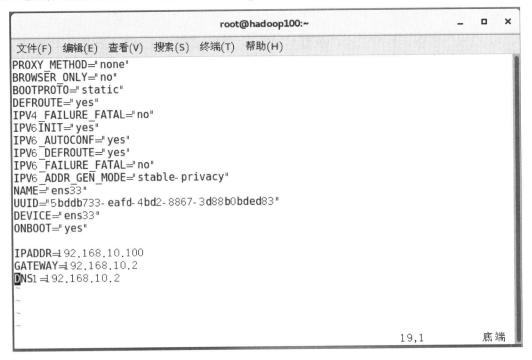

图 2-42　修改 IP 地址后的结果

执行 systemctl restart network 命令，重启网络服务。如果报错，则执行 reboot 命令，重启虚拟机。

2.4.3 主机名配置

将主机名修改为一系列有规律的主机名,并且修改主机映射文件(hosts 文件),添加我们需要的主机名和 IP 地址映射,方便管理,以及方便节点服务器之间通过主机名进行通信。

1. 修改 Linux 操作系统的主机映射文件

(1)进入 Linux 操作系统,执行 hostname 命令,查看本机的主机名。

```
[root@hadoop100 桌面]# hostname
hadoop100
```

(2)如果感觉当前主机名不合适,则可以通过编辑/etc/hostname 文件进行修改。

```
[root@hadoop100 桌面]# vim /etc/hostname
hadoop100
```

注意:主机名中不要有"_"(下画线)。

(3)打开/etc/hostname 文件,可以看到主机名,可以对主机名进行修改,这里不做修改,仍为 hadoop100。

(4)保存并退出。

(5)打开/etc/hosts 文件。

```
[root@hadoop100 桌面]# vim /etc/hosts
```

添加以下内容。

```
192.168.10.100 hadoop100
192.168.10.101 hadoop101
192.168.10.102 hadoop102
192.168.10.103 hadoop103
192.168.10.104 hadoop104
192.168.10.105 hadoop105
192.168.10.106 hadoop106
192.168.10.107 hadoop107
192.168.10.108 hadoop108
```

(6)重启设备,查看主机名,可以看到主机名修改成功。

2. 修改 Windows 操作系统的主机映射文件

(1)进入路径 C:\Windows\System32\drivers\etc。

(2)将 hosts 文件复制到桌面上。

(3)打开桌面上的 hosts 文件并添加以下内容。

```
192.168.10.100 hadoop100
192.168.10.101 hadoop101
192.168.10.102 hadoop102
192.168.10.103 hadoop103
192.168.10.104 hadoop104
192.168.10.105 hadoop105
192.168.10.106 hadoop106
192.168.10.107 hadoop107
192.168.10.108 hadoop108
```

(4)用桌面上的 hosts 文件覆盖 C:\Windows\System32\drivers\etc 路径下的 hosts 文件。

2.4.4 防火墙配置

为了使 Windows 或其他操作系统可以访问 Linux 虚拟机内的服务,有时需要关闭虚拟机的防火墙,以下是常见的防火墙开/关命令。

1. 临时关闭防火墙

（1）查看防火墙状态。

```
[root@hadoop100 桌面]# systemctl status firewalld
```

（2）临时关闭防火墙。

```
[root@hadoop100 桌面]# systemctl stop firewalld
```

2. 在开机时关闭防火墙

（1）查看开机时的防火墙状态。

```
[root@hadoop100 桌面]# systemctl list-unit-files|grep firewalld
```

（2）设置在开机时不自动启动防火墙。

```
[root@hadoop100 桌面]# systemctl disable firewalld.service
```

2.4.5 一般用户配置

root 用户具有非常高的操作权限，而在实际操作中需要对用户权限进行限制，所以我们需要创建一般用户。

创建一般用户 atguigu，配置 atguigu 用户具有 root 权限，接下来的所有操作都在一般用户 atguigu 身份下完成。

（1）添加 atguigu 用户，并且给其设置密码。

```
[root@hadoop100 ~]#useradd atguigu
[root@hadoop100 ~]#passwd atguigu
```

（2）修改配置文件。

```
[root@hadoop100 ~]#vi /etc/sudoers
```

修改/etc/sudoers 文件，找到第 91 行，在 root 下面添加一行。

```
## Allow root to run any commands anywhere
root      ALL=(ALL)     ALL
atguigu   ALL=(ALL)     ALL
```

或者配置成在执行 sudo 命令时，不需要输入密码。

```
## Allow root to run any commands anywhere
root      ALL=(ALL)     ALL
atguigu   ALL=(ALL)     NOPASSWD:ALL
```

在修改完毕后，用户使用 atguigu 账号或执行 sudo 命令进行登录，即可获得 root 操作权限。

2.4.6 克隆虚拟机

Hadoop 是一个分布式的大数据框架，在企业的实际生产环境中，通常是集群部署的，为了模拟真实的生产环境，我们需要对虚拟机进行克隆，得到 3 台服务器，形成一个微型的服务器集群。克隆虚拟机的过程如下。

（1）关闭要被克隆的虚拟机，右击虚拟机名称，在弹出的快捷菜单中选择"管理"→"克隆"命令，如图 2-43 所示。

（2）在欢迎界面中单击"下一步"按钮，弹出"克隆虚拟机向导"对话框，设置"克隆自"为"虚拟机中的当前状态"，单击"下一步"按钮，如图 2-44 所示。

（3）设置"克隆方法"为"创建完整克隆"，单击"下一步"按钮，如图 2-45 所示。

（4）设置克隆的虚拟机名称和位置，如图 2-46 所示。

图 2-43 选择"克隆"命令　　　　　图 2-44 "克隆虚拟机向导"对话框

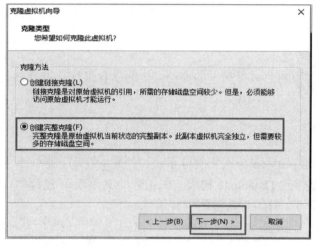 

图 2-45 设置"克隆方法"为"创建完整克隆"　　　图 2-46 设置克隆的虚拟机名称和位置

（5）单击"完成"按钮，开始克隆，需要等待一段时间，在克隆完成后，单击"关闭"按钮。
（6）打开 IP 地址配置文件 ifcfg-ens33，修改克隆后的虚拟机 IP 地址。

```
[root@hadoop100 /]#vim /etc/sysconfig/network-scripts/ifcfg-ens33
```

将 IPADDR 修改为想要设置的 IP 地址。

```
IPADDR=192.168.10.102        #IP 地址
```

按照 2.4.3 节中主机名的配置方法修改虚拟机的主机名。

重新启动服务器，按照上述操作克隆 3 台虚拟机，将其分别命名为 hadoop102、hadoop103、hadoop104，主机名和 IP 地址分别与 2.4.3 节中的 hosts 文件设置一一对应。

2.5 本章总结

本章主要介绍了大数据开发过程中经常使用的环境搭建方法，为后续内容的学习打下环境基础。本章以实践为主，读者要严格按照步骤执行，细心为上。在搭建环境的过程中，熟悉 Linux 操作系统的命令行和文件系统结构，是学习大数据相关知识的第一步。

第3章 Hadoop 快速上手

Hadoop 的诞生，在存储和计算方面为大数据领域提供了可大规模扩展的可能性。Hadoop 是以集群的形式对外提供优秀的存储和计算性能的。本章主要讲解如何搭建一个 Hadoop 集群。在搭建 Hadoop 集群前，读者首先要对 Hadoop 的集群角色有所了解，前面已经简单讲解过 Hadoop 的主要组成部分，我们了解到，HDFS 和 YARN 是 Hadoop 的核心组成部分，那么 HDFS 和 YARN 具体是如何运作的呢？通过对本章内容的学习，读者对此就可见一斑了。

3.1 集群角色

在搭建 Hadoop 集群前，首先对 Hadoop 集群的主要角色进行简单的介绍，帮助读者进行集群规划、理解配置文件的含义。

3.1.1 Hadoop 集群的主要角色

HDFS，全称 Hadoop Distributed File System，意为 Hadoop 分布式文件系统。HDFS 整体上是一个主从式的体系结构，由一个主节点 NameNode 和若干个从节点 DataNode 构成。在实际的文件读/写过程中，还会存在一个客户端，用于在用户与 HDFS 之间进行交互。HDFS 集群的主要角色如下。

1. NameNode

NameNode 简称 NN，主要用于存储文件的元数据。元数据是指描述数据的数据，如文件名、文件目录结构、文件属性（生成时间、副本数量、文件权限），以及每个文件的块列表和块所在的 DataNode 等信息。

2. DataNode

DataNode 简称 DN，主要用于存储文件数据。DataNode 实际上实现了 HDFS 的文件存储功能，在本地文件系统中存储数据块及数据块的校验和。

3. SecondaryNameNode

SecondaryNameNode 简称 2NN，每隔一段时间会对 NameNode 中的元数据进行备份，在 NameNode 发生故障时，可以辅助 NameNode 进行故障恢复，但是在实际运行过程中并不承担 NameNode 的工作任务。

4. Client

Client（客户端）主要用于在用户与 NameNode 和 DataNode 之间进行交互，开发人员实际上是面向 Client 进行编程的。

3.1.2 YARN 的主要组成部分

YARN（Yet Another Resource Negotiator，另一种资源协调者）是 Hadoop 的资源管理器，主要负责 Hadoop

数据计算过程中的资源调度工作，也可以为其他部署在 Hadoop 集群上的计算引擎提供资源调度服务。YARN 的基本架构如图 3-1 所示。

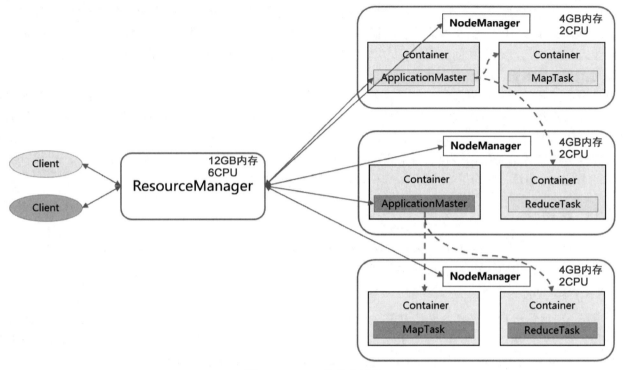

图 3-1　YARN 的基本架构

1. ResourceManager

ResourceManager 简称 RM，主要负责管理和分配全局资源（内存、CPU 等）。ResourceManager 主要由调度器（Scheduler）和应用程序管理器（Applications Manager）构成。调度器主要负责根据限定条件（资源量、队列等）将系统的资源分配给各个正在运行的应用程序。应用程序管理器主要负责管理整个系统中的所有应用程序，包括应用程序的提交、与调度器协调资源、启动 ApplicationMaster 等。

2. NodeManager

NodeManager 简称 NM，是每个节点服务器中的资源和任务管理器。NodeManager 一方面需要定时向 ResourceManager 汇报本节点上的资源使用情况和各个 Container（容器）的运行状态，另一方面需要接收并处理来自 ApplicationMaster 的启动 Container、停止 Container 等请求。

3. ApplicationMaster

ApplicationMaster 简称 AM，主要负责管理用户提交的具体作业。用户每提交一个作业，都会开启一个 ApplicationMaster，负责与 RM 调度器协商，从而获取 Container 资源，划分并分配任务；还负责与 NM 通信，从而启动和停止任务，监控任务运行状态。集群上可以运行多个 ApplicationMaster。

4. Container

Container 是 YARN 中的资源抽象，主要用于封装节点服务器中的内存、CPU 等资源。当 AM 向 RM 申请资源时，RM 为 AM 返回的资源就是用 Container 表示的。每个 NodeManager 中都可以有多个 Container。

5. Client

Client 主要负责向 YARN 集群提交任务。Client 也可以有多个。

3.2 本地模式

Hadoop 为用户提供了灵活的运行模式，主要包括本地模式、伪分布式模式和完全分布式模式。

本地模式是指 Hadoop 在本地单机运行，主要用于进行程序的简单调试和示例程序的演示，在企业的实际生产环境中，没有太高的应用价值。本节主要介绍本地模式的相关知识，读者对本地模式有简单的了解即可，不要求掌握。

伪分布式模式是指单机运行，但是具备 Hadoop 集群的所有角色和功能，相当于使用一台服务器模拟一个分布式环境。在企业的实际生产环境中，同样没有太高的应用价值，只能做简单的程序测试。本书不对此模式进行具体讲解。

完全分布式模式是指由多台服务器构成的分布式环境。多台服务器担任不同的集群角色，构成一组分布式的 Hadoop 集群，是企业在实际生产环境中主要使用的模式。

3.2.1 安装

本地模式的安装非常简便，直接将安装包解压即可。Hadoop 的安装包可以从 Hadoop 的官方网站获取，从本书附赠的资料包中也可找到。本书采用的是 Hadoop 3.1.3。

1．安装 JDK

在安装 Hadoop 前，需要先安装 JDK。JDK 是 Java 的开发工具箱，是整个 Java 的核心，包括 Java 运行环境、Java 工具和 Java 基础类库。JDK 是学习大数据技术的基础，即将搭建的 Hadoop 分布式集群的安装程序就是用 Java 开发的，要使 Hadoop 分布式集群正常运行，必须安装 JDK。

（1）卸载现有的 JDK。

① 检查计算机中是否已经安装了 JDK。

```
[atguigu@hadoop102 opt]$ rpm -qa | grep java
```

② 如果安装的 JDK 版本低于 JDK 1.7，则卸载该 JDK。

```
[atguigu@hadoop102 opt]$ sudo rpm -e 软件包
```

（2）在下载 JDK 的安装包后，将其上传至服务器中，可以直接将安装包拖曳至目标文件夹/opt/software 的对话框中，如图 3-2 所示。

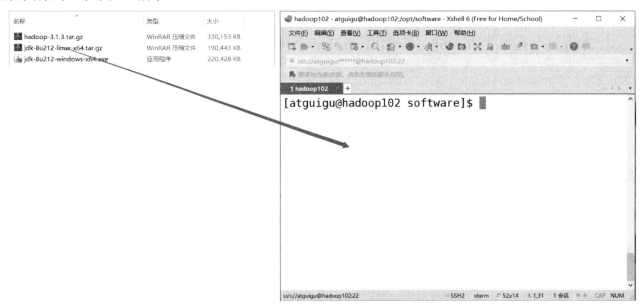

图 3-2　拖曳安装包至指定路径下

① 在 Linux 的 opt 目录下查看软件包是否导入成功。
```
[atguigu@hadoop102 opt]$ cd software/
[atguigu@hadoop102 software]$ ls
jdk-8u212-linux-x64.tar.gz
```
② 将 JDK 解压缩到/opt/module 目录下，tar 命令主要用于解压缩.tar 或.tar.gz 格式的压缩包，-z 选项主要用于指定解压缩.tar.gz 格式的压缩包，-f 选项主要用于指定解压缩文件，-x 选项主要用于指定解压缩操作，-v 选项主要用于显示解压缩过程，-C 选项主要用于指定解压缩路径。
```
[atguigu@hadoop102 software]$ tar -zxvf jdk-8u144-linux-x64.tar.gz -C /opt/module/
```
（3）配置 JDK 环境变量，方便使用 JDK 的程序调用 JDK。
① 获取 JDK 路径。
```
[atgui@hadoop102 jdk1.8.0_144]$ pwd
/opt/module/jdk1.8.0_212
```
② 新建/etc/profile.d/my_env.sh 文件。需要注意的是，/etc/profile.d 路径属于 root 用户，需要使用 sudo vim 命令对其进行编辑。
```
[atguigu@hadoop102 software]$ sudo vim /etc/profile.d/my_env.sh
```
在 profile 文件末尾添加 JDK 路径，添加的内容如下：
```
#JAVA_HOME
export JAVA_HOME=/opt/module/jdk1.8.0_144
export PATH=$PATH:$JAVA_HOME/bin
```
保存文件并退出。
```
:wq
```
③ 在修改环境变量后，执行 source 命令，使修改后的文件生效。
```
[atguigu@hadoop102 jdk1.8.0_144]$ source /etc/profile.d/my_env.sh
```
（4）通过执行 java -version 命令，测试 JDK 是否安装成功。
```
[atguigu@hadoop102 jdk1.8.0_144]# java -version
java version "1.8.0_144"
```
如果在执行 java -version 命令后无法显示 Java 版本，则执行以下命令，重启服务器。
```
[atguigu@hadoop102 jdk1.8.0_144]$ sync
[atguigu@hadoop102 jdk1.8.0_144]$ sudo reboot
```

2. 安装 Hadoop 本地模式

（1）首先需要将 Hadoop 的安装包上传至服务器的安装包存储路径/opt/software 下，具体操作方式与上传 JDK 安装包相同。

（2）进入 Hadoop 安装包的存储路径，将 Hadoop 安装包解压缩至安装路径/opt/module 下。
```
[atguigu@hadoop102 ~]$ cd /opt/software/
[atguigu@hadoop102 software]$ tar -zxvf hadoop-3.1.3.tar.gz -C /opt/module/
```
（3）查看是否解压缩成功。
```
[atguigu@hadoop102 software]$ ls /opt/module/
hadoop-3.1.3
```
（4）将 Hadoop 的安装路径添加至环境变量中。
① 获取 Hadoop 的安装路径。
```
[atguigu@hadoop102 hadoop-3.1.3]$ pwd
/opt/module/hadoop-3.1.3
```
② 打开/etc/profile.d/my_env.sh 文件。
```
[atguigu@hadoop102 hadoop-3.1.3]$ sudo vim /etc/profile.d/my_env.sh
```

③ 在 my_env.sh 文件末尾添加以下内容。
```
#HADOOP_HOME
export HADOOP_HOME=/opt/module/hadoop-3.1.3
export PATH=$PATH:$HADOOP_HOME/bin
export PATH=$PATH:$HADOOP_HOME/sbin
```
④ 执行:wq 命令，保存文件并退出。

（5）执行以下命令，使修改后的环境变量文件生效。
```
[atguigu@hadoop102 hadoop-3.1.3]$ source /etc/profile
```
（6）执行以下命令，测试是否安装成功，以及环境变量配置是否成功。
```
[atguigu@hadoop102 hadoop-3.1.3]$ hadoop version
Hadoop 3.1.3
```
（7）如果上述测试命令不能成功执行，则执行以下命令，尝试重启虚拟机。
```
[atguigu@hadoop102 hadoop-3.1.3]$ sudo reboot
```
（8）执行以下命令，查看 Hadoop 的目录结构。
```
[atguigu@hadoop102 hadoop-3.1.3]$ ll
总用量 52
drwxr-xr-x. 2 atguigu atguigu  4096 5月  22 2017 bin
drwxr-xr-x. 3 atguigu atguigu  4096 5月  22 2017 etc
drwxr-xr-x. 2 atguigu atguigu  4096 5月  22 2017 include
drwxr-xr-x. 3 atguigu atguigu  4096 5月  22 2017 lib
drwxr-xr-x. 2 atguigu atguigu  4096 5月  22 2017 libexec
-rw-r--r--. 1 atguigu atguigu 15429 5月  22 2017 LICENSE.txt
-rw-r--r--. 1 atguigu atguigu   101 5月  22 2017 NOTICE.txt
-rw-r--r--. 1 atguigu atguigu  1366 5月  22 2017 README.txt
drwxr-xr-x. 2 atguigu atguigu  4096 5月  22 2017 sbin
drwxr-xr-x. 4 atguigu atguigu  4096 5月  22 2017 share
```
目录说明如下。

① bin 目录：存储对 Hadoop 相关服务（HDFS、YARN、MapReduce）进行操作的脚本。

② etc 目录：Hadoop 的配置文件目录，存储 Hadoop 的配置文件。

③ lib 目录：存储 Hadoop 的本地库（对数据进行压缩及解压缩的功能包）。

④ sbin 目录：存储启动或停止 Hadoop 相关服务的脚本。

⑤ share 目录：存储 Hadoop 的依赖 jar 包、文档和官方示例程序。

3.2.2 运行官方示例程序

Hadoop 安装包提供了多个官方示例程序，方便用户在未编写 Hadoop 代码的情况下进行测试，本节带领大家运行一下 Hadoop 官方的词频统计程序 WordCount。

（1）在 hadoop-3.1.3 文件夹中创建一个文件夹 wcinput。
```
[atguigu@hadoop102 hadoop-3.1.3]$ mkdir wcinput
```
（2）进入 wcinput 文件夹。
```
[atguigu@hadoop102 hadoop-3.1.3]$ cd wcinput
```
（3）创建一个 word.txt 文件，并且编辑该文件。
```
[atguigu@hadoop102 wcinput]$ vim word.txt
```
① 在 word.txt 文件中输入以下内容。
```
hadoop yarn
hadoop mapreduce
```

```
atguigu
atguigu
```

② 执行:wq 命令，保存文件并退出。

（4）返回 Hadoop 的/opt/module/hadoop-3.1.3 目录，执行以下命令，运行官方提供的示例程序 jar 包中的 WordCount 程序。

```
[atguigu@hadoop102 hadoop-3.1.3]$ hadoop jar share/hadoop/mapreduce/hadoop-mapreduce-examples-3.1.3.jar wordcount wcinput wcoutput
```

（5）查看 WordCount 程序的运行结果。

```
[atguigu@hadoop102 hadoop-3.1.3]$ cat wcoutput/part-r-00000
```

如果看到以下结果，则表示运行成功。

```
atguigu    2
hadoop     2
mapreduce  1
yarn       1
```

3.3 完全分布式模式

完全分布式模式是指由多台服务器构成分布式环境。要搭建一组 Hadoop 完全分布式集群，至少需要 3 台节点服务器，这 3 台节点服务器要满足以下条件。

- 关闭防火墙。
- 配置静态 IP 地址。
- 配置统一的主机名与 IP 地址映射文件。
- 安装 JDK 并配置 JAVA_HOME 环境变量。

关于虚拟机的配置工作，在第 2 章中已经做过详细讲解。本节主要讲解 Hadoop 完全分布式模式的完整配置过程。

3.3.1 SSH 免密登录

为什么需要配置免密登录呢？这与 Hadoop 分布式集群的架构有关。我们搭建的 Hadoop 分布式集群属于主从架构，在配置节点服务器之间免密登录后，即可通过主节点服务器启动从节点服务器，无须手动输入用户名和密码。

第 1 步：配置 SSH。

（1）基本语法：假设要以用户名 user 登录远程主机 host，那么只需输入"ssh user@host"，如输入"ssh atguigu@192.168.10.100"，如果本地用户名与远程用户名一致，那么在登录时可以省略用户名，即输入"ssh host"。

（2）如果在进行 SSH 连接时出现错误提示"Host key verification failed"，那么直接输入"yes"即可。

```
[atguigu@hadoop102 opt] $ ssh 192.168.10.103
The authenticity of host '192.168.10.103 (192.168.10.103)' can't be established.
RSA key fingerprint is cf:1e:de:d7:d0:4c:2d:98:60:b4:fd:ae:b1:2d:ad:06.
Are you sure you want to continue connecting (yes/no)?
Host key verification failed.
```

第 2 步：无密钥配置。

（1）免密登录的原理如图 3-3 所示。

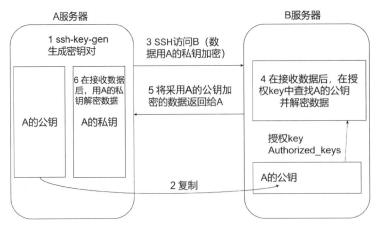

图 3-3 免密登录的原理

（2）生成公钥和私钥。
```
[atguigu@hadoop102 .ssh]$ ssh-keygen -t rsa
```
然后连续按 3 次回车键，会生成两个文件：id_rsa（私钥）、id_rsa.pub（公钥）。

（3）将公钥复制到要免密登录的目标节点服务器中。
```
[atguigu@hadoop102 .ssh]$ ssh-copy-id hadoop102
[atguigu@hadoop102 .ssh]$ ssh-copy-id hadoop103
[atguigu@hadoop102 .ssh]$ ssh-copy-id hadoop104
```

注意：在 hadoop102 节点服务器上对 root 账号进行配置，使其可以免密登录到 hadoop102、hadoop103 和 hadoop104 节点服务器上；在 hadoop103 节点服务器上对 atguigu 账号进行配置，使其可以免密登录到 hadoop102、hadoop103 和 hadoop104 节点服务器上。

.ssh 文件夹中文件的功能如下。
- known_hosts：记录 SSH 访问过的计算机公钥。
- id_rsa：生成的私钥。
- id_rsa.pub：生成的公钥。
- authorized_keys：存储授权过的免密登录服务器公钥。

3.3.2 shell 脚本准备

有时，直接使用 Linux 自带的命令行会比较烦琐，可能需要重复输入一些命令参数。为了执行命令更加方便、快捷，可以编写一些通用脚本。本节带领读者编写两个非常实用的开发脚本，分别是集群间文件分发脚本和集群命令同时执行脚本。

1．集群间文件分发脚本 xsync

集群间数据复制的两个通用命令是 scp 和 rsync，其中，使用 rsync 命令可以只对差异文件进行更新，非常方便，但在使用时需要操作者频繁输入各种命令参数，为了能够更方便地使用该命令，下面编写一个集群间文件分发脚本，用于实现目前集群间的数据分发功能。

第 1 步：脚本需求分析。循环复制文件到所有节点服务器的相同目录下。
（1）原始复制。
```
rsync -rv /opt/module root@hadoop103:/opt/
```
（2）期望脚本效果。
```
xsync path/filename #要同步的文件路径或文件名
```
（3）在/home/atguigu/bin 目录下存储的脚本，atguigu 用户可以在操作系统的任何地方直接执行。

第 2 步：脚本实现。
（1）在/home/atguigu 目录下创建 bin 文件夹，并且使用 vim 命令在 bin 文件夹中创建一个脚本文件

xsync，该文件中的内容如下：
```
[atguigu@hadoop102 ~]$ mkdir bin
[atguigu@hadoop102 ~]$ cd bin/
[atguigu@hadoop102 bin]$ touch xsync
[atguigu@hadoop102 bin]$ vim xsync
#!/bin/bash
#获取输入参数的个数，如果没有参数，则直接退出
pcount=$#
if((pcount==0)); then
echo no args;
exit;
fi

#获取文件名称
p1=$1
fname=`basename $p1`
echo fname=$fname

#获取上级目录的绝对路径
pdir=`cd -P $(dirname $p1); pwd`
echo pdir=$pdir

#获取当前用户名
user=`whoami`

#循环
for((host=103; host<105; host++)); do
        echo -------------------- hadoop$host ----------------
        rsync -rvl $pdir/$fname $user@hadoop$host:$pdir
done
```

（2）修改脚本文件 xsync，使其具有执行权限。
```
[atguigu@hadoop102 bin]$ chmod 777 xsync
```
（3）调用脚本文件 xsync 的形式为该文件的路径，具体代码如下：
```
[atguigu@hadoop102 bin]$ xsync /home/atguigu/bin
```

2．集群命令同时执行脚本

在启动集群后，用户需要使用 jps 命令查看各节点服务器进程的启动情况，操作起来比较麻烦，所以我们通过编写一个集群命令同时执行脚本，达到使用一个脚本查看所有节点上的所有进程的目的。使用该脚本，还可以执行一些需要同时在集群不同节点上运行的命令。

（1）在/home/atguigu/bin 目录下创建 xcall.sh 脚本。
```
[atguigu@hadoop102 bin]$ vim xcall.sh
```
（2）xcall.sh 脚本的思路：通过 i 变量在 hadoop102、hadoop103 和 hadoop104 节点服务器之间遍历，分别使用 ssh 命令进入 3 台节点服务器，执行传入指定参数的命令。

在 xcall.sh 脚本中编写以下内容。
```
#! /bin/bash

for i in hadoop102 hadoop103 hadoop104
do
        echo --------- $i ----------
```

```
        ssh $i "$*"
done
```

（3）增加 xcall.sh 脚本的执行权限。

```
[atguigu@hadoop102 bin]$ chmod 777 xcall.sh
```

（4）启动 xcall.sh 脚本。

```
[atguigu@hadoop102 bin]$ xcall.sh jps
```

3.3.3 集群配置

1．集群部署规划

在搭建一个集群前，需要对集群的角色进行恰当的规划——哪些角色需要部署在哪些节点服务器上。在 3.1 节中，我们已经对 Hadoop 的集群角色做过简单介绍，HDFS 的主要角色包括 NameNode、DataNode、SecondaryNameNode 和 Client。其中，DataNode 主要负责数据的存储工作，所以需要部署在每一台节点服务器上，SecondaryNameNode 主要负责在集群遇到故障时，协助 NameNode 进行故障恢复，所以不推荐与 NameNode 位于同一台节点服务器上。YARN 的主要角色包括 ResourceManager 和 NodeManager。其中，NodeManager 是单个节点服务器上的资源和任务管理器，所以需要将其部署在每一台节点服务器上；ResourceManager 主要负责集群整体的资源调度工作，是非常耗费内存资源的，所以不要将其与同样耗费内存资源的 NameNode 配置在同一台节点服务器上。综上所示，集群的整体部署规划如表 3-1 所示。

表 3-1 集群的整体部署规划

节点服务器	hadoop102	hadoop103	hadoop104
HDFS	NameNode DataNode	DataNode	SecondaryNameNode DataNode
YARN	NodeManager	ResourceManager NodeManager	NodeManager

2．配置文件说明

Hadoop 配置文件分为两类：默认配置文件和自定义配置文件，默认配置文件是 Hadoop 源码中自带的，提供了所有参数的默认值，如果用户要修改默认值，则需要在自定义配置文件中修改。自定义配置文件的优先级高于默认配置文件。默认配置文件包括 core-default.xml、hdfs-default.xml、yarn-default.xml 和 mapred-default.xml，可以在官网主页通过 Getting Started→Configuration 命令找到，官网对配置文件中的所有参数均给出了详细解释。自定义配置文件包括 core-site.xml、hdfs-site.xml、yarn-site.xml 和 mapred-site.xml。在生产环境中，通常会根据用户需求在 4 个自定义配置文件中调整默认参数值。

3．配置集群

1）core-site.xml 文件。

core-site.xml 文件是核心配置文件，该文件属于 Hadoop 的全局配置文件，主要用于将分布式文件系统 HDFS 的 NameNode 的入口地址和分布式文件系统中的数据存储于服务器本地磁盘中进行配置。

找到 core-site.xml 文件。

```
[atguigu@hadoop102 ~]$ cd $HADOOP_HOME/etc/hadoop
[atguigu@hadoop102 hadoop]$ vim core-site.xml
```

修改配置如下：

```
<?xml version="1.0" encoding="UTF-8"?>
<?xml-stylesheet type="text/xsl" href="configuration.xsl"?>


<configuration>
    <!-- 指定 NameNode 的地址 -->
```

```xml
    <property>
        <name>fs.defaultFS</name>
        <value>hdfs://hadoop102:8020</value>
    </property>

    <!-- 指定 Hadoop 数据的存储目录 -->
    <property>
        <name>hadoop.tmp.dir</name>
        <value>/opt/module/hadoop-3.1.3/data</value>
    </property>

    <!-- 配置 HDFS 网页登录使用的静态用户为 atguigu -->
    <property>
        <name>hadoop.http.staticuser.user</name>
        <value>atguigu</value>
    </property>
</configuration>
```

2）hdfs-site.xml 文件。

hdfs-site.xml 文件是关于 HDFS 的配置文件，在该文件中，我们主要对 HDFS 的属性进行配置。

找到 hdfs-site.xml 文件。

```
[atguigu@hadoop102 hadoop]$ vim hdfs-site.xml
```

修改配置如下：

```xml
<?xml version="1.0" encoding="UTF-8"?>
<?xml-stylesheet type="text/xsl" href="configuration.xsl"?>

<configuration>
    <!-- NN Web 端访问地址-->
    <property>
        <name>dfs.namenode.http-address</name>
        <value>hadoop102:9870</value>
    </property>
    <!-- 2NN Web 端访问地址-->
    <property>
        <name>dfs.namenode.secondary.http-address</name>
        <value>hadoop104:9868</value>
    </property>
</configuration>
```

3）yarn-site.xml 文件。

yarn-site.xml 文件是关于 YARN 的配置文件。

找到 yarn-site.xml 文件。

```
[atguigu@hadoop102 hadoop]$ vim yarn-site.xml
```

修改配置如下：

```xml
<?xml version="1.0" encoding="UTF-8"?>
<?xml-stylesheet type="text/xsl" href="configuration.xsl"?>

<configuration>
    <!— NodeManager 辅助服务 -->
    <property>
        <name>yarn.nodemanager.aux-services</name>
        <value>mapreduce_shuffle</value>
```

```xml
        </property>

        <!-- 指定ResourceManager的地址-->
        <property>
            <name>yarn.resourcemanager.hostname</name>
            <value>hadoop103</value>
        </property>

        <!-- 环境变量的继承 -->
        <property>
            <name>yarn.nodemanager.env-whitelist</name>
            <value>JAVA_HOME,HADOOP_COMMON_HOME,HADOOP_HDFS_HOME,HADOOP_CONF_DIR,CLASSPATH_PREPEND_DISTCACHE,HADOOP_YARN_HOMEID/value>
        </property>
</configuration>
```

4) mapred-site.xml 文件。

mapred-site.xml 文件是关于 MapReduce 的配置文件,主要配置一个参数,用于指定 MapReduce 的运行框架为 YARN。

找到 mapred-site.xml 文件。

```
[atguigu@hadoop102 hadoop]$ vim mapred-site.xml
```

修改配置如下:

```xml
<?xml version="1.0" encoding="UTF-8"?>
<?xml-stylesheet type="text/xsl" href="configuration.xsl"?>

<configuration>
    <!-- 指定MapReduce程序运行在YARN上 -->
    <property>
        <name>mapreduce.framework.name</name>
        <value>yarn</value>
    </property>
</configuration>
```

5) workers 文件。

主节点服务器 NameNode 和 ResourceManager 的角色已经在配置文件中进行了配置,下面配置从节点服务器的角色。配置文件 workers 主要用于配置 Hadoop 分布式集群中各个从节点服务器的角色。对 workers 文件进行修改,将 3 台节点服务器全部指定为从节点服务器,启动 DataNode 和 NodeManager 进程。

```
[atguigu@hadoop102 hadoop]$ vim workers
```

在 workers 文件中增加以下内容。

```
hadoop102
hadoop103
hadoop104
```

注意:workers 文件中添加的内容结尾不允许有空格,文件中不允许有空行。

4. 分发安装包

(1) 使用分发脚本将已经配置好的 Hadoop 安装包同步至所有节点服务器上。

```
[atguigu@hadoop102 ~]$ xsync /opt/module/hadoop-3.1.3
```

在分发完成后,在 hadoop103 和 hadoop104 节点服务器上查看 Hadoop 安装包的分发情况。

```
[atguigu@hadoop103 ~]$ cd /opt/module/hadoop-3.1.3
[atguigu@hadoop104 ~]$ cd /opt/module/hadoop-3.1.3
```

（2）将 JDK 安装包分发至所有节点服务器上。JDK 是 Hadoop 程序能够运行的必要条件。
```
[atguigu@hadoop102 ~]$ xsync /opt/module/jdk1.8.0_212
```
在分发完成后，在 hadoop103 和 hadoop104 节点服务器上查看 JDK 安装包的分发情况。
```
[atguigu@hadoop103 ~]$ cd /opt/module/jdk1.8.0_212
[atguigu@hadoop104 ~]$ cd /opt/module/jdk1.8.0_212
```
（3）将环境变量文件分发至其他节点服务器上。
```
[atguigu@hadoop102 hadoop]$ sudo /home/atguigu/bin/xsync /etc/profile.d/my_env.sh
```
（4）执行以下命令，使环境变量生效。
```
[atguigu@hadoop103 ~]$ source /etc/profile
[atguigu@hadoop104 ~]$ source /etc/profile
```

5．启动集群

（1）如果第一次启动集群，则需要在 hadoop102 节点服务器上格式化 NameNode。
```
[atguigu@hadoop102 hadoop-3.1.3]$ hdfs namenode -format
```
（2）在格式化 NameNode 后，执行 start-dfs.sh 命令启动 HDFS，即可同时启动所有的 DataNode 和 SecondaryNameNode。
```
[atguigu@hadoop102 hadoop-3.1.3]$ sbin/start-dfs.sh
```
在上述命令执行成功后，查看所有节点服务器的进程情况。
```
[atguigu@hadoop102 hadoop-3.1.3]$ xcall.sh jps
--------- hadoop102 ----------
7349 Jps
6919 NameNode
7087 DataNode
--------- hadoop103 ----------
5856 Jps
5776 DataNode
--------- hadoop104 ----------
5922 SecondaryNameNode
5794 DataNode
5972 Jps
```
（3）通过执行 start-yarn.sh 命令启动 YARN，可以同时启动 ResourceManager 和所有的 NodeManager。需要注意的是，如果 NameNode 和 ResourceManager 不在同一台节点服务器上，则不能在 NameNode 上启动 YARN，应该在 ResourceManager 所在的节点服务器上启动 YARN。
```
[atguigu@hadoop103 hadoop-3.1.3]$ sbin/start-yarn.sh
```
在上述命令执行成功后，查看所有节点服务器的进程情况，可以发现，目前的进程情况与本节开始所做的集群规划是一致的。
```
[atguigu@hadoop102 hadoop-3.1.3]$ xcall.sh jps
--------- hadoop102 ----------
8210 NodeManager
6919 NameNode
8350 Jps
7087 DataNode
--------- hadoop103 ----------
6352 ResourceManager
5776 DataNode
6856 Jps
6490 NodeManager
--------- hadoop104 ----------
6416 NodeManager
```

```
5922 SecondaryNameNode
5794 DataNode
6552 Jps
```

（4）在 Web 端输入之前配置的 NameNode 的节点服务器地址和端口 9870，可以查看 HDFS 的相关信息。例如，在浏览器中输入"http://hadoop102:9870"，可以检查 NameNode 和 DataNode 是否正常。NameNode 的 Web 端如图 3-4 所示。

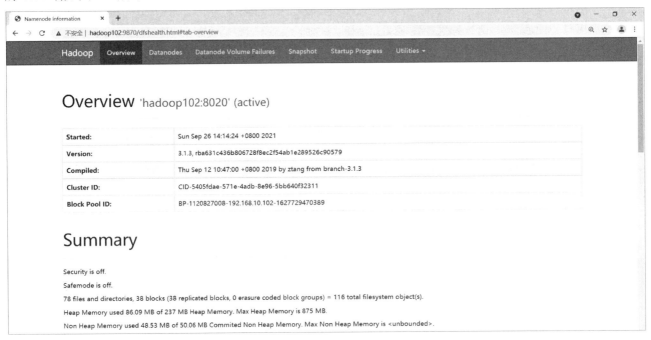

图 3-4　NameNode 的 Web 端

（5）通过在 Web 端输入 ResourceManager 的 IP 地址和端口 8088，可以查看 YARN 上的任务运行情况。例如，在浏览器中输入"http://hadoop103:8088"，可以查看本集群 YARN 上的任务运行情况。YARN 的 Web 端如图 3-5 所示。

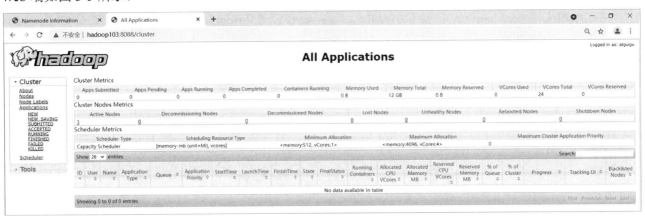

图 3-5　YARN 的 Web 端

6．集群基本测试

在集群初步搭建完成后，需要对 Hadoop 集群的基本功能进行简单的测试。Hadoop 的基本功能包括数据存储和数据计算，测试主要围绕文件上传、文件下载和简单计算三方面展开。

1）文件上传测试。

① 小文件上传测试。使用 Hadoop 的 shell 操作命令将 3.2.2 节中创建的小文件 word.txt 上传至集群

中。关于 Hadoop 的 shell 操作命令，将在第 4 章进行详细介绍。

执行以下命令，在 Hadoop 集群中创建 input 文件夹。

```
[atguigu@hadoop102 ~]$ hadoop fs -mkdir /input
```

执行以下命令，将服务器本地的 word.txt 文件上传至集群的 input 文件夹中。

```
[atguigu@hadoop102 ~]$ hadoop fs -put $HADOOP_HOME/wcinput/word.txt /input
```

② 大文件上传测试。Hadoop 集群中通常会存储一些很大的文件，所以需要对大文件的上传进行测试，下面以上传 JDK 安装包为例进行讲解。

执行以下命令，将 JDK 安装包上传至 Hadoop 的根目录下。

```
[atguigu@hadoop102 ~]$ hadoop fs -put /opt/software/jdk-8u212-linux-x64.tar.gz /
```

在上传 JDK 安装包后，登录 NameNode 的 Web 端，查看 JDK 安装包的上传情况，如图 3-6 所示，可以看到 JDK 安装包已经成功上传到了集群中。

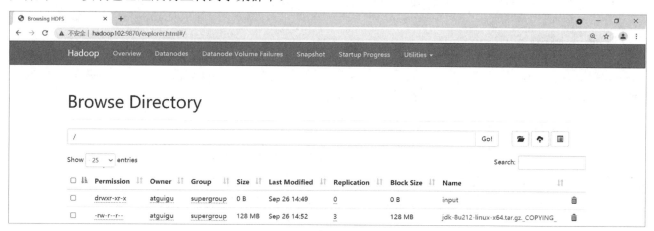

图 3-6 在 NameNode 的 Web 端查看 JDK 安装包的上传情况

③ 上传文件的存储位置。

执行以下命令，进入 HDFS 文件的存储路径。

```
[atguigu@hadoop102 subdir0]$ cd
/opt/module/hadoop-3.1.3/data/dfs/data/current/BP-1436128598-192.168.10.102-1610603650062/current/finalized/subdir0/subdir0
```

在进入 HDFS 文件的存储路径后，执行以下命令，查看该路径下的文件列表，其中的 blk_********* 文件就是存储于 Hadoop 中的数据块文件（文件在 HDFS 中是分块存储的）。

```
[atguigu@hadoop102 subdir0]$ ll
总用量 1364
-rw-rw-r--. 1 atguigu atguigu       330 7月  31 23:20 blk_1073741825
-rw-rw-r--. 1 atguigu atguigu       300 7月  31 23:20 blk_1073741825_1028.meta
-rw-rw-r--. 1 atguigu atguigu 134217728 7月  31 23:21 blk_1073741836
-rw-rw-r--. 1 atguigu atguigu   1048583 7月  31 23:21 blk_1073741836_1040.meta
-rw-rw-r--. 1 atguigu atguigu  63439959 7月  31 23:21 blk_1073741837
-rw-rw-r--. 1 atguigu atguigu    495635 7月  31 23:21 blk_1073741837_1051.meta
```

④ 使用 cat 命令查看 HDFS 在磁盘中存储的文件内容，其中，较小的数据块 blk_1073741825 即之前上传的 word.txt 文件，可以直接使用 cat 命令查看该文件中的内容。

```
[atguigu@hadoop102 subdir0]$ cat blk_1073741825
hadoop yarn
hadoop mapreduce
atguigu
atguigu
```

⑤ 两个较大的数据块 blk_1073741836 和 blk_1073741837 即之前上传的 JDK 安装包，由于 JDK 安装

包体积较大，因此 HDFS 将其切分成了两个数据块进行存储（具体以何种标准切分，将在后面的内容中详细讲解）。将数据块 blk_1073741836 和 blk_1073741837 重新拼接成一个压缩包，将拼接的压缩包解压缩，即可得到原来的 JDK 安装包。

```
-rw-rw-r--. 1 atguigu atguigu 134217728 5月  23 16:01 blk_1073741836
-rw-rw-r--. 1 atguigu atguigu   1048583 5月  23 16:01 blk_1073741836_1012.meta
-rw-rw-r--. 1 atguigu atguigu  63439959 5月  23 16:01 blk_1073741837
-rw-rw-r--. 1 atguigu atguigu    495635 5月  23 16:01 blk_1073741837_1013.meta
[atguigu@hadoop102 subdir0]$ cat blk_1073741836>>tmp.tar.gz
[atguigu@hadoop102 subdir0]$ cat blk_1073741837>>tmp.tar.gz
[atguigu@hadoop102 subdir0]$ tar -zxvf tmp.tar.gz
```

2）文件下载测试。

执行以下命令，将已经上传至 HDFS 中的文件下载至本地。

```
[atguigu@hadoop104 software]$ hadoop fs -get /jdk-8u212-linux-x64.tar.gz ./
```

3）简单计算测试。

运行官方提供的示例程序 jar 包中的 WordCount 程序。

```
[atguigu@hadoop102 hadoop-3.1.3]$ hadoop jar share/hadoop/mapreduce/hadoop-mapreduce-examples-3.1.3.jar wordcount /input /output
```

3.3.4 NameNode 格式化问题

如果集群安装出错，则需要重新格式化集群。首先停止进程，然后删除各节点上的 data 和 logs 目录，最后对 NameNode 进行格式化。这是因为 NameNode 在第一次格式化并启动后，会产生一个集群 id，DataNode 在启动后，会在/tmp 路径下生成相同的集群 id，如图 3-7 所示。

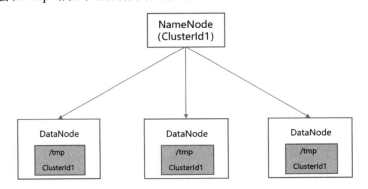

图 3-7　NameNode 与 DataNode 中的集群 id

NameNode 在重新格式化后，会产生新的集群 id，如果 DataNode 上的历史数据未删除，则会与 NameNode 新生成的集群 id 不一致，导致集群不能正常运行，如图 3-8 所示。

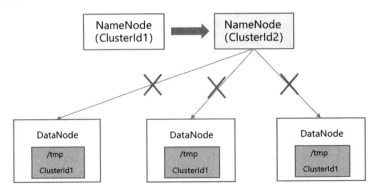

图 3-8　重新格式化后的集群 id 不一致

删除 DataNode 中的旧数据，重新启动 DataNode，会自动生成新的集群 id，如图 3-9 所示。这个操作要谨慎，因为会丢失集群上的数据。

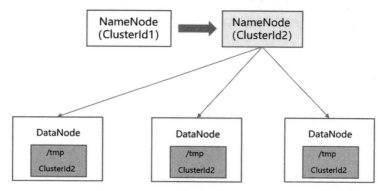

图 3-9　删除 DataNode 中的旧数据并生成新的集群 id

综上所述，正确的格式化集群并重启的操作步骤如下。

（1）停止集群的所有进程。

```
[atguigu@hadoop102 hadoop-3.1.3]$ stop-dfs.sh
[atguigu@hadoop103 hadoop-3.1.3]$ stop-yarn.sh
```

（2）删除所有节点上的 data 和 logs 目录。

```
[atguigu@hadoop102 hadoop-3.1.3]$ rm -rf ./logs ./data
[atguigu@hadoop103 hadoop-3.1.3]$ rm -rf ./logs ./data
[atguigu@hadoop104 hadoop-3.1.3]$ rm -rf ./logs ./data
```

（3）格式化 NameNode。

```
[atguigu@hadoop102 hadoop-3.1.3]$ hdfs namenode -format
```

（4）启动 HDFS 和 YARN。

```
[atguigu@hadoop102 hadoop-3.1.3]$ sbin/start-dfs.sh
[atguigu@hadoop103 hadoop-3.1.3]$ sbin/start-yarn.sh
```

（5）在 Web 端查看 HDFS 中 NameNode 的运行情况，如图 3-10 所示，运行正常。

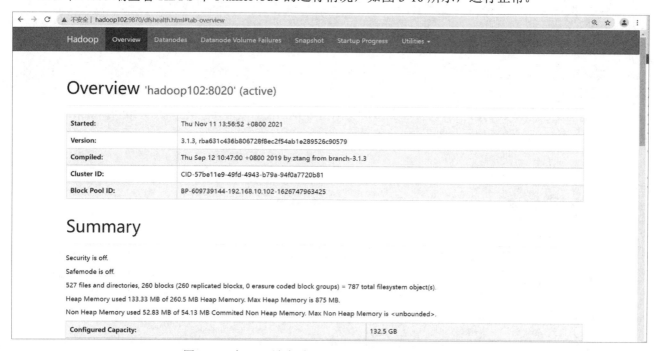

图 3-10　在 Web 端查看 HDFS 中 NameNode 的运行情况

3.3.5 配置历史服务器与日志聚集功能

在 Hadoop 集群上执行完一个计算任务后，所有的运行信息和相关日志都会被清除，用户无法回溯查看历史任务的运行情况。为了方便查看历史任务的运行情况，以及对历史任务的运行情况进行分析，可以配置历史服务器与日志聚集功能。

1．配置历史服务器

为了查看历史任务的运行情况，需要配置历史服务器。历史服务器可以配置在任意一台节点服务器上，我们选择 hadoop102 节点服务器进行配置，具体配置步骤如下。

（1）打开配置文件 mapred-site.xml。

```
[atguigu@hadoop102 hadoop]$ vim mapred-site.xml
```

在 mapred-site.xml 文件中添加以下配置。

```xml
<!-- 历史服务器端地址 -->
<property>
    <name>mapreduce.jobhistory.address</name>
    <value>hadoop102:10020</value>
</property>

<!-- 历史服务器Web端地址 -->
<property>
    <name>mapreduce.jobhistory.webapp.address</name>
    <value>hadoop102:19888</value>
</property>
```

（2）分发配置文件至其他节点服务器中。

```
[atguigu@hadoop102 hadoop]$ xsync $HADOOP_HOME/etc/hadoop/mapred-site.xml
```

（3）在 hadoop102 节点服务器上启动历史服务器。

```
[atguigu@hadoop102 hadoop]$ mapred --daemon start historyserver
```

（4）查看历史服务器是否启动。

```
[atguigu@hadoop102 hadoop]$ jps
10357 DataNode
10824 JobHistoryServer
10986 Jps
10666 NodeManager
10188 NameNode
```

（5）在历史服务器成功启动后，可以登录历史服务器的 Web 端（访问地址为 http://hadoop102:19888/jobhistory），查看历史任务列表，如图 3-11 所示。

图 3-11　查看历史任务列表（一）

2. 配置日志聚集功能

在程序运行完毕后，运行程序的节点服务器上会产生一些本地的日志文件。为了方便查看程序运行情况，进行程序调试，可以在程序运行完毕后，将程序运行的日志信息上传至 HDFS 中，如图 3-12 所示。

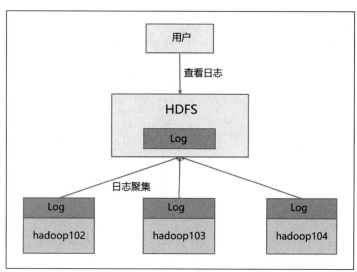

图 3-12　日志聚集功能示意图

需要注意的是，如果要启用日志聚集功能，则需要重新启动 NodeManager、ResourceManager 和 HistoryServer 进程。

启用日志聚集功能的具体步骤如下。

（1）打开配置文件 yarn-site.xml。

```
[atguigu@hadoop102 hadoop]$ vim yarn-site.xml
```

在 yarn-site.xml 文件中添加以下配置。

```xml
<!-- 启用日志聚集功能 -->
<property>
    <name>yarn.log-aggregation-enable</name>
    <value>true</value>
</property>
<!-- 设置日志聚集服务器地址 -->
<property>
    <name>yarn.log.server.url</name>
    <value>http://hadoop102:19888/jobhistory/logs</value>
</property>
<!-- 设置日志保留时间为 7 天 -->
<property>
    <name>yarn.log-aggregation.retain-seconds</name>
    <value>604800</value>
</property>
```

（2）分发配置文件。

```
[atguigu@hadoop102 hadoop]$ xsync $HADOOP_HOME/etc/hadoop/yarn-site.xml
```

（3）关闭 NodeManager、ResourceManager 和 HistoryServer 进程。

```
[atguigu@hadoop103 hadoop-3.1.3]$ sbin/stop-yarn.sh
[atguigu@hadoop103 hadoop-3.1.3]$ mapred --daemon stop historyserver
```

（4）启动 NodeManager、ResourceManager 和 HistoryServer 进程。
```
[atguigu@hadoop103 ~]$ start-yarn.sh
[atguigu@hadoop102 ~]$ mapred --daemon start historyserver
```
（5）删除 HDFS 中已经存在的输出文件。
```
[atguigu@hadoop102 ~]$ hadoop fs -rm -r /output
```
（6）重新运行 WordCount 程序。
```
[atguigu@hadoop102 hadoop-3.1.3]$ hadoop jar share/hadoop/mapreduce/hadoop-mapreduce-examples-3.1.3.jar wordcount /input /output
```
（7）查看日志。

① 访问历史服务器的 Web 端（访问地址为 http://hadoop102:19888/jobhistory），查看历史任务列表，如图 3-13 所示。

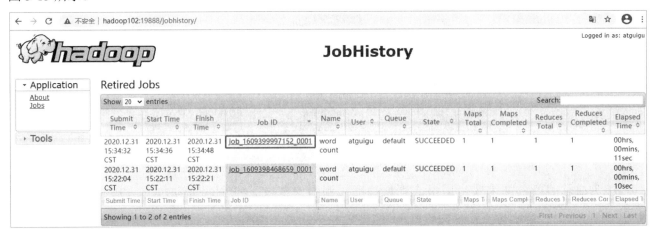

图 3-13　查看历史任务列表（二）

② 在历史任务列表中单击 Job ID，可以查看相应的历史任务详情，如图 3-14 所示。

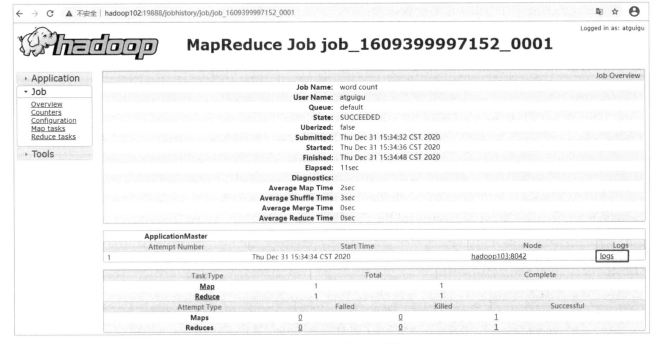

图 3-14　查看历史任务详情

③ 在历史任务详情中单击 logs 超链接，可以查看历史任务的运行日志，如图 3-15 所示。

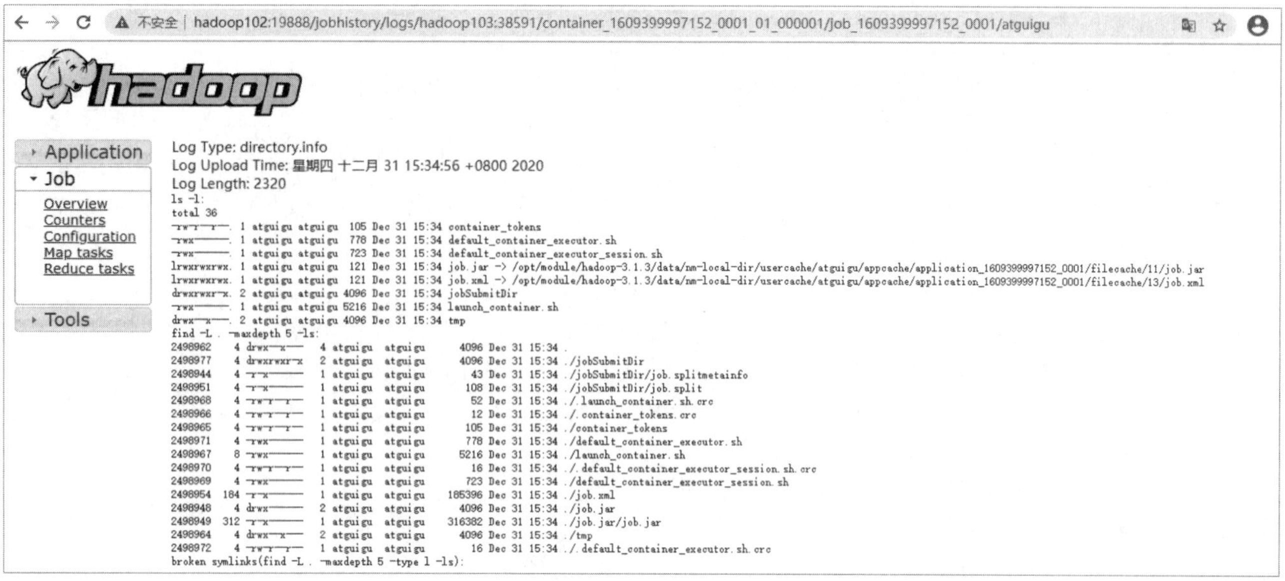

图 3-15　查看历史任务的运行日志

3.3.6　Hadoop 集群启停脚本

通过对前面内容的学习，读者可以了解到，要想完整地启动 Hadoop 集群，需要执行多个启动命令，分别启动 HDFS、YARN 和 HistoryServer 等相关进程。本节编写一个脚本，用于一次性执行这些命令，从而更快捷地启动和关闭 Hadoop 集群。

（1）在/home/atguigu/bin 路径下创建脚本文件 myhadoop.sh。

```
[atguigu@hadoop102 ~]$ cd /home/atguigu/bin
[atguigu@hadoop102 bin]$ vim myhadoop.sh
```

（2）在 myhadoop.sh 脚本文件中输入以下内容。

```
#!/bin/bash

if [ $# -lt 1 ]
then
    echo "No Args Input..."
    exit ;
fi

case $1 in
"start")
        echo " =================== 启动 hadoop集群 ==================="

        echo " --------------- 启动 hdfs ---------------"
        ssh hadoop102 "/opt/module/hadoop-3.1.3/sbin/start-dfs.sh"
        echo " --------------- 启动 yarn ---------------"
        ssh hadoop103 "/opt/module/hadoop-3.1.3/sbin/start-yarn.sh"
        echo " --------------- 启动 historyserver ---------------"
        ssh hadoop102 "/opt/module/hadoop-3.1.3/bin/mapred --daemon start historyserver"
;;
"stop")
        echo " =================== 关闭 hadoop集群 ==================="

        echo " --------------- 关闭 historyserver ---------------"
```

```
        ssh hadoop102 "/opt/module/hadoop-3.1.3/bin/mapred --daemon stop historyserver"
        echo " --------------- 关闭 yarn ---------------"
        ssh hadoop103 "/opt/module/hadoop-3.1.3/sbin/stop-yarn.sh"
        echo " --------------- 关闭 hdfs ---------------"
        ssh hadoop102 "/opt/module/hadoop-3.1.3/sbin/stop-dfs.sh"
;;
*)
    echo "Input Args Error..."
;;
esac
```

(3) 执行:wq 命令，保存文件并退出，然后执行以下命令，赋予脚本执行权限。

```
[atguigu@hadoop102 bin]$ chmod +x myhadoop.sh
```

(4) 脚本测试。

① 使用 myhadoop.sh 脚本文件启动 Hadoop 集群。

```
[atguigu@hadoop102 bin]$ myhadoop.sh start
```

在启动 Hadoop 集群后，查看所有节点服务器上的进程情况。

```
[atguigu@hadoop102 ~]$ xcall.sh jps
--------- hadoop102 ----------
10357 DataNode
10824 JobHistoryServer
10666 NodeManager
11371 Jps
10188 NameNode
--------- hadoop103 ----------
8100 DataNode
8437 NodeManager
8299 ResourceManager
9342 Jps
--------- hadoop104 ----------
8017 NodeManager
8595 Jps
7785 DataNode
7919 SecondaryNameNode
```

② 使用 myhadoop.sh 脚本文件关闭 Hadoop 集群。

```
[atguigu@hadoop102 bin]$ myhadoop.sh stop
```

在关闭 Hadoop 集群后，查看所有节点服务器上的进程情况。

```
[atguigu@hadoop102 ~]$ xcall.sh jps
--------- hadoop102 ----------
12033 Jps
--------- hadoop103 ----------
9751 Jps
--------- hadoop104 ----------
8843 Jps
```

(5) 分发 /home/atguigu/bin 目录，保证所有自定义脚本在 3 台节点服务器上都可以使用。

```
[atguigu@hadoop102 ~]$ xsync /home/atguigu/bin/
```

3.3.7 集群时间同步

即将搭建的 Hadoop 分布式集群需要解决两个问题：数据的存储和数据的计算。

Hadoop 采用分块的方法存储大型文件，它将大型文件切分成多块，以块为单位，将其分发到多台节点服务器中进行存储。当再次访问这个大型文件时，需要从 3 台节点服务器中分别拿出数据，然后进行计算。计算机之间的通信和数据的传输一般是以时间为约定条件的，如果 3 台节点服务器的时间不一致，那么在读取块数据时会出现时间延迟，可能导致访问文件的时间过长，甚至访问文件失败，因此需要配置节点服务器，使不同节点服务器之间的时间同步，具体操作如下。

第 1 步：配置时间服务器（必须是 root 用户）。

（1）检查所有节点服务器的 ntp 服务状态和开机自启动状态。

```
[root@hadoop102 ~]# systemctl status ntpd
[root@hadoop102 ~]# systemctl is-enabled ntpd
```

（2）在所有节点服务器上关闭 ntp 服务和开机自启动功能。

```
[root@hadoop102 ~]# systemctl stop ntpd
[root@hadoop102 ~]# systemctl disable ntpd
```

（3）修改 ntp 服务的配置文件。

```
[root@hadoop102 ~]# vim /etc/ntp.conf
```

修改内容如下。

① 设置本地网络中的主机不受限制，将以下配置前的"#"符号删除，解开此行注释。

```
#restrict 192.168.10.0 mask 255.255.255.0 nomodify notrap
```

② 设置为不采用公共服务器。

```
server 0.centos.pool.ntp.org iburst
server 1.centos.pool.ntp.org iburst
server 2.centos.pool.ntp.org iburst
server 3.centos.pool.ntp.org iburst
```

将上述内容修改如下：

```
#server 0.centos.pool.ntp.org iburst
#server 1.centos.pool.ntp.org iburst
#server 2.centos.pool.ntp.org iburst
#server 3.centos.pool.ntp.org iburst
```

③ 添加一个默认的内部时钟数据，用于为局域网用户提供服务。

```
server 127.127.1.0
fudge 127.127.1.0 stratum 10
```

（4）修改/etc/sysconfig/ntpd 文件。

```
[root@hadoop102 ~]# vim /etc/sysconfig/ntpd
```

增加以下内容（让硬件时间与系统时间同步）。

```
SYNC_HWCLOCK=yes
```

重新启用 ntp 服务。

```
[root@hadoop102 ~]# systemctl status ntpd
ntpd 已停
[root@hadoop102 ~]# systemctl start ntpd
正在启动 ntpd:                                              [确定]
```

执行以下命令，重新启用 ntp 服务的开机自启动功能。

```
[root@hadoop102 ~]# systemctl enable ntpd
```

第 2 步：配置其他服务器（必须是 root 用户）。

（1）配置其他服务器与时间服务器每 10 分钟同步一次。

```
[root@hadoop103 ~]# crontab -e
```

（2）编写脚本。

```
*/10 * * * * /usr/sbin/ntpdate hadoop102
```

（3）修改 hadoop103 节点服务器的时间，使其与另外两台节点服务器的时间不同步。
```
[root@hadoop103 hadoop]# date -s "2017-9-11 11:11:11"
```
（4）在 10 分钟后，查看 hadoop103 节点服务器是否与时间服务器同步。
```
[root@hadoop103 hadoop]# date
```

3.4 本章总结

本章从 Hadoop 的集群角色入手，主要讲解了 Hadoop 的不同模式，读者需要重点掌握的是完全分布式模式，该模式是企业实际开发过程中用到的模式。在完全分布式模式的配置过程中，需要注意的问题很多，如 SSH 免密登录、集群配置、NameNode 格式化问题、Hadoop 集群启停脚本的编写、集群时间同步等。

第4章

分布式文件系统 HDFS

HDFS 是 Hadoop 为大数据领域提供的一套分布式文件系统，使用 HDFS 可以在廉价的硬件设备上构建一套稳健、可扩展的文件存储系统，在极大程度上解决了企业的海量数据存储问题。前面已经初步介绍过 HDFS，本章将深度解读 HDFS 的使用方法和底层运行机制，揭开 HDFS 运行的神秘面纱。

4.1 HDFS 概述

4.1.1 HDFS 背景及定义

随着互联网行业的飞速发展，产生的数据量越来越大，当产生的数据量远远超过一台独立的物理计算机的存储能力时，人们理所应当地想到了使用多台计算机分别存储海量数据中一部分的解决方案。但是这时新的问题出现了，如何管理这些分别存储于不同节点上的同一份数据呢？一个超大的文件要以什么标准进行切分？分别存储的文件在需要使用时要如何获取？如果其中一份数据文件丢失，那么是否可以恢复？面对这些问题，就需要一个强大的分布式文件系统，这个文件系统可以综合管理所有的文件存储节点、容忍节点故障、为海量数据提供可靠的上传和下载通道，Hadoop 为用户提供了这样的分布式文件系统——HDFS。

HDFS（Hadoop Distributed File System，Hadoop 分布式文件系统）主要用于存储文件，其存储结构与普通文件存储系统的存储结构相似，都是通过目录树定位文件的，但是与普通文件存储系统不同的是，HDFS 的底层是分布式的，由多台服务器联合起来对外提供文件存储服务。

HDFS 通常在一次写入、多次读取的场景中使用。HDFS 中存储的数据通常是从数据源采集来的海量数据集，这些数据集需要长期存储于 HDFS 中，用于进行分析计算。例如，一天内产生的用户访问记录，在计算每天活跃用户时会用到，在计算每周活跃用户时也可能会用到。对海量数据集进行分析计算的特别之处在于，每次计算的着眼点都是数据集的整体，而不是某条数据，因此读取数据集中第 1 条记录的延迟并不重要，重要的是获取整体数据集的延迟。

HDFS 作为一个分布式文件系统，具有很多优点。

- 高容错性。HDFS 以多副本的形式提高容错性。数据文件在被上传后，可以自动存储为多个副本，在其中一个副本丢失后，可以通过其他副本自动恢复。
- 适合处理大数据。数据的存储规模从 KB 级到 GB 级、TB 级，甚至 PB 级，并且支持对海量数据的处理计算。
- Hadoop 不需要运行在高可靠的硬件上，在廉价的硬件上即可构建高可靠性的集群。

HDFS 不是万能的，其缺点如下。

- HDFS 并不适合低时间延迟的数据访问。要求低时间延迟数据访问的应用程序是不适合在 HDFS 上运行的。HDFS 是为高数据量、高吞吐的数据应用程序设计的，而这必然会以提高时间延迟为代价。
- 无法存储大量小文件。HDFS 会将文件的元数据存储于 NameNode 中，所以其文件存储量是受限于

NameNode 的内存量的。如果集群中存储了大量的小文件，则会对 HDFS 的存储能力造成极大的损害。在实际应用中，小文件问题应该尽量避免。
- 不支持并发写入操作，不允许多个线程同时写文件。文件只能同时由一个写入者写入，不允许多个线程同时写文件。对于一个文件的写入操作，要以"仅添加"的模式在文件末尾写入数据，不支持在文件的任意位置进行修改。

4.1.2 HDFS 的基本架构

HDFS 的基本架构如图 4-1 所示。根据图 4-1 可知，HDFS 主要由 NameNode、DataNode、Client、SecondaryNameNode 组成。

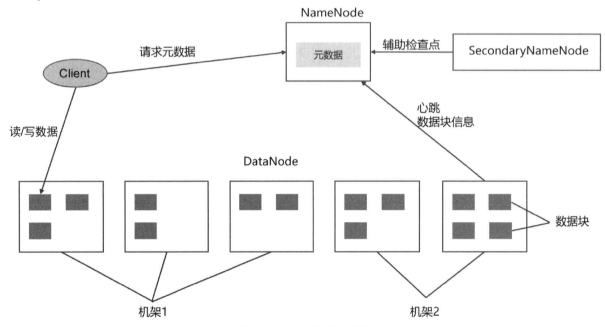

图 4-1 HDFS 的基本架构

1．NameNode

NameNode 是 HDFS 中的主节点，是 HDFS 集群中的 Master 角色，是整个 HDFS 集群的管理者，负责的任务如下。
- 管理 HDFS 的名称空间，这个名称空间是指 HDFS 提供的层次化的文件组织形式。
- 配置副本策略，副本策略是指每一份文件产生副本的存储位置策略，合理的副本策略可以最大限度地保证系统可用性。
- 管理数据块（Block）的映射信息。NameNode 维护着所有数据块到具体 DataNode 的映射关系，对文件的读取操作都需要通过 NameNode 获取数据块的位置信息。
- 处理客户端的读/写请求。

2．DataNode

DataNode 是 HDFS 中的从节点，是 HDFS 集群中的 Slave 角色，是整个 HDFS 集群的功能节点，通常运行在每一台节点服务器上，负责管理所在节点服务器的存储空间。在系统内部，文件在被切分成数据块后，实际上是存储于 DataNode 所在的节点服务器中的。DataNode 最终负责按照 NameNode 的指令执行数据块的创建、删除、复制等工作。

3. Client

Client 就是客户端，用户每提交一个文件的读/写请求，都会生成一个 Client。Client 主要负责以下工作。
- 文件切分。将需要上传至 HDFS 集群中的文件按照规则切分成数据块（Block），然后进行上传。
- 与 NameNode 交互，获取文件的位置信息。
- 与 DataNode 交互，读取或写入数据。
- 执行一些管理 HDFS 的命令，如格式化 NameNode、查看块信息等。
- 访问 HDFS 的名称空间，查看 HDFS 内部的文件层次结构信息。

4. SecondaryNameNode

SecondaryNameNode 从名称上看似乎是 NameNode 的备份机，但实际上并不是 NameNode 的热备。当 NameNode 发生故障宕机时，SecondaryNameNode 并不能马上替换 NameNode 并提供服务。SecondaryNameNode 主要负责辅助 NameNode，分担其工作量。例如，定期合并 FsImage 文件和 EditLog 文件，并且将其推送给 NameNode。在 NameNode 发生故障宕机时，可以辅助恢复 NameNode 进程。此部分内容将在 4.5 节详细讲解。

4.2 HDFS 的 shell 操作

HDFS 是一个分布式文件系统，其内部的文件以传统的层次性结构进行组织。Hadoop 为内部的文件系统提供了多种访问接口，包括 Web 端页面访问接口、shell 命令行访问接口、Java API 访问接口等。本节主要介绍通过 shell 命令行命令操作 HDFS。

4.2.1 命令大全

HDFS 为我们提供了两个 shell 命令行命令，分别是 hadoop fs 和 hdfs fs，这两个命令的使用方式是完全相同的。执行 hadoop fs -help 或 hdfs dfs -help 命令，可以查看所有的 HDFS 操作命令及其详细介绍，如下所示。

```
[atguigu@hadoop102 hadoop-3.1.3]$ hadoop fs -help
Usage: hadoop fs [generic options]
    [-appendToFile <localsrc> ... <dst>]
    [-cat [-ignoreCrc] <src> ...]
    [-checksum <src> ...]
    [-chgrp [-R] GROUP PATH...]
    [-chmod [-R] <MODE[,MODE]... | OCTALMODE> PATH...]
    [-chown [-R] [OWNER][:[GROUP]] PATH...]
    [-copyFromLocal [-f] [-p] [-l] [-d] [-t <thread count>] <localsrc> ... <dst>]
    [-copyToLocal [-f] [-p] [-ignoreCrc] [-crc] <src> ... <localdst>]
    [-count [-q] [-h] [-v] [-t [<storage type>]] [-u] [-x] [-e] <path> ...]
    [-cp [-f] [-p | -p[topax]] [-d] <src> ... <dst>]
    [-createSnapshot <snapshotDir> [<snapshotName>]]
    [-deleteSnapshot <snapshotDir> <snapshotName>]
    [-df [-h] [<path> ...]]
    [-du [-s] [-h] [-v] [-x] <path> ...]
    [-expunge]
    [-find <path> ... <expression> ...]
    [-get [-f] [-p] [-ignoreCrc] [-crc] <src> ... <localdst>]
    [-getfacl [-R] <path>]
    [-getfattr [-R] {-n name | -d} [-e en] <path>]
    [-getmerge [-nl] [-skip-empty-file] <src> <localdst>]
```

```
[-head <file>]
[-help [cmd ...]]
[-ls [-C] [-d] [-h] [-q] [-R] [-t] [-S] [-r] [-u] [-e] [<path> ...]]
[-mkdir [-p] <path> ...]
[-moveFromLocal <localsrc> ... <dst>]
[-moveToLocal <src> <localdst>]
[-mv <src> ... <dst>]
[-put [-f] [-p] [-l] [-d] <localsrc> ... <dst>]
[-renameSnapshot <snapshotDir> <oldName> <newName>]
[-rm [-f] [-r|-R] [-skipTrash] [-safely] <src> ...]
[-rmdir [--ignore-fail-on-non-empty] <dir> ...]
[-setfacl [-R] [{-b|-k} {-m|-x <acl_spec>} <path>]|[--set <acl_spec> <path>]]
[-setfattr {-n name [-v value] | -x name} <path>]
[-setrep [-R] [-w] <rep> <path> ...]
[-stat [format] <path> ...]
[-tail [-f] [-s <sleep interval>] <file>]
[-test -[defsz] <path>]
[-text [-ignoreCrc] <src> ...]
[-touch [-a] [-m] [-t TIMESTAMP ] [-c] <path> ...]
[-touchz <path> ...]
[-truncate [-w] <length> <path> ...]
[-usage [cmd ...]]
```

根据上面的命令列表可知，HDFS 的命令行命令与 Linux 的普通文件命令有很多相似之处，都可以执行上传文件、下载文件、查看文件列表、创建文件目录、删除文件、重命名文件等操作。

4.2.2 命令行命令实操

本节主要介绍常用的命令行命令。

1．准备工作

（1）启动 Hadoop 集群，包括 HDFS 和 YARN。需要注意的是，HDFS 和 YARN 的启动命令要分别在 NameNode 和 ResourceManager 所在的节点上执行。

```
[atguigu@hadoop102 hadoop-3.1.3]$ sbin/start-dfs.sh
[atguigu@hadoop103 hadoop-3.1.3]$ sbin/start-yarn.sh
```

（2）使用-help 命令可以输出所查询命令的具体参数，遇到不会使用的命令，可以通过以下命令获取帮助。

```
[atguigu@hadoop102 hadoop-3.1.3]$ hadoop fs -help rm
```

（3）在 HDFS 的根目录下创建测试文件夹/sanguo。

```
[atguigu@hadoop102 hadoop-3.1.3]$ hadoop fs -mkdir /sanguo
```

2．上传

1）-moveFromLocal 命令。

-moveFromLocal 命令主要用于将文件从本地文件系统中剪切到 HDFS 中。

在 Hadoop 的安装目录下创建一个文件 shuguo.txt，并且编写该文件中的内容（本节中的所有命令均默认在 Hadoop 的安装路径/opt/module/hadoop3.1.3 下执行）。

```
[atguigu@hadoop102 hadoop-3.1.3]$ vim shuguo.txt
shuguo
```

执行-moveFromLocal 命令，将 shuguo.txt 文件上传至/sanguo 路径下。

```
[atguigu@hadoop102 hadoop-3.1.3]$ hadoop fs -moveFromLocal ./shuguo.txt /sanguo
```

在-moveFromLocal 命令执行完毕后，再次查看本地路径，可以发现，shuguo.txt 文件已经被剪切走了。
```
[atguigu@hadoop200 hadoop-3.1.3]$ ll
总用量 208
drwxr-xr-x.  2 atguigu atguigu    4096 7月  29 2020 bin
drwxrwxr-x.  4 atguigu atguigu    4096 7月   7 2020 data
drwxr-xr-x.  3 atguigu atguigu    4096 9月  12 2019 etc
drwxr-xr-x.  2 atguigu atguigu    4096 9月  12 2019 include
drwxr-xr-x.  3 atguigu atguigu    4096 9月  12 2019 lib
drwxr-xr-x.  4 atguigu atguigu    4096 7月  29 2020 libexec
-rw-rw-r--.  1 atguigu atguigu  147145 9月   4 2019 LICENSE.txt
drwxrwxr-x.  3 atguigu atguigu    4096 9月  28 14:45 logs
-rw-rw-r--.  1 atguigu atguigu   21867 9月   4 2019 NOTICE.txt
-rw-rw-r--.  1 atguigu atguigu    1366 9月   4 2019 README.txt
drwxr-xr-x.  3 atguigu atguigu    4096 9月  12 2019 sbin
drwxr-xr-x.  4 atguigu atguigu    4096 9月  12 2019 share
```

2）-copyFromLocal 命令。

-copyFromLocal 命令主要用于从本地文件系统中上传文件到 HDFS 中。

创建一个文件 weiguo.txt，并且编写该文件中的内容。
```
[atguigu@hadoop102 hadoop-3.1.3]$ vim weiguo.txt
weiguo
```
执行-copyFromLocal 命令，将 weiguo.txt 文件上传至/sanguo 路径下。
```
[atguigu@hadoop102 hadoop-3.1.3]$ hadoop fs -copyFromLocal weiguo.txt /sanguo
```

3）-put 命令。

-put 命令的功能与-copyFromLocal 命令的功能相同，在生产环境中，-put 命令更常用。

创建一个文件 wuguo.txt，并且编写该文件中的内容。
```
[atguigu@hadoop102 hadoop-3.1.3]$ vim wuguo.txt
wuguo
```
执行-put 命令，将 wuguo.txt 文件上传至/sanguo 路径下。
```
[atguigu@hadoop102 hadoop-3.1.3]$ hadoop fs -put ./wuguo.txt /sanguo
```

4）-appendToFile 命令。

-appendToFile 命令主要用于将一个文件追加到已经存在的文件的末尾。

创建一个文件 liubei.txt，并且编写该文件中的内容。
```
[atguigu@hadoop102 hadoop-3.1.3]$ vim liubei.txt
liubei
```
执行-appendToFile 命令，将 liubei.txt 文件追加到 HDFS 中 shuguo.txt 文件的末尾。
```
[atguigu@hadoop102 hadoop-3.1.3]$ hadoop fs -appendToFile liubei.txt /sanguo/shuguo.txt
```

3．下载

1）-copyToLocal 命令。

-copyToLocal 命令主要用于将文件从 HDFS 中下载到本地文件系统中。
```
[atguigu@hadoop102 hadoop-3.1.3]$ hadoop fs -copyToLocal /sanguo/shuguo.txt ./
```
在之前的操作中，已经将 shuguo.txt 文件从本地文件系统中直接剪切至 HDFS 中了，在-copyToLocal 命令执行完毕后，再次查看本地文件系统，可以发现 shuguo.txt 文件又出现了。
```
[atguigu@hadoop102 hadoop-3.1.3]$ ll
总用量 224
drwxr-xr-x.  2 atguigu atguigu    4096 7月  29 2020 bin
drwxrwxr-x.  4 atguigu atguigu    4096 7月   7 2020 data
drwxr-xr-x.  3 atguigu atguigu    4096 9月  12 2019 etc
```

```
drwxr-xr-x. 2 atguigu atguigu      4096 9月  12 2019 include
drwxr-xr-x. 3 atguigu atguigu      4096 9月  12 2019 lib
drwxr-xr-x. 4 atguigu atguigu      4096 7月  29 2020 libexec
-rw-rw-r--. 1 atguigu atguigu    147145 9月   4 2019 LICENSE.txt
-rw-rw-r--  1 atguigu atguigu         7 10月 15 11:24 liubei.txt
drwxrwxr-x. 3 atguigu atguigu      4096 10月 15 11:09 logs
-rw-rw-r--. 1 atguigu atguigu     21867 9月   4 2019 NOTICE.txt
-rw-rw-r--. 1 atguigu atguigu      1366 9月   4 2019 README.txt
drwxr-xr-x. 3 atguigu atguigu      4096 9月  12 2019 sbin
drwxr-xr-x. 4 atguigu atguigu      4096 9月  12 2019 share
-rw-r--r--  1 atguigu atguigu        15 10月 15 11:25 shuguo.txt
-rw-rw-r--  1 atguigu atguigu         7 10月 15 11:22 weiguo.txt
-rw-rw-r--  1 atguigu atguigu         6 10月 15 11:23 wuguo.txt
```

2）-get 命令。

-get 命令的功能与-copyToLocal 命令的功能相同，在生产环境中，-get 命令更常用。

```
[atguigu@hadoop102 hadoop-3.1.3]$ hadoop fs -get /sanguo/shuguo.txt ./shuguo2.txt
```

4．HDFS 直接操作

1）-ls 命令。

-ls 命令主要用于显示目录信息。

```
[atguigu@hadoop102 hadoop-3.1.3]$ hadoop fs -ls /sanguo
Found 3 items
-rw-r--r--   3 atguigu supergroup         15 2021-10-15 11:24 /sanguo/shuguo.txt
-rw-r--r--   3 atguigu supergroup          7 2021-10-15 11:23 /sanguo/weiguo.txt
-rw-r--r--   3 atguigu supergroup          6 2021-10-15 11:23 /sanguo/wuguo.txt
```

2）-cat 命令。

-cat 命令主要用于显示文件内容。

```
[atguigu@hadoop102 hadoop-3.1.3]$ hadoop fs -cat /sanguo/shuguo.txt
2021-10-15 13:55:26,984 INFO sasl.SaslDataTransferClient: SASL encryption trust check:
localHostTrusted = false, remoteHostTrusted = false
shuoguo
liubei
```

3）-chgrp、-chmod、-chown 命令。

-chgrp、-chmod 和-chown 命令都是修改文件或目录所属权限的相关命令，分别用于修改文件或目录的所属用户组、权限管理模式和所属用户。

HDFS 中文件和目录的权限管理模式与 Linux 中文件和目录的权限管理模式相似。HDFS 为文件和目录提供了 3 种权限，分别是只读权限（r）、写入权限（w）和可执行权限（x）。当读取文件内容或列出目录下的所有内容时，需要只读权限。当写入、追加文件，或者在目录下创建、删除文件或目录时，需要写入权限。在 HDFS 中，对文件来说，可执行权限可以忽略，因为不可以在 HDFS 中执行文件，但是当访问一个目录的子项时，需要该目录具有可执行权限。

与 Linux 的权限管理模型不同，HDFS 权限管理模型中的文件没有 sticky、setuid 或 setgid 位，因为 HDFS 权限管理模型中没有可执行文件的概念。为了便于管理，HDFS 权限管理模型中的目录也没有 sticky、setuid 或 setgid 位。总而言之，文件或目录的权限就是它的模式（mode）。HDFS 采用 UNIX 操作系统表示和显示模式的习惯，如使用八进制数表示权限。当新建一个文件或目录时，它的所有者（owner）是客户进程的用户，它的所属组（group）是父目录的组。

通过讲解 HDFS 中文件的权限管理模式，相信读者可以充分理解-chgrp、-chmod 和-chown 命令的功能。

```
[atguigu@hadoop102 hadoop-3.1.3]$ hadoop fs -chmod 666 /sanguo/shuguo.txt
[atguigu@hadoop102 hadoop-3.1.3]$ hadoop fs -chown atguigu:atguigu /sanguo/shuguo.txt
[atguigu@hadoop102 hadoop-3.1.3]$ hadoop fs -chgrp atguigu /sanguo/weiguo.txt
```

4）-mkdir 命令。

-mkdir 命令主要用于创建路径。

```
[atguigu@hadoop102 hadoop-3.1.3]$ hadoop fs -mkdir /jinguo
```

5）-cp 命令。

-cp 命令主要用于将文件从 HDFS 中的一个路径下复制到 HDFS 中的另一个路径下。

```
[atguigu@hadoop102 hadoop-3.1.3]$ hadoop fs -cp /sanguo/shuguo.txt /jinguo
```

6）-mv 命令。

-mv 命令主要用于在 HDFS 中移动文件，可以移动多个文件，当移动多个文件时，目标路径必须是文件夹。使用该命令也可以给文件重命名。

```
[atguigu@hadoop102 hadoop-3.1.3]$ hadoop fs -mv /sanguo/wuguo.txt /jinguo
[atguigu@hadoop102 hadoop-3.1.3]$ hadoop fs -mv /sanguo/weiguo.txt /jinguo
```

7）-tail 命令。

-tail 命令主要用于显示一个文件末尾 1KB 的数据。

```
[atguigu@hadoop102 hadoop-3.1.3]$ hadoop fs -tail /jinguo/shuguo.txt
```

8）-rm 命令。

-rm 命令主要用于删除文件或文件夹。

```
[atguigu@hadoop102 hadoop-3.1.3]$ hadoop fs -rm /sanguo/shuguo.txt
```

9）-rm -r 命令。

-rm -r 命令主要用于递归删除目录及其中的内容。

```
[atguigu@hadoop102 hadoop-3.1.3]$ hadoop fs -rm -r /sanguo
```

10）-du 命令。

-du 命令主要用于统计文件夹的大小。如果加上-s 参数，则表示只统计文件夹的大小；如果不加-s 参数，则表示统计文件夹中文件的大小。

```
[atguigu@hadoop102 hadoop-3.1.3]$ hadoop fs -du -s -h /sanguo
28  84  /sanguo

[atguigu@hadoop102 hadoop-3.1.3]$ hadoop fs -du  -h /sanguo
15  45  /sanguo/shuguo.txt
7   21  /sanguo/weiguo.txt
6   18  /sanguo/wuguo.txt
```

说明：文件夹或文件前面的两个数字分别表示文件夹或文件的字节数和文件夹或文件的 3 个副本的总字节数。

11）-setrep 命令。

-setrep 命令主要用于设置 HDFS 中文件的副本数量。

```
[atguigu@hadoop102 hadoop-3.1.3]$ hadoop fs -setrep 10 /sanguo/shuguo.txt
```

查看文件详情，可以看到 shuguo.txt 文件的副本数量已经变成了 10 个（文件详情中的第 2 列表示副本数量）。

```
[atguigu@hadoop200 ~]$ hadoop fs -ls /sanguo
Found 3 items
-rw-r--r--  10 atguigu supergroup  15 2021-10-15 11:24 /sanguo/shuguo.txt
-rw-r--r--   3 atguigu supergroup   7 2021-10-15 11:23 /sanguo/weiguo.txt
-rw-r--r--   3 atguigu supergroup   6 2021-10-15 11:23 /sanguo/wuguo.txt
```

通过该命令设置的副本数量只记录在 NameNode 的元数据中，是否真的有这么多副本，取决于

DataNode 的数量。因为目前只有 3 台节点服务器，所以最多存在 3 个副本，只有当节点服务器增加到 10 台时，真正的副本数量才能达到 10 个。

4.3 HDFS 的 API 操作

在学习了如何使用命令行命令操作 HDFS 中的文件后，我们还应学习如何使用 Hadoop 提供的 Java API 操作 HDFS 中的文件，这在实际工作中应用更广泛。Hadoop 底层是使用 Java 语言编写的，使用 Java API 可以进行 Hadoop 文件系统的大部分交互操作，主要使用 FileSystem 类对文件系统内的文件进行操作。

4.3.1 客户端环境准备

HDFS API 的编写工作主要在 Windows 操作系统的 IDE 中进行。在编写代码前，先对需要使用的开发环境和工具进行介绍。

- 操作系统为 Windows 10。
- 需要提前安装 Java 8。
- 集成开发环境（IDE）使用 IntelliJ IDEA，具体安装流程参见 IntelliJ 官方网站。
- 在安装 IntelliJ IDEA 后，还需要安装插件 Maven 和 Git，其中，Maven 主要用于管理项目依赖，通过 Git 可以轻松获取示例代码，并且对本地代码的版本进行控制。

如果对以上运行环境的配置和部署存在疑问，那么欢迎访问尚硅谷 IT 教育官方网站获取全套视频教程。

在 Windows 操作系统中编写 Hadoop 程序前，需要安装 Hadoop 环境，具体步骤如下。

（1）在本书附赠的资料包中可以找到已经编译好的 Hadoop 在 Windows 操作系统中的安装包，打开该安装包文件夹，将 hadoop-3.1.0 文件复制到一个非中文路径下，如 D:\。

（2）在 Windows 桌面上右击"此电脑"图标，在弹出的快捷菜单中选择"属性"命令，在打开的"设置"页面中选择"高级系统设置"选项，然后单击右下角的"环境变量"按钮，开始配置 HADOOP_HOME 环境变量。在系统变量中，增加 HADOOP_HOME 环境变量，并且配置 Hadoop 的安装路径，如图 4-2 所示。

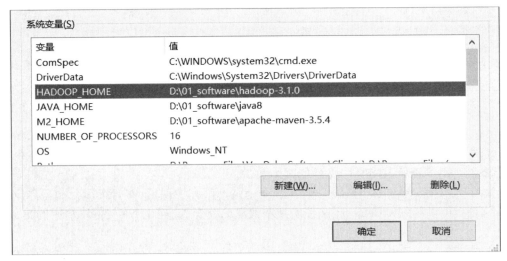

图 4-2　配置系统变量

（3）配置 Path 环境变量，如图 4-3 所示。

在 hadoop-3.1.0 目录下的 bin 文件夹中，双击 winutils.exe 文件，如果报如图 4-4 所示的错误，则说明缺少微软运行库。在本书提供的资料包中找到对应的微软运行库安装包 MSVBCRT_AIO_2018.07.31_

X86+X64.exe，双击安装即可。

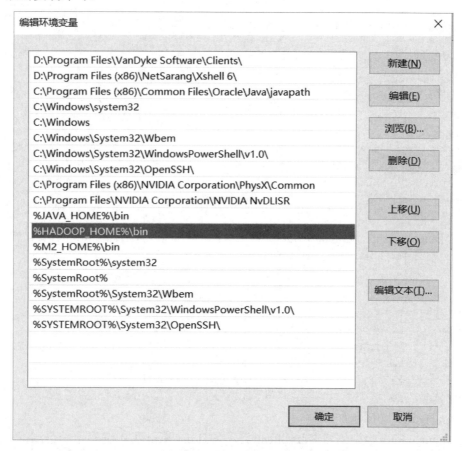

图 4-3　配置 Path 环境变量

图 4-4　运行 winutils.exe 报错

（4）在 IDEA 中创建一个 Maven 工程 HdfsClientDemo，在该工程的 pom.xml 文件中添加以下依赖。其中，hadoop-client 依赖主要用于执行 Hadoop 的相关操作，junit 依赖主要用于进行方法测试，slf4j-log4j12 依赖主要用于打印日志。

```
<dependencies>
    <dependency>
        <groupId>org.apache.hadoop</groupId>
```

```xml
    <artifactId>hadoop-client</artifactId>
    <version>3.1.3</version>
</dependency>
<dependency>
    <groupId>junit</groupId>
    <artifactId>junit</artifactId>
    <version>4.12</version>
</dependency>
<dependency>
    <groupId>org.slf4j</groupId>
    <artifactId>slf4j-log4j12</artifactId>
    <version>1.7.30</version>
</dependency>
</dependencies>
```

在 HdfsClientDemo 工程的 src/main/resources 目录下新建一个文件 log4j.properties，在该文件中添加以下内容，可以在控制台中打印日志。

```
log4j.rootLogger=INFO, stdout
log4j.appender.stdout=org.apache.log4j.ConsoleAppender
log4j.appender.stdout.layout=org.apache.log4j.PatternLayout
log4j.appender.stdout.layout.ConversionPattern=%d %p [%c] - %m%n
log4j.appender.logfile=org.apache.log4j.FileAppender
log4j.appender.logfile.File=target/spring.log
log4j.appender.logfile.layout=org.apache.log4j.PatternLayout
log4j.appender.logfile.layout.ConversionPattern=%d %p [%c] - %m%n
```

（5）创建 com.atguigu.hdfs 包。

（6）在 com.atguigu.hdfs 包中创建 HdfsClient 类，在 HdfsClient 类中编写第一个测试方法 testMkdirs()，用于在 HDFS 中创建一个路径/xiyou/huaguoshan。

编写 HDFS API 的具体思路如下。

① 创建一个 Hadoop 的配置类 Configuration，在该配置类中封装有 Hadoop 的所有默认配置，若用户需要自定义配置，则可以通过该配置类进行修改。

② 创建 FileSystem 对象，FileSystem 是 Hadoop 提供的文件系统的客户端接口。

③ 通过 FileSystem 对象调用具体方法，对文件系统进行操作。

④ 关闭客户端。

HdfsClient 类的相关代码如下：

```java
public class HdfsClient {

    @Test
    public void testMkdirs() throws IOException, URISyntaxException, InterruptedException {

        // 1 获取文件系统
        Configuration configuration = new Configuration();

        // FileSystem fs = FileSystem.get(new URI("hdfs://hadoop102:8020"), configuration);
        FileSystem fs = FileSystem.get(new URI("hdfs://hadoop102:8020"), configuration, "atguigu");

        // 2 创建目录
        fs.mkdirs(new Path("/xiyou/huaguoshan/"));
```

```
        // 3 关闭资源
        fs.close();
    }
}
```

（7）执行程序。由于我们已经添加了 junit 依赖，因此在标有"@Test"的方法上右击，在弹出的快捷菜单中选择"运行"命令，即可测试该方法。

在程序执行完毕后，可以查看 Web 端页面，若出现/xiyou/huaguoshan 文件夹，则说明程序执行成功。

（8）客户端在操作 HDFS 时，是有一个用户身份的。在默认情况下，如果 HDFS 的客户端 API 使用 Windows 的默认用户访问 HDFS，则会报 AccessControlException 异常，如下所示。所以在访问 HDFS 时，一定要配置用户。

```
org.apache.hadoop.security.AccessControlException: Permission denied: user=56576, access=WRITE,
inode="/xiyou/huaguoshan":atguigu:supergroup:drwxr-xr-x
```

用户名可以在创建 FileSystem 对象时传入，将创建 FileSystem 对象的代码替换成以下代码。

```
FileSystem fs = FileSystem.get(new URI("hdfs://hadoop102:8020"), configuration,"atguigu");
```

4.3.2 HDFS 文件上传案例

（1）在 D 盘中新建一个文件 sunwukong.txt，并且在该文件中编写以下内容。

```
sunwukong
```

（2）编写文件上传测试方法 testCopyFromLocalFile()，用于将 sunwukong.txt 文件上传至 HDFS 中的/xiyou/huaguoshan 路径下。在上传 sunwukong.txt 文件前，可以通过 Configuration 对象设置该文件的副本数量。

```
@Test
public void testCopyFromLocalFile() throws IOException, InterruptedException, URISyntaxException {

    // 1 获取文件系统
    Configuration configuration = new Configuration();
    configuration.set("dfs.replication", "2");
    FileSystem fs = FileSystem.get(new URI("hdfs://hadoop102:8020"), configuration, "atguigu");

    // 2 上传文件
    fs.copyFromLocalFile(new Path("d:/sunwukong.txt"), new Path("/xiyou/huaguoshan"));

    // 3 关闭资源
    fs.close();
}
```

（3）将 hdfs-site.xml 文件复制到 HdfsClientDemo 工程的 resources 目录下，并且将 sunwukong.txt 文件的副本数量修改为 1 个。

```
<?xml version="1.0" encoding="UTF-8"?>
<?xml-stylesheet type="text/xsl" href="configuration.xsl"?>

<configuration>
    <property>
        <name>dfs.replication</name>
        <value>1</value>
    </property>
</configuration>
```

（4）执行程序，在程序执行完毕后，访问 HDFS 的 Web 端页面，查看 sunwukong.txt 文件的副本数量，如图 4-5 所示。

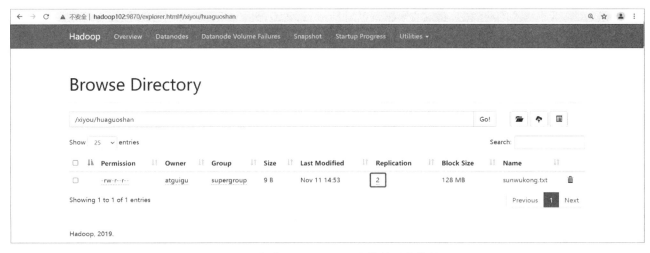

图 4-5　查看 sunwukong.txt 文件的副本数量

根据图 4-5 可知，sunwukong.txt 文件的最终副本数量为 2 个，与代码中配置的参数值一致。Hadoop 服务器可以在不同的位置进行参数配置，具有不同的优先级，优先级由高到低分别如下：

客户端代码中设置的值>ClassPath 中的用户自定义配置文件>服务器中的自定义配置文件（xxx-site.xml）>服务器中的默认配置文件（xxx-default.xml）。

4.3.3　HDFS 文件下载案例

（1）编写文件下载测试方法 testCopyToLocalFile()，将/xiyou/huaguoshan/sunwukong.txt 文件下载至本地文件系统中。HDFS 的文件下载功能主要调用的是 FileSystem 对象的 copyToLocalFile()方法，这个方法要传入以下 4 个参数。

- delSrc：boolean 类型，用于设置是否将原文件删除。
- src：Path 类型，要下载的文件路径。
- dst：Path 类型，存储下载文件的目标路径。
- useRawLocalFileSystem：boolean 类型，用于设置是否启用文件校验功能。

```
@Test
public void testCopyToLocalFile() throws IOException, InterruptedException, URISyntaxException{

    // 1 获取文件系统
    Configuration configuration = new Configuration();
    FileSystem fs = FileSystem.get(new URI("hdfs://hadoop102:8020"), configuration, "atguigu");

    // 2 执行下载操作
    // boolean delSrc : 用于设置是否将原文件删除
    // Path src : 要下载的文件路径
    // Path dst : 存储下载文件的目标路径
    // boolean useRawLocalFileSystem : 用于设置是否启用文件校验功能
    fs.copyToLocalFile(false, new Path("/xiyou/huaguoshan/sunwukong.txt"), new Path("d:/sunwukong2.txt"), true);

    // 3 关闭资源
    fs.close();
}
```

（2）执行程序，在程序执行完毕后，去目标路径下查看文件是否下载成功。

4.3.4 HDFS 文件重命名案例

编写 HDFS 文件重命名的测试代码，具体如下：

```
@Test
public void testRename() throws IOException, InterruptedException, URISyntaxException{

    // 1 获取文件系统
    Configuration configuration = new Configuration();
    FileSystem fs = FileSystem.get(new URI("hdfs://hadoop102:8020"), configuration, "atguigu");

    // 2 修改文件名称
    fs.rename(new Path("/xiyou/huaguoshan/sunwukong.txt"), new Path("/xiyou/huaguoshan/meihouwang.txt"));

    // 3 关闭资源
    fs.close();
}
```

4.3.5 HDFS 文件删除案例

本案例会使用 FileSystem 对象的 delete() 方法删除文件或文件夹。delete() 方法有两个参数，第 1 个参数是需要删除的文件路径，第 2 个参数是 boolean 类型的参数，用于设置是否对文件夹进行递归删除操作。当传入的文件路径为非空文件夹时，boolean 类型的参数值必须为 true，才会执行删除操作，否则会报错。

```
@Test
public void testDelete() throws IOException, InterruptedException, URISyntaxException{

    // 1 获取文件系统
    Configuration configuration = new Configuration();
    FileSystem fs = FileSystem.get(new URI("hdfs://hadoop102:8020"), configuration, "atguigu");

    // 2 执行删除操作
    fs.delete(new Path("/xiyou"), true);

    // 3 关闭资源
    fs.close();
}
```

4.3.6 HDFS 文件详情查看案例

通过 API 可以查看文件名称、权限、长度、块信息等，通过调用 FileSystem 对象的 listFiles() 方法实现，返回值是一个封装了路径下所有文件状态的迭代器。迭代器内封装的对象类型是 FileStatus 类的子类 LocatedFileStatus，通过 LocatedFileStatus 对象可以获取文件的所有信息。

```
@Test
public void testListFiles() throws IOException, InterruptedException, URISyntaxException {

    // 1 获取文件系统
    Configuration configuration = new Configuration();
    FileSystem fs = FileSystem.get(new URI("hdfs://hadoop102:8020"), configuration, "atguigu");
```

```java
// 2 获取文件详情
RemoteIterator<LocatedFileStatus> listFiles = fs.listFiles(new Path("/"), true);

while (listFiles.hasNext()) {
    LocatedFileStatus fileStatus = listFiles.next();

    System.out.println("========" + fileStatus.getPath() + "=========");
    System.out.println(fileStatus.getPermission());
    System.out.println(fileStatus.getOwner());
    System.out.println(fileStatus.getGroup());
    System.out.println(fileStatus.getLen());
    System.out.println(fileStatus.getModificationTime());
    System.out.println(fileStatus.getReplication());
    System.out.println(fileStatus.getBlockSize());
    System.out.println(fileStatus.getPath().getName());

    // 获取块信息
    BlockLocation[] blockLocations = fileStatus.getBlockLocations();
    System.out.println(Arrays.toString(blockLocations));
}
// 3 关闭资源
fs.close();
}
```

4.3.7 HDFS 文件和文件夹判断案例

在上一个案例中，通过调用 listFiles()方法可以获取 FileStatus 集合，遍历集合内的元素（FileStatus 对象），对获取的每个 FileStatus 对象调用 isFile()方法，可以判断该 FileStatus 对象是文件还是文件夹。

```java
@Test
public void testListStatus() throws IOException, InterruptedException, URISyntaxException{

    // 1 获取文件配置信息
    Configuration configuration = new Configuration();
    FileSystem fs = FileSystem.get(new URI("hdfs://hadoop102:8020"), configuration, "atguigu");

    // 2 判断是文件还是文件夹
    FileStatus[] listStatus = fs.listStatus(new Path("/"));

    for (FileStatus fileStatus : listStatus) {

        if (fileStatus.isFile()) {// 如果是文件
            System.out.println("f:"+fileStatus.getPath().getName());
        }else {                    // 如果是文件
            System.out.println("d:"+fileStatus.getPath().getName());
        }
    }

    // 3 关闭资源
    fs.close();
}
```

4.4 HDFS 的读/写流程

4.4.1 HDFS 中的数据块大小

在 HDFS 中，数据块是指 HDFS 中的文件在物理上是分块存储的。数据块的大小可以通过配置参数 dfs.blocksize 进行配置，在 Hadoop 2.x 和 Hadoop 3.x 中默认是 128MB，在 Hadoop 1.x 中默认是 64MB。

HDFS 中的数据块不能太大，也不能太小，原因如下。

- 如果数据块太小，那么一个大文件会被切分成过多个数据块，从而增加整个大文件的寻址时间，也会生成过多的元数据信息，对 NameNode 造成更大的存储负担。
- 如果数据块太大，那么虽然可以降低整个文件的寻址时间占比，但是磁盘传输数据的时间占比会大幅提高。

假设寻址时间是 10ms，磁盘传输速度是 100MB/s，当寻址时间为传输时间的 1%时，状态最佳，那么可以计算得出，最佳数据块大小为 100MB（0.01s÷1%×100MB/s），如图 4-6 所示。通过计算可知，数据块的大小主要受制于磁盘传输速度的大小。在实际工作中，随着磁盘传输速度的提高，很多企业会配置更大的数据块。但是数据块的大小不能过大，因为在执行 MapReduce 任务时，map 任务通常一次只处理一个数据块中的数据，如果数据块过大，那么大文件被切分成的数据块数量会减少，从而减少并行任务数，降低工作效率。MapReduce 任务划分的相关知识将在后续章节中进行详细介绍。

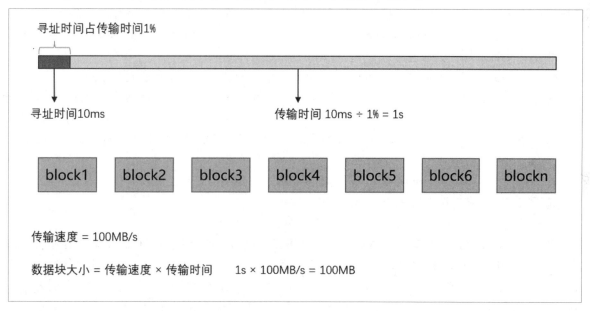

图 4-6　HDFS 中数据块大小的计算方法

根据上述计算过程可知，当磁盘传输速度是 100MB/s 时，将数据块大小设置为 100MB 最佳，但是磁盘的容量一般是 128MB 的倍数，所以将数据块大小设置为 128MB。

数据块的出现大大提升了存储文件的灵活性。有了数据块，HDFS 可以存储大于单节点物理存储空间的大文件。同一个文件的所有数据块不需要存储于一个 DataNode 中，可以分散存储于多个 DataNode 中。

4.4.2 写数据流程

1. 剖析写数据流程

HDFS 中的写数据流程是指创建一个文件，将数据写入文件，最后关闭该文件的过程。下面详细介绍 HDFS 的写数据流程，如图 4-7 所示。

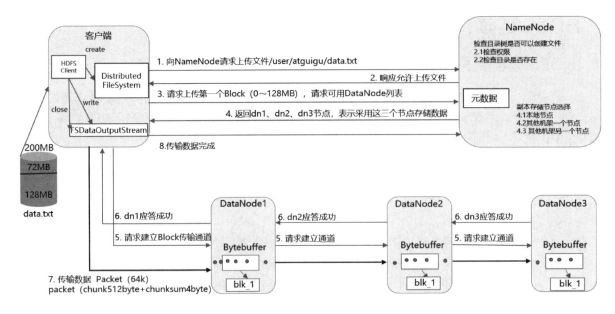

图 4-7 HDFS 中的写数据流程

假设有一个 200MB 的文件 data.txt，通过执行 hadoop fs -put ./data.txt /atguigu 命令将其上传至 HDFS 中。客户端首先将文件切分成两个数据块，一个 128MB，另一个 72MB，然后进行以下操作。

（1）客户端通过 DistributedFileSystem 模块向 NameNode 请求上传文件，NameNode 验证用户权限并检查目标文件及其父目录是否已存在。

（2）NameNode 向 DistributedFileSystem 模块返回一个信号，用于表示是否可以上传。

（3）客户端向 NameNode 请求上传第一个数据块，并且向 NameNode 询问目标 DataNode 服务器地址列表。

（4）NameNode 返回 3 个 DataNode 服务器地址，假设这 3 个 DataNode 服务器地址分别为 dn1、dn2、dn3。

（5）客户端通过 FSDataOutputStream 模块向 dn1 请求上传数据，dn1 在收到请求后会继续请求 dn2，然后 dn2 请求 dn3，最终将这个通信管道建立完成。

（6）在通信管道建立完成后，dn1、dn2、dn3 逐级应答客户端。

（7）客户端开始向 dn1 上传第一个数据块（先从磁盘中读取数据并将其缓存于本地内存中），以 Packet（数据包）为单位发送数据，dn1 每收到一个 Packet，都会将其上传给 dn2，dn2 再将其上传给 dn3。

（8）在一个数据块传输完成后，客户端会向 NameNode 请求上传第二个数据块，并且获取可用的 DataNode 服务器列表。重复执行步骤（3）～（7），直到所有数据块传输完毕，客户端关闭连接。

2．副本放置策略与机架感知

在写数据流程中提到，NameNode 会向客户端提供可用的 DataNode 服务器列表，那么 NameNode 是如何选择期望的目标节点，以供副本块存储的呢？

决定文件副本如何放置的策略称为副本放置策略（Block Placement Policy），HDFS 默认采用的副本放置策略概括起来有以下 3 点。

- 若客户端所处位置为其中一台 DataNode 服务器，那么将第一个文件副本存储于本地文件系统中，否则随机选择一台 DataNode 服务器进行存储。
- 第二个文件副本存储于不同于第一个文件副本所在机架的随机节点中。
- 第三个文件副本存储于第二个文件副本所在机架的随机节点中。

关于副本放置策略的制订，充分考虑了如何最大限度地提高读/写性能及容错性。根据副本放置策略，3 个文件副本的最终存储位置如图 4-8 所示。

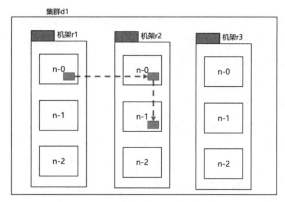

图 4-8　3 个文件副本的最终存储位置

在副本放置策略中，提到了机架的概念。基本上，多个连接到同一个网络交换机的物理存储节点的集合就可以称为机架。可以将机架理解为节点服务器的集合，而 Hadoop 集群是机架的集合。Hadoop 是如何确定任意两个节点是否属于同一个机架的呢？答案是机架感知。在 Hadoop 中，机架感知功能是默认关闭的，在机架感知功能关闭的情况下，副本放置策略会失效，但是并不会失败。也就是说，文件的上传操作可以完整进行，但并不能正确按照副本放置策略判断各个节点服务器是否属于同一个机架，也不能选择副本列表。

启用 Hadoop 机架感知功能的方法非常简单：在 Hadoop 集群的 hadoop-site.xml 配置文件中，对以下属性进行配置。

```
<property>
  <name>topology.script.file.name</name>
  <value>/path/to/RackAware.py</value>
</property>
```

以上属性值为一个可执行程序，如一个脚本，该可执行程序可以接收某个 DataNode 服务器的 IP 地址，并且输出对应服务器所处的机架。

3. 网络拓扑距离与 PipeLine 的形成

前面在剖析写数据流程时曾经提到，NameNode 会将可用的 DataNode 服务器列表返回客户端，这一组 DataNode 服务器会按照一定的顺序进行排序，形成一个 Pipeline。我们已经知道，客户端在上传数据时，首先将数据发送给 dn1，然后 dn1 将数据上传给 dn2，最后 dn2 将数据上传给 dn3。dn1→dn2→dn3 的顺序决定了数据传输的速率，只有当 Pipeline 的"距离"最近时，传输速度才最快。

利用机架感知功能，可以得到如图 4-9 所示的网络拓扑图。

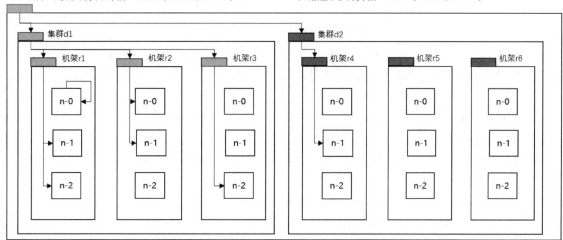

图 4-9　网络拓扑图

根据该网络拓扑图,计算两个节点之间的网络拓扑距离。假设数据中心 d1 的机架 r1 中有一个节点 n1,那么该节点可以表示为/d1/r1/n1,根据这种表示方法,给出 4 种距离描述。

- 同一个节点上的进程:Distance(/d1/r1/n0, /d1/r1/n0)=0。
- 同一个机架中的不同节点:Distance(/d1/r1/n1, /d1/r1/n2)=2。
- 同一个数据中心、不同机架中的节点:Distance(/d1/r2/n0, /d1/r3/n2)=4。
- 不同数据中心中的节点:Distance(/d1/r2/n1, /d2/r4/n1)=6。

根据上述计算节点间距离的规定,对得到的 DataNode 服务器列表进行排序。首先选出一个源节点,然后根据这个源节点遍历其他节点,找出一个距离最短的节点,将其作为下一轮选择的源节点,直至节点选择完毕。这样,每两个节点之间的距离都是最近的,整个 Pipeline 也是距离最近的。

在了解了网络拓扑距离的概念后,我们再回过头来看一下副本放置策略。在选择这 3 个文件副本时,不仅要考虑可靠性、写入带宽和读取带宽,还要考虑网络拓扑距离。第一个文件副本存储于 Client 所在的节点中,可以确保距离最短,第二个文件副本与第三个文件副本存储于同一个机架的不同节点中,可以确保网络拓扑距离最短。

4.4.3 读数据流程

HDFS 中的读数据流程如图 4-10 所示。

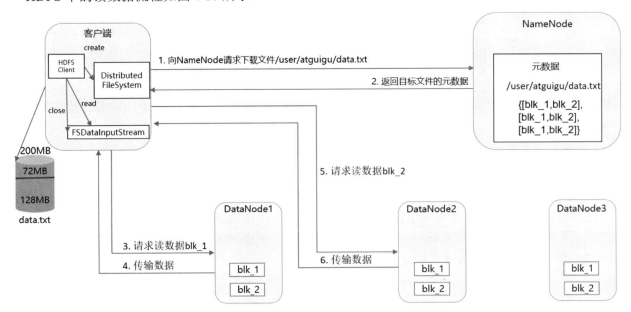

图 4-10 HDFS 中的读数据流程

(1)客户端通过 DistributedFileSystem 模块向 NameNode 请求下载文件,NameNode 通过查询元数据,找到数据块所在的 DataNode 服务器地址,这些 DataNode 服务器按照与客户端所在服务器的网络拓扑距离进行排序。

(2)客户端挑选一台距离最近的 DataNode 服务器,如果所有 DataNode 服务器的距离相同,则随机选择一台 DataNode 服务器,请求读取数据。

(3)DataNode 服务器开始向客户端传输数据,从磁盘中读取数据输入流,并且以 Packet 为单位进行校验。

(4)客户端以 Packet 为单位接收数据,先将其存储于本地缓存中,再将其写入目标文件。

(5)客户端会根据需要向 NameNode 询问下一个数据块的 DataNode 服务器地址。客户端在完成所有数据块的读取工作后,即可关闭连接。

4.5　HDFS 的工作机制

4.5.1　NameNode 和 SecondaryNameNode 的工作机制

我们已经知道，NameNode 中存储的是整个 Hadoop 集群的元数据信息，每次对 HDFS 进行操作，都需要访问元数据或产生新的元数据。下面思考一个问题，NameNode 中的元数据应该存储在哪里？

我们先做一个假设，将元数据存储于 NameNode 所在的磁盘中。在 Hadoop 中，元数据被访问的频率非常高，并且需要低延迟响应，存储于磁盘中显然会大大降低效率。为了能够更快地响应访问请求，元数据需要存储于内存中。但如果只存储于内存中，一旦突发故障断电，元数据就会丢失，那么整个集群都无法工作了，并且无法恢复数据。因此，HDFS 提供了两种文件，用于保障 NameNode 中元数据的可靠性，这两种文件分别是编辑日志文件 EditLog 和镜像文件 FsImage。

文件系统的客户端在执行文件的写操作时，这些操作会首先被记录到编辑日志中，当修改编辑日志时，NameNode 在内存中的元数据信息也会同步更新。但是如果编辑日志一直记录下去，那么编辑日志文件会无限增大，如果集群出现问题，需要重启或从编辑日志中恢复数据，则会是一个巨大的工程。所以，需要定期对编辑日志文件进行合并，形成镜像文件 FsImage。每个 FsImage 文件都是文件系统元数据的完整检查点快照。

完整的工作机制如下：每当元数据有更新或添加时，修改内存中的元数据并将其追加到 EditLog 文件中。定期对 EditLog 和 FsImage 文件进行合并，形成新的 FsImage 文件。如果 NameNode 发生故障或集群重启，那么 NameNode 会将最近的 FsImage 文件载入内存，从而重构元数据的最近状态，再从相关点开始向前执行编辑日志中记录的每个操作。

现在出现一个新的问题，如果 NameNode 不仅负责响应客户端的请求，还要负责 EditLog 和 FsImage 文件的合并工作，那么负担太重了，所以 Hadoop 引入了另一个角色 SecondaryNameNode，用于辅助 NameNode 完成这项工作。

NameNode 和 SecondaryNameNode 的工作机制如图 4-11 所示。

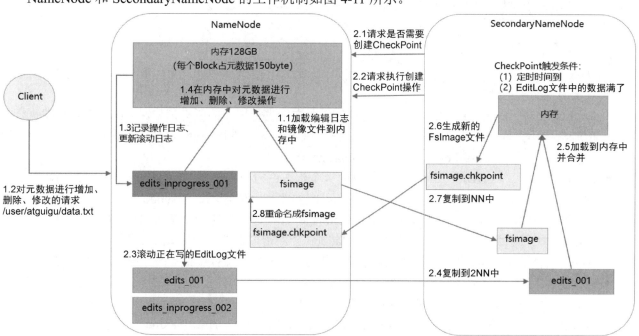

图 4-11　NameNode 与 SecondaryNameNode 的工作机制

1. 第一阶段：启动 NameNode

（1）在第一次格式化并启动 NameNode 后，会创建 EditLog 和 FsImage 文件。如果不是第一次启动，那么直接加载 EditLog 和 FsImage 文件到内存中，重构元数据的最近状态。

（2）客户端向 NameNode 发送对元数据进行增加、删除、修改的请求。

（3）NameNode 记录编辑日志，更新滚动日志。

（4）NameNode 在内存中对元数据进行增加、删除、修改操作。

2．第二阶段：SecondaryNameNode 工作

（1）SecondaryNameNode 询问 NameNode 是否需要创建 CheckPoint（检查点），并且返回 NameNode 的响应结果。

（2）SecondaryNameNode 请求创建 CheckPoint。

（3）NameNode 滚动（roll）正在操作的 EditLog 文件，并且将文件名由 edits_inprogress_001 修改为 edits_001。

（4）将滚动后的 EditLog 和 FsImage 文件复制到 SecondaryNameNode 中。

（5）SecondaryNameNode 将 EditLog 和 FsImage 文件加载到内存中，并且对其进行合并操作。

（6）生成新的镜像文件 fsimage.chkpoint。

（7）将 fsimage.chkpoint 文件复制到 NameNode 中。

（8）NameNode 将 fsimage.chkpoint 重命名为 fsimage。

4.5.2 EditLog 和 FsImage 文件解析

NameNode 在被格式化后，会在/opt/module/hadoop-3.1.3/data/tmp/dfs/name/current 目录下产生以下文件结构。

```
edits_0000000000000000001-0000000000000000002
edits_0000000000000000003-0000000000000000116
edits_inprogress_0000000000000000117
fsimage_0000000000000000002
fsimage_0000000000000000002.md5
fsimage_0000000000000000116
fsimage_0000000000000000116.md5
seen_txid
VERSION
```

对以上文件结构的说明如下。

- EditLog 文件：EditLog 文件中存储了对 HDFS 中文件进行操作的所有信息。文件系统客户端执行的所有事务操作首先会被记录到 EditLog 文件中。EditLog 文件以 edits_开头，后面跟一个事务 ID 范围段，正在使用的 EditLog 文件以 edits_inprogress_开头。
- FsImage 文件：是 HDFS 文件系统元数据的一个永久性检查点，是 NameNode 中元数据的快照。其中包含 HDFS 文件系统的所有目录和文件 inode 的序列化信息。每个 inode 都是一个文件或目录的元数据的内部描述方式。FsImage 文件以 fsimage_开头，后面跟一个事务 ID，这个事务 ID 表示此 FsImage 文件包含的最后一个操作事务。Hadoop 会保留最近的两个 FsImage 文件。
- seen_txid 文件：存储的是一个数字，即正在使用的 EditLog 文件中最后一个操作事务的 ID。
- VERSION 文件：存储的是正在运行的 HDFS 的版本信息、文件系统命名空间的唯一标识符、HDFS 集群作为一个整体被赋予的唯一标识符等重要信息。
- NameNode 在启动时会将最新的 FsImage 文件读入内存，同时加载与 FsImage 文件中的最后一个操作事务 ID 相连的 EditLog 文件中的更新操作，保证内存中的元数据信息是最新且同步的，可以看成 NameNode 在启动时就对 EditLog 和 FsImage 文件进行了合并。

EditLog 和 FsImage 文件都是二进制文件，所以不能直接查看，Hadoop 提供了对应的文件查看命令，可以将文件格式转换成可以直接查看的格式。

1. 使用 oiv 命令查看 FsImage 文件

（1）查看 oiv 命令，如下所示。

```
[atguigu@hadoop102 current]$ hdfs oiv
Usage: bin/hdfs oiv [OPTIONS] -i INPUTFILE -o OUTPUTFILE
Offline Image Viewer
View a Hadoop fsimage INPUTFILE using the specified PROCESSOR,
saving the results in OUTPUTFILE.

The oiv utility will attempt to parse correctly formed image files
and will abort fail with mal-formed image files.

The tool works offline and does not require a running cluster in
order to process an image file.

The following image processors are available:
  * XML: This processor creates an XML document with all elements of
    the fsimage enumerated, suitable for further analysis by XML
    tools.
  * ReverseXML: This processor takes an XML file and creates a
    binary fsimage containing the same elements.
  * FileDistribution: This processor analyzes the file size
    distribution in the image.
    -maxSize specifies the range [0, maxSize] of file sizes to be
     analyzed (128GB by default).
    -step defines the granularity of the distribution. (2MB by default)
    -format formats the output result in a human-readable fashion
     rather than a number of bytes. (false by default)
  * Web: Run a viewer to expose read-only WebHDFS API.
    -addr specifies the address to listen. (localhost:5978 by default)
    It does not support secure mode nor HTTPS.
  * Delimited (experimental): Generate a text file with all of the elements common
    to both inodes and inodes-under-construction, separated by a
    delimiter. The default delimiter is \t, though this may be
    changed via the -delimiter argument.

Required command line arguments:
-i,--inputFile <arg>   FSImage or XML file to process.

Optional command line arguments:
-o,--outputFile <arg>  Name of output file. If the specified
                       file exists, it will be overwritten.
                       (output to stdout by default)
                       If the input file was an XML file, we
                       will also create an <outputFile>.md5 file.
-p,--processor <arg>   Select which type of processor to apply
                       against image file. (XML|FileDistribution|
                       ReverseXML|Web|Delimited)
                       The default is Web.
-delimiter <arg>       Delimiting string to use with Delimited processor.
-t,--temp <arg>        Use temporary dir to cache intermediate result to generate
                  Delimited outputs. If not set, Delimited processor constructs
```

```
                          the namespace in memory before outputting text.
-h,--help                 Display usage information and exit
```

（2）基本语法如下：

```
hdfs oiv -p 文件类型 -i 镜像文件 -o 转换后文件输出路径
```

（3）执行以下命令，将 fsimage_0000000000000000025 文件转换为 fsimage.xml 文件。

```
[atguigu@hadoop102 current]$ pwd
/opt/module/hadoop-3.1.3/data/dfs/name/current

[atguigu@hadoop102 current]$ hdfs oiv -p XML -i fsimage_0000000000000000025 -o /opt/module/hadoop-3.1.3/fsimage.xml

[atguigu@hadoop102 current]$ cat /opt/module/hadoop-3.1.3/fsimage.xml
```

（4）查看 fsimage.xml 文件，部分显示结果如下。可以看到，FsImage 文件是由多个 inode 信息构成的。

```xml
<inode>
    <id>16386</id>
    <type>DIRECTORY</type>
    <name>user</name>
    <mtime>1512722284477</mtime>
    <permission>atguigu:supergroup:rwxr-xr-x</permission>
    <nsquota>-1</nsquota>
    <dsquota>-1</dsquota>
</inode>
<inode>
    <id>16387</id>
    <type>DIRECTORY</type>
    <name>atguigu</name>
    <mtime>1512790549080</mtime>
    <permission>atguigu:supergroup:rwxr-xr-x</permission>
    <nsquota>-1</nsquota>
    <dsquota>-1</dsquota>
</inode>
<inode>
    <id>16389</id>
    <type>FILE</type>
    <name>wc.input</name>
    <replication>3</replication>
    <mtime>1512722322219</mtime>
    <atime>1512722321610</atime>
    <perferredBlockSize>134217728</perferredBlockSize>
    <permission>atguigu:supergroup:rw-r--r--</permission>
    <blocks>
        <block>
            <id>1073741825</id>
            <genstamp>1001</genstamp>
            <numBytes>59</numBytes>
        </block>
    </blocks>
</inode >
```

根据上述文件内容可知，FsImage 文件中并没有记录数据块对应的 DataNode 信息，这是因为在启动集群后，NameNode 会要求 DataNode 上报数据块信息，并且在间隔一段时间后再次上报。

2．使用 oev 命令查看 EditLog 文件

（1）查看 oev 命令，如下所示。

```
[atguigu@hadoop102 current]$ hdfs oev
Usage: bin/hdfs oev [OPTIONS] -i INPUT_FILE -o OUTPUT_FILE
Offline edits viewer
Parse a Hadoop edits log file INPUT_FILE and save results
in OUTPUT_FILE.
Required command line arguments:
-i,--inputFile <arg>   edits file to process, xml (case
                       insensitive) extension means XML format,
                       any other filename means binary format.
                       XML/Binary format input file is not allowed
                       to be processed by the same type processor.
-o,--outputFile <arg>  Name of output file. If the specified
                       file exists, it will be overwritten,
                       format of the file is determined
                       by -p option

Optional command line arguments:
-p,--processor <arg>   Select which type of processor to apply
                       against image file, currently supported
                       processors are: binary (native binary format
                       that Hadoop uses), xml (default, XML
                       format), stats (prints statistics about
                       edits file)
-h,--help              Display usage information and exit
-f,--fix-txids         Renumber the transaction IDs in the input,
                       so that there are no gaps or invalid
                       transaction IDs.
-r,--recover           When reading binary edit logs, use recovery
                       mode. This will give you the chance to skip
                       corrupt parts of the edit log.
-v,--verbose           More verbose output, prints the input and
                       output filenames, for processors that write
                       to a file, also output to screen. On large
                       image files this will dramatically increase
                       processing time (default is false).


Generic options supported are:
-conf <configuration file>     specify an application configuration file
-D <property=value>            define a value for a given property
-fs <file:///|hdfs://namenode:port> specify default filesystem URL to use, overrides
'fs.defaultFS' property from configurations.
-jt <local|resourcemanager:port>  specify a ResourceManager
-files <file1,...>             specify a comma-separated list of files to be copied to
the map reduce cluster
```

```
-libjars <jar1,...>              specify a comma-separated list of jar files to be included
in the classpath
-archives <archive1,...>         specify a comma-separated list of archives to be
unarchived on the compute machines

The general command line syntax is:
command [genericOptions] [commandOptions]
```

（2）基本语法如下：

```
hdfs oev -p 文件类型 -i 编辑日志 -o 转换后文件输出路径
```

（3）执行以下命令，将 edits_0000000000000000012-0000000000000000013 文件转换成 edits.xml 文件。

```
[atguigu@hadoop102 current]$ hdfs oev -p XML -i edits_0000000000000000012-0000000000000000013
-o /opt/module/hadoop-3.1.3/edits.xml

[atguigu@hadoop102 current]$ cat /opt/module/hadoop-3.1.3/edits.xml
```

（4）查看 edits.xml 文件，部分显示结果如下：

```xml
<?xml version="1.0" encoding="UTF-8"?>
<EDITS>
    <EDITS_VERSION>-63</EDITS_VERSION>
    <RECORD>
        <OPCODE>OP_START_LOG_SEGMENT</OPCODE>
        <DATA>
            <TXID>129</TXID>
        </DATA>
    </RECORD>
    <RECORD>
        <OPCODE>OP_ADD</OPCODE>
        <DATA>
            <TXID>130</TXID>
            <LENGTH>0</LENGTH>
            <INODEID>16407</INODEID>
            <PATH>/hello7.txt</PATH>
            <REPLICATION>2</REPLICATION>
            <MTIME>1512943607866</MTIME>
            <ATIME>1512943607866</ATIME>
            <BLOCKSIZE>134217728</BLOCKSIZE>
            <CLIENT_NAME>DFSClient_NONMAPREDUCE_-1544295051_1</CLIENT_NAME>
            <CLIENT_MACHINE>192.168.10.102</CLIENT_MACHINE>
            <OVERWRITE>true</OVERWRITE>
            <PERMISSION_STATUS>
                <USERNAME>atguigu</USERNAME>
                <GROUPNAME>supergroup</GROUPNAME>
                <MODE>420</MODE>
            </PERMISSION_STATUS>
            <RPC_CLIENTID>908eafd4-9aec-4288-96f1-e8011d181561</RPC_CLIENTID>
            <RPC_CALLID>0</RPC_CALLID>
        </DATA>
    </RECORD>
    <RECORD>
        <OPCODE>OP_ALLOCATE_BLOCK_ID</OPCODE>
        <DATA>
```

```xml
            <TXID>131</TXID>
            <BLOCK_ID>1073741839</BLOCK_ID>
        </DATA>
    </RECORD>
    <RECORD>
        <OPCODE>OP_SET_GENSTAMP_V2</OPCODE>
        <DATA>
            <TXID>132</TXID>
            <GENSTAMPV2>1016</GENSTAMPV2>
        </DATA>
    </RECORD>
    <RECORD>
        <OPCODE>OP_ADD_BLOCK</OPCODE>
        <DATA>
            <TXID>133</TXID>
            <PATH>/hello7.txt</PATH>
            <BLOCK>
                <BLOCK_ID>1073741839</BLOCK_ID>
                <NUM_BYTES>0</NUM_BYTES>
                <GENSTAMP>1016</GENSTAMP>
            </BLOCK>
            <RPC_CLIENTID></RPC_CLIENTID>
            <RPC_CALLID>-2</RPC_CALLID>
        </DATA>
    </RECORD>
    <RECORD>
        <OPCODE>OP_CLOSE</OPCODE>
        <DATA>
            <TXID>134</TXID>
            <LENGTH>0</LENGTH>
            <INODEID>0</INODEID>
            <PATH>/hello7.txt</PATH>
            <REPLICATION>2</REPLICATION>
            <MTIME>1512943608761</MTIME>
            <ATIME>1512943607866</ATIME>
            <BLOCKSIZE>134217728</BLOCKSIZE>
            <CLIENT_NAME></CLIENT_NAME>
            <CLIENT_MACHINE></CLIENT_MACHINE>
            <OVERWRITE>false</OVERWRITE>
            <BLOCK>
                <BLOCK_ID>1073741839</BLOCK_ID>
                <NUM_BYTES>25</NUM_BYTES>
                <GENSTAMP>1016</GENSTAMP>
            </BLOCK>
            <PERMISSION_STATUS>
                <USERNAME>atguigu</USERNAME>
                <GROUPNAME>supergroup</GROUPNAME>
                <MODE>420</MODE>
            </PERMISSION_STATUS>
        </DATA>
    </RECORD>
</EDITS >
```

4.5.3 检查点时间设置

创建检查点的触发条件由 hdfs-default.xml 文件中的配置参数 dfs.namenode.checkpoint.period 和 dfs.namenode.checkpoint.txns 控制。在默认配置下，SecondaryNameNode 每隔一个小时创建一次检查点。但是，从上一个检查点开始，当 EditLog 文件中的操作次数达到 100 万次时，即使不到一个小时，也会创建检查点。检查操作次数是否达到标准（100 万次）的频率为每分钟一次，由 dfs.namenode.checkpoint.check.period 参数决定。这 3 个参数的具体配置如下：

```xml
<property>
    <name>dfs.namenode.checkpoint.period</name>
    <value>3600s</value>
</property>
<property>
    <name>dfs.namenode.checkpoint.txns</name>
    <value>1000000</value>
    <description>操作次数</description>
</property>
<property>
    <name>dfs.namenode.checkpoint.check.period</name>
    <value>60s</value>
    <description>1分钟检查一次操作次数</description>
</property>
```

4.5.4 DataNode 的工作机制

HDFS 中的文件数据实际是存储于 DataNode 中的，数据块存储于以 blk_为前缀的文件中，文件名中包含该文件存储的数据块的原始字节数。每个数据块文件都有一个名字相同但是带有 .meta 后缀的元数据文件，在元数据文件中存储了数据块的长度、校验和、时间戳等信息。

DataNode 的工作机制如图 4-12 所示。

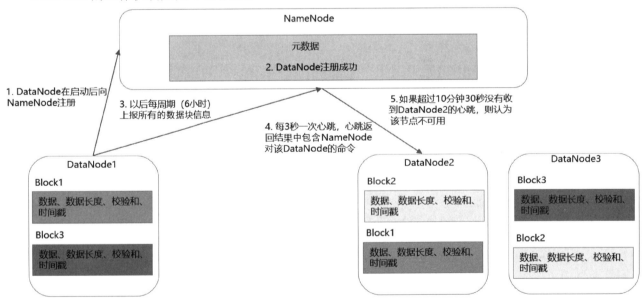

图 4-12 DataNode 的工作机制

DataNode 在启动后会向 NameNode 注册，在注册通过后，每隔一段时间都会向 NameNode 上报所有的数据块信息，时间间隔默认是 6 小时，由 dfs.blockreport.intervalMsec 参数控制，具体参数配置如下：

```xml
<property>
    <name>dfs.blockreport.intervalMsec</name>
    <value>21600000</value>
    <description>Determines block reporting interval in milliseconds.</description>
</property>
```

DataNode 每隔一段时间都会扫描本节点中的数据块信息列表，时间间隔默认是 6 小时，具体参数配置如下：

```xml
<property>
    <name>dfs.datanode.directoryscan.interval</name>
    <value>21600s</value>
    <description>Interval in seconds for Datanode to scan data directories and reconcile
    the difference between blocks in memory and on the disk.
    Support multiple time unit suffix(case insensitive), as described
    in dfs.heartbeat.interval.
    </description>
</property>
```

DataNode 每隔 3 秒都会向 NameNode 发送心跳信息，心跳返回结果中包含 NameNode 对该 DataNode 的命令，如复制数据块、删除数据块等。如果 DataNode 进程挂掉，或者发生网络故障，导致 DataNode 无法与 NameNode 正常通信，那么 NameNode 不会立即将该节点判定为死亡，它会在一段时间后将该节点判定为死亡，这段时间称为超时时长（TimeOut）。在 HDFS 的默认配置中，超时时长为 10 分钟 30 秒。超时时长的时间长短与参数 dfs.namenode.heartbeat.recheck-interval 和 dfs.heartbeat.interval 有关，这两个参数的默认配置如下：

```xml
<property>
    <name>dfs.namenode.heartbeat.recheck-interval</name>
    <value>300000</value>
</property>

<property>
    <name>dfs.heartbeat.interval</name>
    <value>3</value>
</property>
```

超时时长与以上两个参数的关系如下：

```
TimeOut = 2 * dfs.namenode.heartbeat.recheck-interval + 10 * dfs.heartbeat.interval
```

dfs.namenode.heartbeat.recheck-interval 参数的默认值为 5 分钟，dfs.heartbeat.interval 参数的默认值为 3 秒，所以超时时长默认为 10 分钟 30 秒。

需要注意的是，在 hdfs-site.xml 配置文件中，dfs.namenode.heartbeat.recheck-interval 参数的单位为毫秒，dfs.heartbeat.interval 参数的单位为秒。

4.5.5 数据完整性

对一个文件存储系统来说，如何保证内部存储文件的数据完整性是十分关键的。HDFS 将大型文件切分成多个数据块，并且将其分散存储于集群中，数据块在不同的集群节点之间复制传输，很容易将错误引入数据。

DataNode 检测数据完整性的方法如下：当数据第一次被引入系统时，计算校验和（CheckSum），当数据通过一个不可靠的传输通道进行传输（如复制数据块、读取数据块）时，再次计算校验和，如果计算得到的校验和与数据被引入时计算得到的校验和不同，则说明此数据块已经损坏。此时，客户端会尝试读取其他副本数据块中的数据。常见的校验和计算算法有 CRC-32、MD5、SHA1 等。每个 DataNode 都会在文件创建完成后周期性地验证数据的校验和，从而确保存储数据的完整性。

检测数据完整性的思路如图 4-13 所示。一种比较基础的数据校验方法是校验二进制数据中 1 的奇偶个数，如果二进制数据中 1 的个位为奇数，则奇偶校验位为 1，否则为 0。如果数据在传输过程中发生损坏，导致二进制数据中 1 的个数发生变化，则可以通过奇偶校验位发现。但是这种校验方法并不保险，在图 4-13 中的第一种情况下，1 的奇偶个数没有发生变化，但是数据已经损坏。而 CRC 校验是相对可靠的校验方法；在图 4-13 中的第二种情况下，原有数据在经过 CRC 算法计算后，可以得到 3 位校验数字，在数据传输完毕后，重新对其进行计算，并且与原校验位对比，看是否一致，具有更高的可靠性。

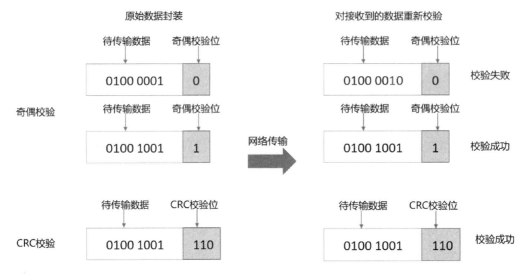

图 4-13　检测数据完整性的思路

4.6　本章总结

HDFS 是 Hadoop 的重要功能模块，可以分布式存储海量数据文件，在大数据体系中具有重要的作用。对 HDFS 的学习，从 shell 命令行操作和 API 操作入手，深入探讨了 HDFS 的文件读/写流程。在这个过程中，读者不仅需要了解如何使用 HDFS，还需要了解 HDFS 是如何提高文件存储可靠性的，包括数据块大小的设置、副本放置策略的制订、NameNode 和 SecondaryNameNode 的工作机制等。本章的重点是 HDFS 的基本架构、命令行命令、API 操作、工作机制解读，希望读者仔细研读。

第 5 章 分布式计算 MapReduce

在讲解完海量数据的存储系统 HDFS 后，下面讲解海量数据的计算。Hadoop 为海量数据提供了一个编程模型，也就是本章将要讲解的 MapReduce。MapReduce 可以充分应用集群的资源，使用大量的节点服务器共同计算同一个海量数据集，突破了单台物理计算机的性能瓶颈。本章将深入 MapReduce 核心，为读者揭开其强大的分布式计算能力的神秘面纱。

5.1 MapReduce 概述

本节从宏观的角度带领读者初步认识 MapReduce，对 MapReduce 进行简单的介绍。

5.1.1 MapReduce 定义

MapReduce 是一个分布式计算程序的编程框架，是用户开发基于 Hadoop 的数据分析应用程序的核心框架。

MapReduce 的核心功能是将用户编写的业务逻辑代码和自带的默认组件整合成一个完整的分布式计算程序，并且将其并发运行在一个 Hadoop 集群上。

1. 优点

1）易于编程。

MapReduce 简单地实现一些接口，就可以完成一个分布式计算程序，这个分布式计算程序可以分布到大量廉价的 PC 上运行。也就是说，编写一个分布式计算程序，与编写一个简单的串行程序是一样的，因此 MapReduce 编程变得非常流行。

2）良好的扩展性。

当计算资源不能满足海量数据的计算需求时，MapReduce 可以简单地通过增加节点数量增强计算能力。

3）高容错性。

MapReduce 的设计初衷是使程序能够部署在廉价的 PC 上，这要求它具有很高的容错性。例如，一个节点宕机，它可以将该节点上的计算任务转移到另一个节点上运行，不至于导致这个任务运行失败，并且这个过程不需要人工参与，完全是由 Hadoop 内部完成的。

4）适合 PB 级以上数据的离线处理。

MapReduce 可以实现上千台服务器集群并发工作，提供数据处理能力。

2. 缺点

1）不擅长实时计算。

MapReduce 无法像 MySQL 一样，在毫秒级或秒级内返回结果。

2）不擅长流式计算。

流式计算的输入数据是动态的，而 MapReduce 的输入数据是静态的，不能动态变化。MapReduce 的设计特点决定了其数据源必须是静态的。

3）不擅长 DAG（有向无环图）计算。

多个应用程序之间存在依赖关系，后一个应用程序的输入数据是前一个应用程序的输出数据。在这种情况下，MapReduce 不是不能运行，而是在运行后，每个 MapReduce 程序的输出结果都会被写入磁盘，会造成大量的磁盘 I/O，导致性能非常低下。

5.1.2 MapReduce 核心思想

下面通过一个简单的词频统计程序讲解 MapReduce 程序是如何工作的，如图 5-1 所示。

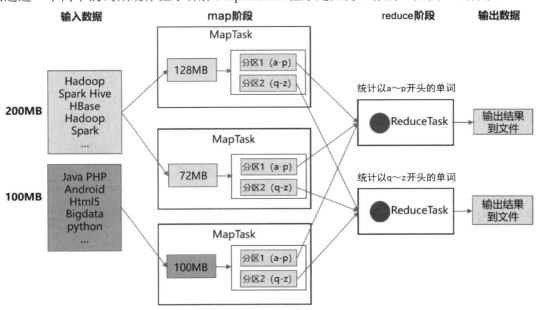

图 5-1　词频统计程序

这个词频统计程序主要用于统计一个数据集（由 2 个文件构成，其大小分别是 200MB 和 100MB）中每个单词出现的次数，并且将统计结果按照单词首字母分为 2 个文件，将以 a～p 开头的单词存储于一个文件中，将以 q～z 开头的单词存储于另一个文件中。当数据量较少时，这是一个很容易完成的工作，但是当数据量达到 300MB 时，MapReduce 是如何完成工作的？

首先思考一下这个数据集在 Hadoop 中是如何存储的。根据前面介绍过的 HDFS 存储流程，可以将其中 200MB 的文件分成 128MB 和 72MB 的 2 个数据块进行存储，将另外 100MB 的文件单独作为 1 个数据块进行存储。针对这 3 个数据块，Hadoop 会为每个数据块构建一个 map 任务，并且由该任务运行用户自定义的 map() 方法，从而处理数据块中的每条记录。

将数据集切分成多个数据块，不仅在进行 HDFS 存储时具有重大意义，在进行分布式计算时显然也有很多好处。对数据集进行合理切分，分析计算每一个数据块的时间都明显少于处理整个数据集的时间。此外，Hadoop 会将 map 任务尽量运行在数据块所在的节点服务器上，从而获得最佳的计算性能，从而大幅节省宝贵的集群带宽资源。

回到词频统计程序，每个 map 任务都会将其负责的数据块按行进行处理，并且按空格划分单词，将每个单词都处理成键/值对（word，1）的形式，并且按照单词首字母分成 2 个分区溢写至本地磁盘中。

在所有的 map 任务运行完毕并将运行结果溢写至本地磁盘中后，启动 2 个 reduce 任务，分别拉取 map 任务的输出结果文件，将相同的单词合并起来统计其出现次数，并且将最终计算结果输出至指定位置（一般为 HDFS 中）。

根据以上计算过程，我们可以总结得出以下几点。
- 分布式计算程序通常需要分成至少 2 个阶段——map 阶段和 reduce 阶段。
- map 阶段的 MapTask 并发实例完全并行运行，互不相关。
- reduce 阶段的 ReduceTask 并发实例互不相关，但是它们的数据依赖于 map 阶段的所有 MapTask 并发实例的输出结果。
- MapReduce 编程模型只能包含一个 map 阶段和一个 reduce 阶段，如果用户的业务逻辑非常复杂，那么多个 MapReduce 程序只能串行运行。

在运行一个完整的 MapReduce 程序时，在集群中会发现以下 3 类实例进程。
- MRAppMaster：负责整个程序的过程调度及状态协调。
- MapTask：负责 map 阶段的整个数据处理流程。
- ReduceTask：负责 reduce 阶段的整个数据处理流程。

在后面的章节中，我们会自己编写 MapReduce 程序并将其提交到集群上运行，读者可以自行查看。进程的运行情况可以进一步印证 MapReduce 程序会将计算过程分为两个阶段——map 阶段和 reduce 阶段。

5.2　MapReduce 编程入门

在了解了 MapReduce 的基本概念后，本节从官方提供的 WordCount 程序入手，步步拆解，带领读者编写第一个 MapReduce 程序。

5.2.1　官方示例程序 WordCount 源码

在第 3 章，我们在安装 Hadoop 后，会运行 Hadoop 提供的示例程序 WordCount。在 Hadoop 的安装目录下，打开 share/hadoop/mapreduce 文件夹，找到 hadoop-mapreduce-examples-3.1.3.jar，使用解压缩工具打开压缩包，发现 WordCount 程序中有 Map 类、Reduce 类和驱动类，如图 5-2 所示。

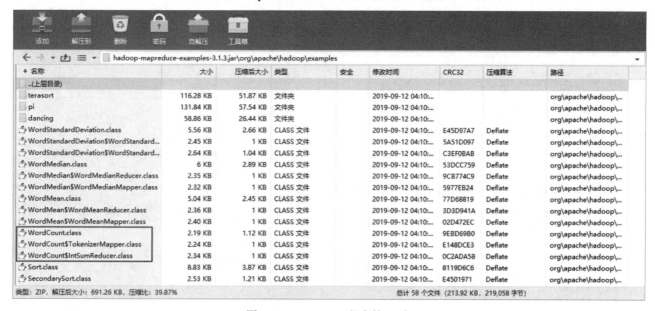

图 5-2　WordCount 程序的 jar 包

使用反编译工具进一步解析以上字节码文件，可以看到更详细的实现逻辑。

5.2.2 编程规范

用户编写的 MapReduce 程序主要分为 3 部分：Mapper、Reducer 和 Driver。

1．Mapper 组件

- 用户自定义类会继承 Mapper 类，并且给出 4 个泛型，即 map()方法的输入键、输入值、输出键和输出值的数据类型。
- Mapper 组件的输入数据是键/值对的形式，输入键是一个长整型的偏移量，输入值是一行文本。
- Mapper 组件的业务逻辑写在 map()方法中，map()方法的实现逻辑由用户自行编写。
- Mapper 组件的输出数据也是键/值对的形式，输出键和输出值的数据类型由 map()方法的处理逻辑决定。
- map()方法对每个输入数据键/值对都只调用一次。

2．Reducer 组件

- 用户自定义类会继承 Reducer 类，并且给出 4 个泛型，即 reduce()方法的输入键、输入值、输出键和输出值的数据类型。
- Reducer 组件的输入数据是键/值对的形式，其中，输入键和输入值的数据类型与 Mapper 的输出键和输出值的数据类型必须匹配。
- Reducer 组件的业务逻辑写在 reduce()方法中，reduce()方法的实现逻辑由用户自行编写。
- ReduceTask 进程对每组相同键的键/值对都只调用一次 reduce()方法。

3．Driver 驱动器

这部分代码主要负责运行整个 MapReduce 程序、指定 MapReduce 程序的执行规范，以及控制整个 MapReduce 程序的运行。在这部分代码中，我们首先构建一个 Job 对象，通过 Job 对象指定输入数据和输出数据的路径，指定 MapReduce 程序的 Mapper 组件和 Reducer 组件，指定输出键和输出值的数据类型。以上都是必需的设置，用户还可以进行各种自定义设置，在后面的章节中会进行讲解。

5.2.3 WordCount 案例实操

1．需求分析

1）统计数据。

统计数据如下：

```
atguigu atguigu
ss ss
cls cls
jiao
banzhang
xue
hadoop
```

2）具体需求分析。

统计指定的文本文件中每个单词出现的总次数，并且将其输出至文件中。

3）期望输出数据。

期望输出数据如下：

```
atguigu 2
banzhang    1
cls 2
hadoop  1
```

```
jiao    1
ss      2
xue 1
```

4）实现思路分析。

WordCount 程序的实现思路如图 5-3 所示。在 Mapper 组件中，输入数据是一行一行的文本，将一行文本按照空格切分成多个单词，然后将单词转换成(word,1)的形式输出。例如，将第一行文本"atguigu atguigu"按照空格切分成两个单词"atguigu"，并且将其转换为两个键/值对(atguigu,1)。

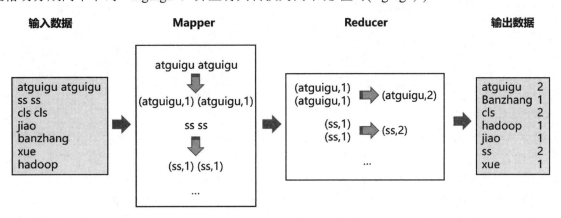

图 5-3　WordCount 程序的实现思路

在 Reducer 组件中，输入数据是(word,1)形式的数据，每次获得的是同一个单词的一组键/值对数据，汇总这组键/值对的 value 和，如将 2 个(atguigu,1)键/值对汇总成(atguigu,2)。

在编写完 Mapper 组件和 Reducer 组件的代码后，编写 Driver 驱动器的代码，进行必要的配置和类型指定，完成作业的提交。

2．本地测试

1）环境准备。

（1）使用 IDEA 创建一个 Maven 工程，将其命名为"MapReduceDemo"。

（2）在 pom.xml 文件中添加以下依赖。

```
<dependencies>
    <dependency>
        <groupId>org.apache.hadoop</groupId>
        <artifactId>hadoop-client</artifactId>
        <version>3.1.3</version>
    </dependency>
    <dependency>
        <groupId>org.slf4j</groupId>
        <artifactId>slf4j-log4j12</artifactId>
        <version>1.7.30</version>
    </dependency>
</dependencies>
```

（3）在 MapReduceDemo 工程的 src/main/resources 目录下新建一个文件 log4j.properties，在该文件中添加以下内容。

```
log4j.rootLogger=INFO, stdout
log4j.appender.stdout=org.apache.log4j.ConsoleAppender
log4j.appender.stdout.layout=org.apache.log4j.PatternLayout
log4j.appender.stdout.layout.ConversionPattern=%d %p [%c] - %m%n
log4j.appender.logfile=org.apache.log4j.FileAppender
```

```
log4j.appender.logfile.File=target/spring.log
log4j.appender.logfile.layout=org.apache.log4j.PatternLayout
log4j.appender.logfile.layout.ConversionPattern=%d %p [%c] - %m%n
```

（4）创建一个包，将其命名为"com.atguigu.mapreduce.wordcount"。

（5）在 MapReduceDemo 工程的 src/main/resources 目录下创建输入数据文件 input.txt，在该文件中输入以下内容。

```
atguigu atguigu
ss ss
cls cls
jiao
banzhang
xue
hadoop
```

2）编写程序。

（1）编写 WordCountMapper 类，使其继承 Mapper 类。Mapper 类是一个泛型类，有 4 个形参，分别是 map()方法的输入键、输入值、输出键和输出值的数据类型。在 WordCount 程序中，输入键是一个长整型偏移量，输入值是一行文本，输出键是文本，输出值是一个整数。Hadoop 为常见的数据类型提供了序列化传输类型，长整型对应 LongWritable 类型，文本对应 Text 类型，整数对应 IntWritable 类型。

继承 Mapper 类的关键是要重写 map()方法。在 map()方法中，首先将获取的一行文本转换成 String 类型的数据，然后将其按照空格进行切分，再将切分得到的单词转换成键/值对，最后通过 Context 对象将数据写入输出数据。

```java
package com.atguigu.mapreduce.wordcount;
import java.io.IOException;
import org.apache.hadoop.io.IntWritable;
import org.apache.hadoop.io.LongWritable;
import org.apache.hadoop.io.Text;
import org.apache.hadoop.mapreduce.Mapper;

public class WordCountMapper extends Mapper<LongWritable, Text, Text, IntWritable>{

    Text k = new Text();
    IntWritable v = new IntWritable(1);

    @Override
    protected void map(LongWritable key, Text value, Context context) throws IOException, InterruptedException {

        // 1 获取一行
        String line = value.toString();

        // 2 切割
        String[] words = line.split(" ");

        // 3 输出
        for (String word : words) {

            k.set(word);
            context.write(k, v);
        }
```

        }
    }

（2）编写 WordCountReducer 类，使其继承 Reducer 类，需要指定 4 个泛型，用于规范输入键、输入值、输出键和输出值的数据类型。输入键和输入值的数据类型必须匹配 map()方法的输出键和输出值的数据类型，分别为 Text 类型和 IntWritable 类型；输出键和输出值分别是单词和单词出现次数，所以也定为 Text 类型和 IntWritable 类型。

reduce()方法每次获取的都是相同输入键的一组键/值对，所以值以迭代器的形式给出，遍历迭代器，对值进行累加求和，即可得到每个单词出现的总次数。

```java
package com.atguigu.mapreduce.wordcount;
import java.io.IOException;
import org.apache.hadoop.io.IntWritable;
import org.apache.hadoop.io.Text;
import org.apache.hadoop.mapreduce.Reducer;

public class WordCountReducer extends Reducer<Text, IntWritable, Text, IntWritable>{

int sum;
IntWritable v = new IntWritable();

    @Override
    protected void reduce(Text key, Iterable<IntWritable> values,Context context) throws IOException, InterruptedException {

        // 1 累加求和
        sum = 0;
        for (IntWritable count : values) {
            sum += count.get();
        }

        // 2 输出
        v.set(sum);
        context.write(key,v);
    }
}
```

（3）编写 Driver 类。在 Driver 类中，首先获取默认配置信息及 Job 对象。通过 Job 对象进行以下操作。
① 关联当前 Driver 程序的主类。
② 关联 Mapper 组件和 Reducer 组件。
③ 设置 Mapper 组件输出键和输出值的数据类型。
④ 设置 Reducer 组件输出键和输出值的数据类型。
⑤ 设置输入路径和输出路径。
⑥ 提交任务。

```java
package com.atguigu.mapreduce.wordcount;
import java.io.IOException;
import org.apache.hadoop.conf.Configuration;
import org.apache.hadoop.fs.Path;
import org.apache.hadoop.io.IntWritable;
import org.apache.hadoop.io.Text;
import org.apache.hadoop.mapreduce.Job;
import org.apache.hadoop.mapreduce.lib.input.FileInputFormat;
```

```java
import org.apache.hadoop.mapreduce.lib.output.FileOutputFormat;

public class WordCountDriver {

    public static void main(String[] args) throws IOException, ClassNotFoundException, InterruptedException {

        // 1 获取配置信息及Job对象job
        Configuration conf = new Configuration();
        Job job = Job.getInstance(conf);

        // 2 关联当前Driver程序的jar包
        job.setJarByClass(WordCountDriver.class);

        // 3 关联Mapper组件和Reducer组件的jar包
        job.setMapperClass(WordCountMapper.class);
        job.setReducerClass(WordCountReducer.class);

        // 4 设置Mapper组件输出KV的数据类型
        job.setMapOutputKeyClass(Text.class);
        job.setMapOutputValueClass(IntWritable.class);

        // 5 设置最终输出KV的数据类型
        job.setOutputKeyClass(Text.class);
        job.setOutputValueClass(IntWritable.class);

        // 6 设置输入路径和输出路径
        FileInputFormat.setInputPaths(job, new Path("input.txt"));
        FileOutputFormat.setOutputPath(job, new Path("output"));

        // 7 提交Job
        boolean result = job.waitForCompletion(true);
        System.exit(result ? 0 : 1);
    }
}
```

3）本地测试。

（1）确保在计算机本地已经按照第 4 章中的内容配置好 HADOOP_HOME 环境变量及 Windows 运行依赖。

（2）运行 Driver 类的 main()方法，观察运行结果，在工程的根目录下出现输出文件夹，打开输出文件夹，可以看到输出结果，如图 5-4 所示。

名称	修改日期	类型	大小
._SUCCESS.crc	2021/11/11 15:11	CRC 文件	1 KB
.part-r-00000.crc	2021/11/11 15:11	CRC 文件	1 KB
_SUCCESS	2021/11/11 15:11		0 KB
part-r-00000	2021/11/11 15:11		1 KB

图 5-4 输出结果

输出文件夹中有 4 个文件，其中真正的数据结果文件是 part-r-00000，该文件中的内容如下。可以发现，已经对原文件进行了词频统计。

```
atguigu 2
banzhang 1
cls 2
hadoop 1
jiao 1
ss 2
xue 1
```

3．集群测试

（1）在 pom.xml 文件中添加打包插件依赖，具体如下：

```xml
<build>
    <plugins>
        <plugin>
            <artifactId>maven-compiler-plugin</artifactId>
            <version>3.6.1</version>
            <configuration>
                <source>1.8</source>
                <target>1.8</target>
            </configuration>
        </plugin>
        <plugin>
            <artifactId>maven-assembly-plugin</artifactId>
            <configuration>
                <descriptorRefs>
                    <descriptorRef>jar-with-dependencies</descriptorRef>
                </descriptorRefs>
            </configuration>
            <executions>
                <execution>
                    <id>make-assembly</id>
                    <phase>package</phase>
                    <goals>
                        <goal>single</goal>
                    </goals>
                </execution>
            </executions>
        </plugin>
    </plugins>
</build>
```

因为程序最终要被提交到集群上运行，所以输入路径和输出路径不是在代码中设置的，而是通过参数传入的，将输入路径和输出路径的设置代码修改如下：

```
// 6 设置输入路径和输出路径
FileInputFormat.setInputPaths(job, new Path(args[0]));
FileOutputFormat.setOutputPath(job, new Path(args[1]));
```

（2）将程序打包成 jar 包，具体操作如图 5-5 所示。

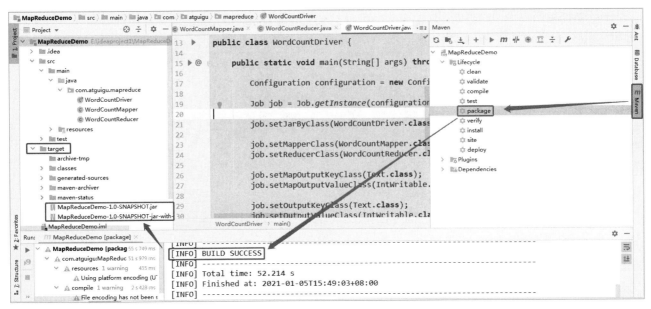

图 5-5　将程序打包成 jar 包

（3）将不带依赖的 jar 包名称修改为"wc.jar"，并且将该 jar 包复制到 Hadoop 集群的 /opt/module/hadoop-3.1.3 路径下。

（4）启动 Hadoop 集群，代码如下：

```
[atguigu@hadoop102 hadoop-3.1.3]$ sbin/start-dfs.sh
[atguigu@hadoop103 hadoop-3.1.3]$ sbin/start-yarn.sh
```

（5）创建输入文件 input，并且输入以下内容。

```
[atguigu@hadoop102 hadoop-3.1.3]$ vim input
atguigu atguigu
ss ss
cls cls
jiao
banzhang
xue
hadoop
```

（6）将 input 文件上传至 HDFS 的 /user/atguigu/input 路径下。

```
[atguigu@hadoop102 hadoop-3.1.3]$ hadoop fs -put ./input /user/atguigu/input
```

（7）执行 WordCount 程序进行测试。

```
[atguigu@hadoop102 hadoop-3.1.3]$ hadoop jar wc.jar
com.atguigu.mapreduce.wordcount.WordCountDriver /user/atguigu/input /user/atguigu/output
```

（8）查看 HDFS 的 /user/atguigu/ 路径下有没有出现输出文件 output。执行以下命令，查看 output 文件中的内容。

```
[atguigu@hadoop102 hadoop-3.1.3]$ hadoop fs -cat /user/atguigu/output
atguigu 2
banzhang 1
cls 2
hadoop 1
jiao 1
ss 2
xue 1
```

5.3 Hadoop 的序列化

在编写 MapReduce 程序时，我们使用了 Hadoop 提供的序列化数据类型，如 Text、LongWritable、IntWritable 等，本节详细介绍 Hadoop 的序列化。

5.3.1 序列化概述

1．什么是序列化

序列化（serialization）是指将内存中的对象转换成字节序列（或其他数据传输协议），以便将其存储于磁盘中（持久化存储）或进行网络传输的过程。

反序列化是指将收到的字节序列（或其他数据传输协议）或磁盘中的持久化数据转换成内存中对象的过程。

2．为什么要进行序列化

在一般情况下，"活的"对象只生存在内存中，在关机、断电后就没有了。而且"活的"对象只能由本地的进程使用，不能被发送到网络上的另一台计算机中。通过序列化可以将对象在进程之间进行通信，以及使对象持久化存储。

3．为什么不用 Java 的序列化机制

Java 具有一套序列化机制（Writable），但是 Java 的序列化机制是一个重量级序列化框架（Serializable）。一个对象在被序列化后，会附带很多额外的信息（各种校验信息、Header、继承体系等），不便于在网络中进行高效传输。所以，Hadoop 自己开发了一套序列化机制。

4．Hadoop 序列化机制的特点

- 紧凑：紧凑的格式有助于高效使用存储空间，充分利用网络带宽。
- 快速：序列化和反序列化的性能开销小，可以实现进程之间的快速通信。
- 互操作：统一的序列化框架可以支持多语言与服务器的交互。

5.3.2 Writable 接口

Hadoop 使用的序列化机制是 Writable。Writable 是 Hadoop 提供的一个接口，Writable 接口的定义代码如下。在 Writable 接口中定义了 2 个方法，分别是 write()方法和 readFields()方法。write()方法主要用于将类中的信息写入 DataOutput 二进制流，readFields()方法主要用于从 DataInput 二进制流中读取信息。

```
package org.apache.hadoop.io;

import java.io.DataOutput;
import java.io.DataInput;
import java.io.IOException;

public interface Writable {
  void write(DataOutput out) throws IOException;
  void readFields(DataInput in) throws IOException;
}
```

Hadoop 自带的序列化数据类型非常多，并且为 Java 中的基本数据类型提供了 Writable 接口的实现类，如表 5-1 所示，Writable 接口的所有实现类都提供了 get()方法和 set()方法，分别用于获取和存储所封装的值。

表 5-1 Java 中基本数据类型的 Writable 接口实现类

Java 中的基本数据类型	Hadoop Writable 接口的实现类
Boolean	BooleanWritable
Byte	ByteWritable
Int	IntWritable
Float	FloatWritable
Long	LongWritable
Double	DoubleWritable
String	Text
Map	MapWritable
Array	ArrayWritable
Null	NullWritable

用户不仅可以使用 Hadoop 提供的序列化数据类型，还可以自定义 Writable 接口的实现类，用于在 Hadoop 环境中实现序列化，在下一节将进行详细讲解。

5.3.3 序列化案例实操

1. 自定义 bean 对象，实现序列化接口 Writable

JavaBean 是一种遵循某套开发规范的类，在开发过程中可以存储具有整体性的实体信息。例如，一个学生的姓名、年龄、性别、学号、班级等信息，一个商品的名称、编号、价格等信息，都可以存储于一个 JavaBean 中。JavaBean 要符合以下规则。

- 必须有一个无参构造方法。
- 所有属性必须私有化。
- 私有化属性必须提供 public 类型的 getter()方法和 setter()方法。

在开发过程中，使用 JavaBean 封装信息的对象称为 bean 对象。在 Hadoop 框架内部，一个未经序列化的 bean 对象不能直接进行进程之间的传输，需要实现一些序列化的接口。下面通过 Hadoop 提供的 Writable 接口，实现对象的序列化，具体流程如下。

（1）创建一个封装流量信息的 JavaBean，将其命名为 FlowBean，使其继承 Writable 接口，其中有 3 个重要属性，分别是上行流量、下行流量和总流量。

```
public class FlowBean implements Writable {
    private long upFlow;
    private long downFlow;
    private long sumFlow;
}
```

（2）在进行反序列化时，需要通过反射技术调用无参构造方法，所以必须有无参构造方法。

```
public FlowBean() {
    super();
}
```

（3）重写序列化方法，代码如下。通过 DataOutput 的实例化对象 out 将属性值写入二进制流，针对不同的数据类型调用不同的方法。例如，长整型参数调用 writeLong()方法，整型参数调用 writeInt()方法。

```
@Override
public void write(DataOutput out) throws IOException {
    out.writeLong(upFlow);
    out.writeLong(downFlow);
    out.writeLong(sumFlow);
}
```

（4）重写反序列化方法，代码如下。通过 DataInput 的实例化对象 in 读取二进制流中的值。需要注意的是，反序列化的顺序和序列化的顺序完全一致，并且读取方法的调用方式与参数的数据类型也需要一一对应。

```java
@Override
public void readFields(DataInput in) throws IOException {
    upFlow = in.readLong();
    downFlow = in.readLong();
    sumFlow = in.readLong();
}
```

（5）要将结果显示在文件中，需要重写 toString()方法，字段与字段之间用"\t"进行分隔，方便后续使用。

```java
@Override
public String toString() {
    return "FlowBean\t" +
            "\t" + upFlow +
            "\t" + downFlow +
            "\t" + sumFlow ;
}
```

（6）如果需要将 FlowBean 类放在 key 中传输，则需要实现 Comparable 接口，因为 MapReduce 框架中的 shuffle 过程要求 key 必须可排序（详见后面的排序案例）。

```java
@Override
public int compareTo(FlowBean o) {
    // 按照 sumflow 倒序（从大到小）排列
    return this.sumFlow > o.getSumFlow() ? -1 : 1;
}
```

2．序列化案例实操

1）需求分析。

有一个数据文件，存储了大量手机号码、IP 地址、浏览页面、上行流量、下行流量等信息，要求统计每个手机号码耗费的总上行流量、总下行流量、总流量。

在 inputflow.txt 文件中输入以下数据，字段与字段之间使用 tab 制表符进行分隔。

```
1   13736230513 192.196.100.1   www.atguigu.com 2481    24681   200
2   13846544121 192.196.100.2                   264     0       200
3   13956435636 192.196.100.3                   132     1512    200
4   13966251146 192.168.100.1                   240     0       404
5   18271575951 192.168.100.2   www.atguigu.com 1527    2106    200
6   84188413    192.168.100.3   www.atguigu.com 4116    1432    200
7   13590439668 192.168.100.4                   1116    954     200
8   15910133277 192.168.100.5   www.hao123.com  3156    2936    200
9   13729199489 192.168.100.6                   240     0       200
10  13630577991 192.168.100.7   www.shouhu.com  6960    690     200
11  15043685818 192.168.100.8   www.baidu.com   3659    3538    200
12  15959002129 192.168.100.9   www.atguigu.com 1938    180     500
13  13560439638 192.168.100.10                  918     4938    200
14  13470253144 192.168.100.11                  180     180     200
15  13682846555 192.168.100.12  www.qq.com      1938    2910    200
16  13992314666 192.168.100.13  www.gaga.com    3008    3720    200
17  13509468723 192.168.100.14  www.qinghua.com 7335    110349  404
18  18390173782 192.168.100.15  www.sogou.com   9531    2412    200
19  13975057813 192.168.100.16  www.baidu.com   11058   48243   200
```

```
20  13768778790 192.168.100.17                     120    120 200
21  13568436656 192.168.100.18  www.alibaba.com 2481    24681   200
22  13568436656 192.168.100.19                    1116    954 200
```

输入数据中每个字段的含义如下：

7	13560436666	120.196.100.99	1116	954	200
id	手机号码	网络 IP	上行流量	下行流量	网络状态码

期望的输出数据格式如下：

13560436666	1116	954	2070
手机号码	上行流量	下行流量	总流量

2）实现思路分析。

序列化案例的实现思路如图 5-6 所示。在 map 阶段对接收的每行数据都进行切分，抽取关键字段（手机号码、上行流量、下行流量、总流量），将其封装成 FlowBean 对象，并且将手机号码字段设置为 key，然后将数据输出。在 reduce 阶段，每次接收到同一个手机号码的一组 FlowBean 对象，都对这组 FlowBean 对象进行汇总，计算得到总流量，将手机号码字段设置为 key，然后将数据输出。

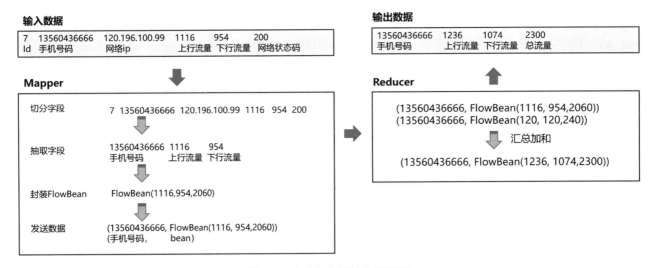

图 5-6　序列化案例的实现思路

3）编写 MapReduce 程序。

（1）按照前面的要求编写统计流量的 FlowBean 类，使其继承 Writable 接口，代码如下：

```java
package com.atguigu.mapreduce.writable;

import org.apache.hadoop.io.Writable;
import java.io.DataInput;
import java.io.DataOutput;
import java.io.IOException;

//1 继承Writable接口
public class FlowBean implements Writable {

    private long upFlow;    //上行流量
    private long downFlow;  //下行流量
    private long sumFlow;   //总流量

    //2 提供无参构造方法
```

```java
public FlowBean() {
}

//3 提供3个参数的getter()方法和setter()方法
public long getUpFlow() {
    return upFlow;
}

public void setUpFlow(long upFlow) {
    this.upFlow = upFlow;
}

public long getDownFlow() {
    return downFlow;
}

public void setDownFlow(long downFlow) {
    this.downFlow = downFlow;
}

public long getSumFlow() {
    return sumFlow;
}

public void setSumFlow(long sumFlow) {
    this.sumFlow = sumFlow;
}

public void setSumFlow() {
    this.sumFlow = this.upFlow + this.downFlow;
}

//4 实现序列化和反序列化方法，注意顺序一定要保持一致
@Override
public void write(DataOutput dataOutput) throws IOException {
    dataOutput.writeLong(upFlow);
    dataOutput.writeLong(downFlow);
    dataOutput.writeLong(sumFlow);
}

@Override
public void readFields(DataInput dataInput) throws IOException {
    this.upFlow = dataInput.readLong();
    this.downFlow = dataInput.readLong();
    this.sumFlow = dataInput.readLong();
}

//5 重写toString()方法
@Override
public String toString() {
    return upFlow + "\t" + downFlow + "\t" + sumFlow;
```

```
    }
}
```

（2）编写 Mapper 组件，代码如下：

```java
package com.atguigu.mapreduce.writable;

import org.apache.hadoop.io.LongWritable;
import org.apache.hadoop.io.Text;
import org.apache.hadoop.mapreduce.Mapper;
import java.io.IOException;

public class FlowMapper extends Mapper<LongWritable, Text, Text, FlowBean> {
    private Text outK = new Text();
    private FlowBean outV = new FlowBean();

    @Override
    protected void map(LongWritable key, Text value, Context context) throws IOException, InterruptedException {

        //1 获取一行数据，将其转换成字符串
        String line = value.toString();

        //2 切割数据
        String[] split = line.split("\t");

        //3 抓取我们需要的数据:手机号码、上行流量、下行流量
        String phone = split[1];
        String up = split[split.length - 3];
        String down = split[split.length - 2];

        //4 封装 outK、outV
        outK.set(phone);
        outV.setUpFlow(Long.parseLong(up));
        outV.setDownFlow(Long.parseLong(down));
        outV.setSumFlow();

        //5 写出 outK、outV
        context.write(outK, outV);
    }
}
```

思考：为什么要提前定义 key 和 value？

（3）编写 Reducer 组件，代码如下：

```java
package com.atguigu.mapreduce.writable;

import org.apache.hadoop.io.Text;
import org.apache.hadoop.mapreduce.Reducer;
import java.io.IOException;

public class FlowReducer extends Reducer<Text, FlowBean, Text, FlowBean> {
    private FlowBean outV = new FlowBean();
    @Override
```

```java
protected void reduce(Text key, Iterable<FlowBean> values, Context context) throws
IOException, InterruptedException {

    long totalUp = 0;
    long totalDown = 0;

    //1 遍历values，将其中的上行流量、下行流量分别累加
    for (FlowBean flowBean : values) {
        totalUp += flowBean.getUpFlow();
        totalDown += flowBean.getDownFlow();
    }

    //2 封装outV
    outV.setUpFlow(totalUp);
    outV.setDownFlow(totalDown);
    outV.setSumFlow();

    //3 写出key和outV
    context.write(key,outV);
  }
}
```

（4）编写 Driver 驱动类，代码如下：

```java
package com.atguigu.mapreduce.writable;

import org.apache.hadoop.conf.Configuration;
import org.apache.hadoop.fs.Path;
import org.apache.hadoop.io.Text;
import org.apache.hadoop.mapreduce.Job;
import org.apache.hadoop.mapreduce.lib.input.FileInputFormat;
import org.apache.hadoop.mapreduce.lib.output.FileOutputFormat;
import java.io.IOException;

public class FlowDriver {
    public static void main(String[] args) throws IOException, ClassNotFoundException,
InterruptedException {

        //1 获取Job对象job
        Configuration conf = new Configuration();
        Job job = Job.getInstance(conf);

        //2 关联当前Driver类
        job.setJarByClass(FlowDriver.class);

        //3 关联Mapper组件和Reducer组件
        job.setMapperClass(FlowMapper.class);
        job.setReducerClass(FlowReducer.class);

        //4 设置map端输出KV的数据类型
        job.setMapOutputKeyClass(Text.class);
        job.setMapOutputValueClass(FlowBean.class);
```

```java
        //5 设置程序最终输出KV的数据类型
        job.setOutputKeyClass(Text.class);
        job.setOutputValueClass(FlowBean.class);

        //6 设置程序的输入路径和输出路径
        FileInputFormat.setInputPaths(job, new Path("D:\\inputflow"));
        FileOutputFormat.setOutputPath(job, new Path("D:\\flowoutput"));

        //7 提交Job
        boolean b = job.waitForCompletion(true);
        System.exit(b ? 0 : 1);
    }
}
```

（5）执行 Driver 类中的 main()方法，查看输出文件中的内容，如下所示。可以发现计算结果正常。根据计算结果可知，FlowBean 对象实现了进程之间的通信。由此可知，JavaBean 对象通过实现 Writable 接口，可以实现进程之间的通信。

```
13470253144 180 180 360
13509468723 7335 110349 117684
13560439638 918 4938 5856
13568436656 3597 25635 29232
13590439668 1116 954 2070
13630577991 6960 690 7650
13682846555 1938 2910 4848
13729199489 240 0 240
13736230513 2481 24681 27162
13768778790 120 120 240
13846544121 264 0 264
13956435636 132 1512 1644
13966251146 240 0 240
13975057813 11058 48243 59301
13992314666 3008 3720 6728
15043685818 3659 3538 7197
15910133277 3156 2936 6092
15959002129 1938 180 2118
18271575951 1527 2106 3633
18390173782 9531 2412 11943
84188413 4116 1432 5548
```

5.4 MapReduce 框架原理之 InputFormat 数据输入

通过前面的学习，我们已经知道，MapReduce 计算框架将数据的计算过程分为数据输入阶段、map 阶段、reduce 阶段和数据输出阶段，本节主要介绍数据输入阶段。在数据输入阶段，我们需要考虑如何划分数据，可以在提高任务处理并发度的同时提高集群性能。

5.4.1 切片与 MapTask 并行度决定机制

1．问题引出

MapTask 的并行度决定 map 阶段的任务处理并发度，进而影响整个 Job 的处理速度。

思考：对于 1GB 的数据，如果启动 8 个 MapTask，那么可以提高集群的并发处理能力。那么对于 1KB 的数据，如果也启动 8 个 MapTask，那么可以提高集群的并发处理能力吗？MapTask 并行任务是否越多越好呢？MapTask 的并行度应该如何决定？

2．MapTask 的并行度决定机制

在讲解 MapTask 的并行度决定机制前，需要先讲解两个概念——数据块和数据切片。

HDFS 可以在物理层面上将数据分成若干个数据块（Block）。数据块是 HDFS 的数据存储单位。

数据切片是指在逻辑上对输入数据进行切分，并不会在物理层面上将输入数据切分成片。数据切片是 MapReduce 程序计算输入数据的单位，一个数据切片可以对应启动一个 MapTask。

下面通过一个案例讲解数据切片大小是如何影响 MapTask 的并行度的。假设一个数据集由两个文件构成，这两个文件的大小分别为 300MB 和 100MB。假设将数据切片大小设置为 128MB，那么会将这两个文件切分成 4 片，切片信息如下。

- split1：0～128MB。
- split2：129～256MB。
- split3：257～300MB。
- split4：0～100MB。

具体切片情况如图 5-7 所示。根据图 5-7 可知，数据切片大小与数据块大小正好吻合，每个数据块都是一个数据切片。

假设将数据切片大小设置为128MB

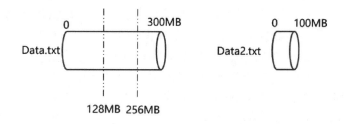

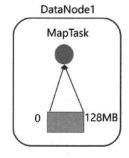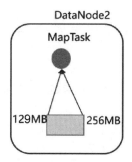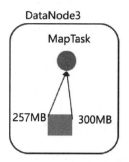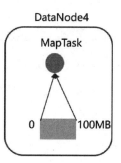

图 5-7 数据切片大小为 128MB 时的具体切片情况

假设将数据切片大小设置为 100MB，那么同样会将这两个文件切分成 4 片，切片信息如下。

- split1：0～100MB。
- spilt2：101～200MB。
- split3：201～300MB。
- split4：0～100MB。

具体切片情况如图 5-8 所示。根据图 5-8 可知，数据切片大小与数据块大小不吻合。

假设将数据切片大小设置为100MB

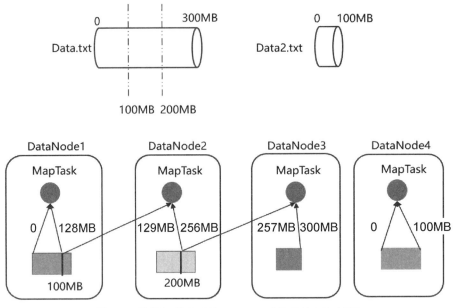

图 5-8 数据切片大小为 100MB 时的具体切片情况

根据以上案例可知，一个 MapReduce 任务的 map 阶段并行度由客户端在提交任务时的切片情况决定，如果设置的数据切片较小，则可以将任务切分成更多个数据切片。在默认情况下，数据切片大小与数据块大小相同，或者数据切片大小是数据块大小的倍数，即 128MB 或 128MB 的倍数。这是因为，与数据块大小相同的数据切片，可以保证一个 MapTask 处理的任务正好是一个数据块，Hadoop 可以将该 MapTask 发送给数据块所在的节点服务器，实现计算的数据本地化，提升计算性能，节省宽带资源。

客户端在将数据按照要求切分成数据切片后，将每个数据切片都分配给一个 MapTask 进行并行处理。可以看到，在切分数据时，并不考虑数据集整体，只会针对数据集中的每个文件单独进行切片。Hadoop 的底层具体是如何执行的，会在后面进行详细的源码讲解。

5.4.2 Job 提交流程源码和 FileInputFormat 切片源码详解

1．Job 提交流程源码详解

在编写 WordCount 程序的代码时，在 Driver 类中，最后会通过 Job 对象 job 调用 waitForCompletion()方法提交任务。

```
boolean result = job.waitForCompletion(true);
```

在 IDEA 中，按住 Ctrl 键并单击 waitForCompletion()方法，查看其底层调用，可以发现，最终调用 submit()方法提交任务。

```
submit();
```

在 submit()方法中，首先通过调用 connect()方法建立连接。在建立连接时，首先通过现有配置文件获取 Cluster 对象，同时判断是本地运行环境，还是 YARN 集群运行环境。

```
// 1 建立连接
connect();
    // 1）创建提交 Job 的代理
    new Cluster(getConfiguration());
        // ①判断是本地运行环境，还是 YARN 集群运行环境
        initialize(jobTrackAddr, conf);
```

在建立连接后，创建 JobSubmiter 的实例化对象 submiter，submitter 对象最终调用 submitJobInternal() 方法完成任务。

```
// 2 提交 Job
submitter.submitJobInternal(Job.this, cluster)
```

进一步追溯 submitJobInternal()方法，该方法中关键步骤的源码如下：

```
// 1 创建给集群提交数据的 Stag 路径
Path jobStagingArea = JobSubmissionFiles.getStagingDir(cluster, conf);

// 2 获取 jobid，并且创建 Job 路径
JobID jobId = submitClient.getNewJobID();

// 3 将 jar 包复制到集群中
copyAndConfigureFiles(job, submitJobDir);
rUploader.uploadFiles(job, jobSubmitDir);

// 4 计算数据切片大小，生成数据切片规划文件
writeSplits(job, submitJobDir);
maps = writeNewSplits(job, jobSubmitDir);
input.getSplits(job);

// 5 向 Stag 路径下写入 XML 配置文件
writeConf(conf, submitJobFile);
conf.writeXml(out);

// 6 提交 Job，返回提交状态
status = submitClient.submitJob(jobId, submitJobDir.toString(), job.getCredentials());
```

Job 提交流程的源码顺序图如图 5-9 所示。通过分析 Job 提交流程的源码可以发现，在最后提交任务时，会使用 getSplits()方法对文件进行切片操作。

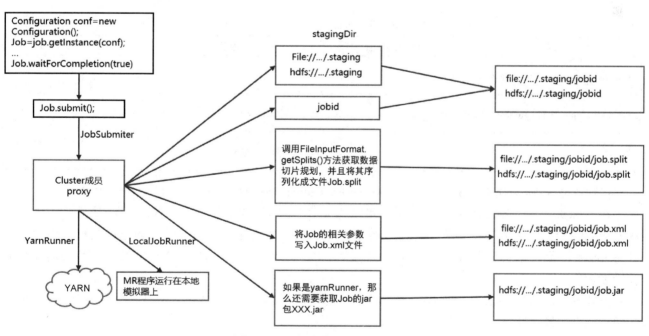

图 5-9 Job 提交流程的源码顺序图

2. FileInputFormat 切片源码解析

在分析 Job 提交流程的源码时可以发现,最后会使用 InputFormat 类的 getSplits()方法对文件进行切分。但是 InputFormat 类是一个抽象类,并没有提供 getSplits()方法的具体实现,因此我们需要一个 InputFormat 的实现类,用于查看在 Hadoop 中具体是如何实现文件切分的。

本节主要分析 FileInputFormat 类的源码。FileInputFormat 类是所有使用文件作为其数据源的 InputFormat 实现类的基类,该类主要实现了两个功能,一个是指定作业的输入文件位置,另一个是将输入文件切分成数据分片的代码实现。更细分的功能,如将数据分片分割成记录的功能,由 FileInputFormat 类的子类实现。

在提交任务时,使用以下代码指定输入文件路径。

```
FileInputFormat.setInputPaths(job, new Path(args[0]));
```

在 FileInputFormat 类中,搜索 getSplits()方法,对文件的分片逻辑主要体现在这个方法的实现中。继续使用 5.4.1 节中数据切分的案例,其逻辑总结如下。

(1)程序先找到数据存储的目录。

```
List<FileStatus> files = listStatus(job);
```

(2)开始遍历处理(规划切片)数据存储目录下的每个文件。

```
for (FileStatus file: files) {
...
}
```

(3)遍历第一个文件 ss.txt,执行以下步骤。

① 获取文件大小。

```
long length = file.getLen();
```

② 计算数据切片大小,计算公式如下:

```
computeSplitSize(Math.max(minSize,Math.min(maxSize,blocksize)))=blocksize=128M
```

以上计算公式中有 3 个关键值,分别是 minsize、maxsize、blocksize,控制这 3 个值的属性如表 5-2 所示。

表 5-2 控制数据切片大小的关键属性

属　性	数 据 类 型	默认值(单位:字节)	描　述
mapreduce.input.fileinputformat.split.minsize	int	1	一个文件分片中最小的有效字节数
mapreduce.input.fileinputformat.split.maxsize	long	Long.MAX_VALUE,即 9 223 372 036 854 775 807	一个文件分片中最大的有效字节数
dfs.blocksize	long	134 217 728,即 128MB	HDFS 中数据块的大小

在默认情况下,minsize<blocksize<maxsize,所以 blocksize 属性值表示默认的数据切片大小。

在实际执行一个 MapReduce 程序时,可以将 minsize 属性的值设置成比 blocksize 属性值大的值,从而得到比数据块大的数据切片。这样会增加对 MapTask 来说不是本地文件的数据块数量。用户也可以将 maxsize 属性的值设置成比 blocksize 属性值小的值,从而得到比数据块小的数据切片,提高数据并行度。

在理论上,我们可以通过调整这 3 个关键属性的值调整数据切片的大小,但是与数据块大小相同的数据切片大小是最合理的。

③ 对 ss.txt 文件进行切分,得到以下数据切片。

- ss.txt.split1:0~128MB。

- ss.txt.split2：128～256MB。
- ss.txt.split3：256～300MB。

在每次进行切片操作时，都要判断剩余部分是否大于数据切片大小的 1.1 倍，如果剩余部分不大于数据切片大小的 1.1 倍，则将剩余部分划分为一个数据切片。

④ 将数据切片信息写入一个数据切片规划列表 InputSplits，InputSplits 中只记录数据切片的元数据信息，如起始位置、长度、所在的节点列表等。

（4）在切片操作完成后，将数据切片规划文件提交到 YARN 中，YARN 中的 MRAppMaster 进程可以根据数据切片规划文件计算需要开启的 MapTask 数量。

5.4.3 FileInputFormat 切片机制总结

1．切片机制

- 简单地按照文件中内容的长度进行切片操作，只记录数据切片的元数据信息，不对文件进行物理层面上的划分。
- 数据切片大小默认等于数据块大小，理论上可以通过调整 3 个关键属性的值进行调节。
- 在进行切片操作时，无须考虑数据集整体，只需逐个针对每个文件单独进行切片操作。

2．案例分析

假设输入数据有以下两个文件。
- file1.txt：320MB。
- file2.txt：10MB。

在经过 FileInputFormat 切片机制运算后，形成的数据切片信息如下。
- file1.txt.split1：0～128MB。
- file1.txt.split2：128～256MB。
- file1.txt.split3：256～320MB。
- file2.txt.split1：0～10MB。

3．参数设置

源码中数据切片大小的计算公式如下：

```
Math.max(minSize, Math.min(maxSize, blockSize));
```

- mapreduce.input.fileinputformat.split.minsize 属性的默认值为 1。
- mapreduce.input.fileinputformat.split.maxsize 属性的默认值为 Long.MAXValue。

因此，在默认情况下，数据切片大小=blocksize。

切片大小的相关设置如下。

- maxsize（数据切片最大值）：如果该属性的值比 blocksize 属性的值小，则会让数据切片变小，相当于配置这个属性的值。
- minsize（数据切片最小值）：如果该属性的值比 blocksize 属性的值大，则会让数据切片大小变得比 blockSize 属性的值还大。

5.4.4 TextInputFormat

1．FileInputFormat 接口的实现类

在运行 MapReduce 程序时，输入的文件格式包括基于行的日志文件、二进制格式文件、数据库表等。

针对不同的数据类型，MapReduce 是如何读取这些数据的呢？Hadoop 提供了 FileInputFormat 的多种实现类，用于读取不同格式的输入数据。FileInputFormat 接口的实现类如图 5-10 所示。

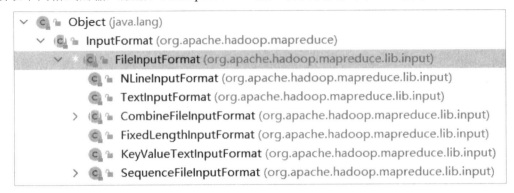

图 5-10　FileInputFormat 接口的实现类

FileInputFormat 接口的常见实现类包括 TextInputFormat、KeyValueTextInputFormat、NLineInputFormat、CombineTextInputFormat 等，用户也可以自定义 InputFormat。

2．TextInputFormat

TextInputFormat 是进行普通文件输入的默认 FileInputFormat 接口实现类，可以按行读取每条记录，读取的键是该行在整个文件中的起始字节偏移量，数据类型为 LongWritable 类型；读取的值是该行内容，不包括任何行终止符（换行符和回车符），数据类型为 Text 类型。

下面来看一个案例，一个数据分片中包含以下 4 条文本记录。

```
Rich learning form
Intelligent learning engine
Learning more convenient
From the real demand for more close to the enterprise
```

每条记录都以键/值对的形式表示，具体如下：

```
(0,Rich learning form)
(20,Intelligent learning engine)
(49,Learning more convenient)
(74,From the real demand for more close to the enterprise)
```

5.4.5　CombineTextInputFormat 切片机制

框架默认的 TextInputFormat 切片机制是对任务按文件进行切片操作，不管文件多小，都会是一个单独的数据切片，并且被交给一个 MapTask 处理。如果输入数据有大量小文件，就会产生大量的 MapTask，处理效率极其低下。

CombineTextInputFormat 主要应用于这种小文件过多的场景，它可以在逻辑上将多个小文件规划到一个数据切片中，然后将多个小文件交给一个 MapTask 处理，从而大幅提升处理效率。

通过以下代码设置数据切片的最大值。

```
CombineTextInputFormat.setMaxInputSplitSize(job, 4194304);// 4m
```

注意：建议根据实际的小文件大小情况设置数据切片最大值。

CombineTextInputFormat 切片机制的具体切片过程分为两部分，分别为虚拟存储过程和切片过程，如图 5-11 所示。

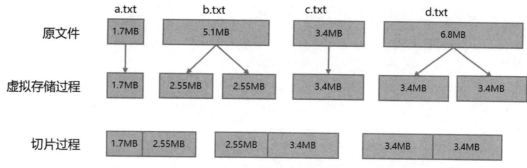

图 5-11 CombineTextInputFormat 切片机制的具体切片过程

1. 虚拟存储过程

将输入目录下所有文件的大小依次与设置的 setMaxInputSplitSize 值进行比较，如果输入目录下所有文件的大小不大于设置的 setMaxInputSplitSize 值，那么在逻辑上将其划分为一个虚拟存储块；如果输入目录下所有文件的大小大于设置的 setMaxInputSplitSize 值，并且大于 setMaxInputSplitSize 值的 2 倍，那么以 setMaxInputSplitSize 值划分第一个虚拟存储块；如果剩余的文件大小大于设置的 setMaxInputSplitSize 值，并且不大于 setMaxInputSplitSize 值的 2 倍，那么将剩余的文件均分成 2 个虚拟存储块（防止出现太小的数据切片）。例如，将 setMaxInputSplitSize 的值设置为 4MB，输入文件的大小为 8.02MB，那么先在逻辑上划分一个 4MB 的虚拟存储块，剩余的文件大小为 4.02MB，大于 4MB，小于 8MB，所以将其均分为两个大小为 2.01MB 的虚拟存储块。

2. 切片过程

判断虚拟存储块的大小是否大于 setMaxInputSplitSize 的值，如果虚拟存储块的大小大于或等于 setMaxInputSplitSize 的值，那么将其单独划分为一个数据切片；如果虚拟存储块的大小不大于 setMaxInputSplitSize 的值，则将与其下一个虚拟存储块进行合并，共同划分为一个数据切片。例如，有 4 个小文件，其大小分别为 1.7MB、5.1MB、3.4MB 和 6.8MB，在进行虚拟存储后，会形成 6 个虚拟存储块，其大小分别为 1.7MB、(2.55MB、2.55MB)、3.4MB 和 (3.4MB、3.4MB)，最终会形成 3 个数据切片，其大小分别为（1.7+2.55）MB、（2.55+3.4）MB、（3.4+3.4）MB。

5.4.6 CombineTextInputFormat 案例实操

下面使用 CombineTextInputFormat 作为 MapReduce 的 InputFormat，测试其对小文件的处理效果。

1. 需求分析

准备 4 个小文件作为输入文件，期望将 4 个小文件合并成一个数据切片进行统一处理。4 个输入文件的大小如下。输入文件在本书附赠的资料包中可以找到。

```
a.txt 1.7M
b.txt 5.1M
c.txt 3.4M
d.txt 6.8M
```

2. 实现过程

（1）不进行任何处理，直接将这 4 个文件作为 5.2 节中 WordCount 程序的输入数据，在 WordCount 程序的运行过程中，观察控制台中打印的日志，可以发现产生了 4 个数据切片。

```
number of splits:4
```

（2）测试一。

① 在 WordCountDriver 中添加以下代码。

```
// 将 InputFormat 设置为 CombineTextInputFormat
```

```
job.setInputFormatClass(CombineTextInputFormat.class);

// 将虚拟存储切片最大值设置为 4MB
CombineTextInputFormat.setMaxInputSplitSize(job, 4194304);
```

② 运行 WordCount 程序，观察控制台中打印的日志，可以发现产生了 3 个数据切片。

```
number of splits:3
```

（3）测试二。

① 在 WordCountDriver 中添加以下代码。

```
// 将 InputFormat 设置为 CombineTextInputFormat
job.setInputFormatClass(CombineTextInputFormat.class);

// 将虚拟存储切片最大值设置为 20MB
CombineTextInputFormat.setMaxInputSplitSize(job, 20971520);
```

② 运行 WordCount 程序，观察控制台中打印的日志，可以发现产生了 1 个数据切片。

```
number of splits:1
```

5.5 MapReduce 框架原理之 shuffle 机制

在 5.1.2 节讲解 MapReduce 的核心思想时，我们提到 map 阶段输出的数据会被 ReduceTask 拉取，用于进行 reduce 阶段的计算。实际上，这个过程是十分复杂的。MapReduce 需要确保 reduce 阶段的输入数据都是按照键进行排序的，MapReduce 执行排序操作并将 map 阶段输出的数据传递给 reduce 阶段的过程称为 shuffle。shuffle 阶段是 MapReduce 程序的核心阶段，了解这个阶段的底层机制有助于开发人员优化程序，提升工作效率。

5.5.1 shuffle 机制

shuffle 机制如图 5-12 所示。shuffle 阶段是指在 map()方法之后、reduce()方法之前的数据处理过程，因此我们将 shuffle 阶段分为两部分，分别为 map 端和 reduce 端。

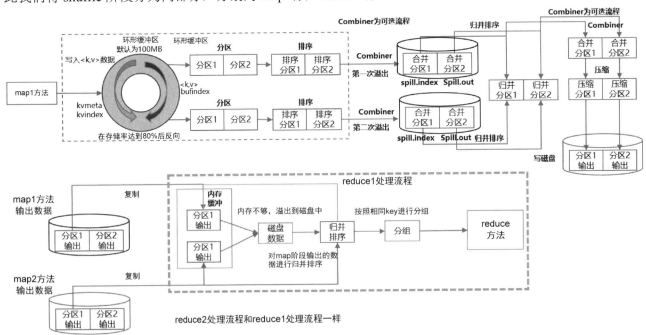

图 5-12 shuffle 机制

1. map 端

当 map() 方法开始产生输出数据时，输出数据会首先被发送至一个环形缓冲区中。环形缓冲区的默认大小为 100MB，可以通过 mapreduce.task.io.sort.mb 属性进行调整。当环形缓冲区内的数据达到某个阈值（默认值为环形缓冲区大小的 80%）时，会将数据溢写至磁盘中。此时 MapTask 数据处理会阻塞，直到溢写过程结束。

在将数据从环形缓冲区溢写至磁盘中前，会根据实际情况对数据进行分区。溢写线程会根据设置的 ReduceTask 的数据分区进行溢写，分区的键即 map 阶段输出数据的键，在每个分区中，数据是按照键进行排序的。在这个过程中，开发者可以自行选择是否配置 Combiner 组件（在 5.5.7 节会详细讲解），使用 Combiner 组件可以使输出数据更加紧凑，降低 reduce 阶段的输入数据量。在溢写完成后，会形成多个分区且有序的小文件。最终，多次溢写形成的小文件会合并成多个分区且有序的文件。在合并文件后，可以再次运行 Combiner 组件，进一步合并数据，使数据更加紧凑，从而减少占用的存储空间。在将合并后的文件写入磁盘前，还可以进行一次压缩操作，进一步节约存储空间。

通过以上操作，最终每个 MapTask 都会输出多个分区且有序的小文件，小文件的数量与 ReduceTask 的数量相同。

2. reduce 端

map 阶段的输出文件最终会存储于 MapTask 所在节点服务器的本地磁盘中。而每个 ReduceTask 执行的计算任务需要的是所有 MapTask 的输出文件对应分区的数据。ReduceTask 通过 HTTP 获取分区文件，将数据拉取到 ReduceTask 中。

ReduceTask 是如何得知所需的数据存储于哪台节点服务器中的呢？在 map 阶段执行完毕后，会使用心跳机制通知任务执行的 ApplicationMaster，而 ReduceTask 中有一个线程会定期询问 ApplicationMaster 获取 map 阶段输出文件的位置。

ReduceTask 在拉取 map 阶段输出的数据后，如果文件很小，则会直接将其复制到 JVM 内存缓冲区中，这个内存缓冲区的大小可以使用 mapreduce.reduce.shuffle.input.buffer.percent 属性进行配置；如果文件大于 JVM 内存缓冲区的大小，则会溢写至磁盘中。在拉取数据的过程中，JVM 内存缓冲区的大小达到阈值（由 mapreduce.reduce.shuffle.merge.percent 属性决定），或者达到 map 阶段的输出阈值（由 mapreduce.reduce.merge.inmem.threshold 属性决定），则将数据合并后溢写至磁盘中。

在 map 阶段输出的数据都被复制完成后，会在磁盘中生成多个文件，这时还会发生一次合并、排序的过程。由于 map 阶段输出的数据是有序的，因此这次排序是归并排序，并且按照 key 进行分组，最终会输出一个整体有序的数据块，用于供 reduce() 方法进行计算。以上就是 shuffle 阶段的整体过程。

在了解了 shuffle 机制后，可以发现，shuffle 机制提到了分区的概念，并且多次提到了排序，那么分区是按照什么标准进行的？排序都涉及哪些算法？在后面的内容中，我们将一一进行解答。

5.5.2 分区

1. 问题引出

要求将统计结果按照条件输出到不同的分区（Partition）中，如将统计结果按照手机归属地输出到不同分区中。

2. 默认分区器 HashPartitioner

默认分区是根据 key 的 hashCode 对 ReduceTask 的数量取模得到的。用户没法控制哪个 key 存储于哪个分区中。默认分区器 HashPartitioner 的源码如下：

```
public class HashPartitioner<K, V> extends Partitioner<K, V> {
  public int getPartition(K key, V value, int numReduceTasks) {
```

```
        return (key.hashCode() & Integer.MAX_VALUE) % numReduceTasks;
    }
}
```

3．自定义分区器

在有些情况下，使用默认分区不能满足用户需求，所以需要自定义分区器，具体步骤如下。

（1）自定义继承 Partitioner 接口的类，需要指定两个泛型，分别是该自定义分区器要应用的 MapReduce 程序中的 Mapper 输出键和输出值的数据类型（以 5.3.3 节中编写的 MapReduce 程序为例，输出键的数据类型为 Text，输出值的数据类型为 FlowBean）；并且需要重写 getPartition()方法，在该方法中编写代码逻辑，实现通过不同的键返回分区值，分区值是 int 类型的数据。例如，如果希望最终得到 5 个分区，则返回 0~4 的整数分区值。

```
public class CustomPartitioner extends Partitioner<Text, FlowBean> {
    @Override
    public int getPartition(Text key, FlowBean value, int numPartitions) {
        // 控制分区代码逻辑
        ...
        return partition;
    }
}
```

（2）在 Job 的 Driver 驱动类中，配置使用自定义分区器，代码如下：

```
job.setPartitionerClass(CustomPartitioner.class);
```

（3）在自定义分区器后，根据自定义分区器的逻辑设置相应数量的 ReduceTask。例如，如果自定义分区器中的逻辑指定返回的分区值为 0~4，则需要将 ReduceTask 的数量设置为 5 个，代码如下：

```
job.setNumReduceTasks(5);
```

4．分区总结

- 如果 ReduceTask 的数量> getPartition()方法的返回值，则会多产生几个空的结果文件 part-r-000xx。
- 如果 1<ReduceTask 的数量<getPartition()方法的返回值，则有一部分分区数据无处安放，会报错。
- 如果 ReduceTask 的数量=1，那么无论 MapTask 端输出多少个分区文件，最终结果都会交给这个 ReduceTask，并且只产生一个结果文件 part-r-00000。
- 分区号必须从 0 开始，逐一累加。

总结：尽量使 ReduceTask 的数量与 getPartition()方法的返回值保持一致。

5．案例分析

假设自定义分区数为 5，则会有以下情况。

- job.setNumReduceTasks(1)：会正常运行，并且只产生一个输出文件。
- job.setNumReduceTasks(2)：会报错。
- job.setNumReduceTasks(6)：大于 5，程序会正常运行，会产生 6 个结果文件，其中 1 个是空文件。

5.5.3 分区案例实操

1．需求分析

在 5.3.3 节统计每个手机号码耗费的总上行流量、总下行流量、总流量的案例基础上，使用自定义分区器，将统计结果按照手机归属地的省份输出到不同的文件中。

1）输入数据。

```
1    13736230513 192.196.100.1   www.atguigu.com 2481    24681   200
2    13846544121 192.196.100.2                   264 0   200
```

```
3    13956435636 192.196.100.3              132    1512    200
4    13966251146 192.168.100.1              240  0       404
5    18271575951 192.168.100.2    www.atguigu.com 1527    2106    200
6    84188413    192.168.100.3    www.atguigu.com 4116    1432    200
7    13590439668 192.168.100.4              1116    954  200
8    15910133277 192.168.100.5    www.hao123.com  3156    2936    200
9    13729199489 192.168.100.6              240  0       200
10   13630577991 192.168.100.7    www.shouhu.com  6960    690  200
11   15043685818 192.168.100.8    www.baidu.com   3659    3538    200
12   15959002129 192.168.100.9    www.atguigu.com 1938    180  500
13   13560439638 192.168.100.10             918  4938    200
14   13470253144 192.168.100.11             180  180  200
15   13682846555 192.168.100.12   www.qq.com      1938    2910    200
16   13992314666 192.168.100.13   www.gaga.com    3008    3720    200
17   13509468723 192.168.100.14   www.qinghua.com 7335    110349  404
18   18390173782 192.168.100.15   www.sogou.com   9531    2412    200
19   13975057813 192.168.100.16   www.baidu.com   11058   48243   200
20   13768778790 192.168.100.17             120  120  200
21   13568436656 192.168.100.18   www.alibaba.com 2481    24681   200
22   13568436656 192.168.100.19             1116    954  200
```

2）期望效果。

将手机号码开头为 136、137、138、139 的数据分别存储于 4 个独立文件中，将其他数据存储于 1 个文件中。

2．需求实现

（1）增加一个自定义分区器 ProvincePartitioner，编写分区逻辑，将手机号码开头为 136 的数据发送至 0 号分区，将手机号码开头为 137 的数据发送至 1 号分区，将手机号码开头为 138 的数据发送至 2 号分区，将手机号码开头为 139 的数据发送至 3 号分区，将其他数据发送至 4 号分区。以 5.3.3 节的自定义 FlowMapper 为基础，设置自定义分区器 ProvincePartitioner 的输入键泛型是 Text，输入值泛型是 FlowBean。具体代码如下：

```
package com.atguigu.mapreduce.partitioner;
import org.apache.hadoop.io.Text;
import org.apache.hadoop.mapreduce.Partitioner;

public class ProvincePartitioner extends Partitioner<Text, FlowBean> {

    @Override
    public int getPartition(Text text, FlowBean flowBean, int numPartitions) {
        //获取手机号码前 3 位 prePhone
        String phone = text.toString();
        String prePhone = phone.substring(0, 3);

        //定义一个分区号变量 partition, 根据 prePhone 的值设置分区号
        int partition;

        if("136".equals(prePhone)){
            partition = 0;
        }else if("137".equals(prePhone)){
            partition = 1;
        }else if("138".equals(prePhone)){
```

```
            partition = 2;
        }else if("139".equals(prePhone)){
            partition = 3;
        }else {
            partition = 4;
        }

        //最后返回分区号 partition
        return partition;
    }
}
```

（2）编写驱动类 FlowDriver，配置自定义分区器，并且设置 ReduceTask 的数量，代码如下：

```
package com.atguigu.mapreduce.partitioner;
import org.apache.hadoop.conf.Configuration;
import org.apache.hadoop.fs.Path;
import org.apache.hadoop.io.Text;
import org.apache.hadoop.mapreduce.Job;
import org.apache.hadoop.mapreduce.lib.input.FileInputFormat;
import org.apache.hadoop.mapreduce.lib.output.FileOutputFormat;
import java.io.IOException;

public class FlowDriver {

    public static void main(String[] args) throws IOException, ClassNotFoundException, InterruptedException {

        //1 获取 Job 对象 job
        Configuration conf = new Configuration();
        Job job = Job.getInstance(conf);

        //2 关联当前 Driver 类
        job.setJarByClass(FlowDriver.class);

        //3 关联 Mapper 组件和 Reducer 组件
        job.setMapperClass(FlowMapper.class);
        job.setReducerClass(FlowReducer.class);

        //4 设置 map 端输出 KV 的数据类型
        job.setMapOutputKeyClass(Text.class);
        job.setMapOutputValueClass(FlowBean.class);

        //5 设置程序最终输出 KV 的数据类型
        job.setOutputKeyClass(Text.class);
        job.setOutputValueClass(FlowBean.class);

        //6 指定自定义分区器
        job.setPartitionerClass(ProvincePartitioner.class);

        //7 同时指定相应数量的 ReduceTask
        job.setNumReduceTasks(5);
```

```
        //8 设置输入路径和输出路径
        FileInputFormat.setInputPaths(job, new Path("D:\\inputflow"));
        FileOutputFormat.setOutputPath(job, new Path("D\\partitionout"));

        //9 提交 Job
        boolean b = job.waitForCompletion(true);
        System.exit(b ? 0 : 1);
    }
}
```

（3）在设置完毕后，重新运行程序，查看运行过程中 ReduceTask 的数量及输出的文件数量，输出文件夹如图 5-13 所示。

名称	修改日期	类型	大小
._SUCCESS.crc	2021/11/11 15:45	CRC 文件	1 KB
.part-r-00000.crc	2021/11/11 15:45	CRC 文件	1 KB
.part-r-00001.crc	2021/11/11 15:45	CRC 文件	1 KB
.part-r-00002.crc	2021/11/11 15:45	CRC 文件	1 KB
.part-r-00003.crc	2021/11/11 15:45	CRC 文件	1 KB
.part-r-00004.crc	2021/11/11 15:45	CRC 文件	1 KB
_SUCCESS	2021/11/11 15:45	.	0 KB
part-r-00000	2021/11/11 15:45	.	1 KB
part-r-00001	2021/11/11 15:45	.	1 KB
part-r-00002	2021/11/11 15:45	.	1 KB
part-r-00003	2021/11/11 15:45	.	1 KB
part-r-00004	2021/11/11 15:45	.	1 KB

图 5-13　输出文件夹

可以看到共输出了 5 个结果文件，分别查看 5 个文件中的内容，可以发现，已经将输出数据按照手机号码前 3 位完成了分区。

（4）将 ReduceTask 的数量设置为 1 个，重新运行程序，再次查看输出文件夹，可以发现只输出了 1 个结果文件，如图 5-14 所示。

名称	修改日期	类型	大小
._SUCCESS.crc	2021/11/11 15:47	CRC 文件	1 KB
.part-r-00000.crc	2021/11/11 15:47	CRC 文件	1 KB
_SUCCESS	2021/11/11 15:47	.	0 KB
part-r-00000	2021/11/11 15:47	.	2 KB

图 5-14　将 ReduceTask 的数量设置为 1 个后的输出文件夹

（5）将 ReduceTask 的数量设置为 2 个，重新运行程序，程序运行报错，如下所示。

```
Caused by: java.io.IOException: Illegal partition for 13846544121 (2)
    at org.apache.hadoop.mapred.MapTask$MapOutputBuffer.collect(MapTask.java:1096)
    at org.apache.hadoop.mapred.MapTask$NewOutputCollector.write(MapTask.java:727)
    at org.apache.hadoop.mapreduce.task.TaskInputOutputContextImpl.write(TaskInputOutputContextImpl.java:89)
    at org.apache.hadoop.mapreduce.lib.map.WrappedMapper$Context.write(WrappedMapper.java:112)
    at com.atguigu.mapreduce.partitioner2.FlowMapper.map(FlowMapper.java:40)
    at com.atguigu.mapreduce.partitioner2.FlowMapper.map(FlowMapper.java:9)
```

（6）将 ReduceTask 的数量设置为 6 个，重新运行程序，再次查看输出文件夹，共输出了 6 个结果文件，但是其中一个是空文件，如图 5-15 所示。

名称	修改日期	类型	大小
._SUCCESS.crc	2021/11/11 15:54	CRC 文件	1 KB
.part-r-00000.crc	2021/11/11 15:54	CRC 文件	1 KB
.part-r-00001.crc	2021/11/11 15:54	CRC 文件	1 KB
.part-r-00002.crc	2021/11/11 15:54	CRC 文件	1 KB
.part-r-00003.crc	2021/11/11 15:54	CRC 文件	1 KB
.part-r-00004.crc	2021/11/11 15:54	CRC 文件	1 KB
.part-r-00005.crc	2021/11/11 15:54	CRC 文件	1 KB
_SUCCESS	2021/11/11 15:54	.	0 KB
part-r-00000	2021/11/11 15:54	.	1 KB
part-r-00001	2021/11/11 15:54	.	1 KB
part-r-00002	2021/11/11 15:54	.	1 KB
part-r-00003	2021/11/11 15:54	.	1 KB
part-r-00004	2021/11/11 15:54	.	1 KB
part-r-00005	2021/11/11 15:54	.	0 KB

图 5-15 将 ReduceTask 的数量设置为 6 个后的输出文件夹

5.5.4 WritableComparable 排序

排序是 MapReduce 框架中最重要的操作之一。在讲解 shuffle 过程时，曾经多次提到排序。

MapTask 和 ReduceTask 均会对数据按照 key 进行排序，该操作属于 Hadoop 的默认行为，任何应用程序中的数据均会被排序，不管在逻辑上是否需要。

默认排序是指按照字典顺序进行排序，实现该排序的方法是快速排序。

对于 MapTask，它会将处理的结果暂时存储于环形缓冲区中，当环形缓冲区使用率达到一定的阈值后，再对缓冲区中的数据进行一次快速排序，并且将这些有序数据溢写到磁盘中，在数据处理完毕后，它会对磁盘中的所有文件进行归并排序。

ReduceTask 会从每个 MapTask 上远程复制相应的数据文件，如果数据文件的大小超过一定的阈值，则会将其溢写至磁盘中，否则将数据文件存储于内存中。如果磁盘中的文件数量达到一定的阈值，那么进行一次归并排序，以便生成一个更大的文件；如果内存中的文件大小或文件数量超过一定的阈值，那么在进行一次合并后将数据溢写至磁盘中。在所有数据复制完毕后，ReduceTask 会统一对内存和磁盘中的所有数据进行一次归并排序。

以上过程中提到的排序都是针对数据的键进行的，在不进行配置的情况下，必须使用 Hadoop 提供的序列化类，如 Text 类、IntWritable 类等。若用户使用自定义 JavaBean 作为键，则需要使其继承 WritableComparable 接口，重写该接口中的 compareTo()方法，定义比较逻辑，代码如下：

```java
@Override
public int compareTo(FlowBean bean) {

    int result;

    // 按照总流量大小进行倒序排列
    if (this.sumFlow > bean.getSumFlow()) {
        result = -1;
    }else if (this.sumFlow < bean.getSumFlow()) {
        result = 1;
    }else {
        result = 0;
    }
```

```
    return result;
}
```

以上提到的是 MapReduce 执行过程中用到的排序，使用 MapReduce 对数据集进行计算，得到的数据结果有时也需要实现一定的排序效果。计算需要实现的排序效果有以下几种。
- 部分排序。MapReduce 根据输入记录的键对数据集进行排序，保证输出的每个文件都内部有序。实际上，MapReduce 程序的输出文件是默认在分区内有序的。
- 全排序。最终输出结果只有一个文件，并且文件内部有序。实现方式是只设置一个 ReduceTask。但该方法在处理大型文件时效率极低，因为一个节点处理所有文件，使 MapReduce 提供的并行架构没有了意义。
- 二次排序。在自定义排序过程中，如果自定义的 compareTo() 方法中的判断条件为两个，则表示此次排序为二次排序。

5.5.5 WritableComparable 排序案例实操（全排序）

1. 需求分析

基于 5.3.3 节的序列化案例的输出结果进行进一步计算，针对总流量对输出结果进行倒序排序。

1）输入数据。

输入数据是 5.3.3 节序列化案例的输出文件夹中的数据文件 part-r-00000，删除其他校验位文件，只保留此数据文件作为本次案例的输入数据。数据文件 part-r-00000 中的数据如下：

```
13470253144 180  180  360
13509468723 7335 110349 117684
13560439638 918  4938 5856
13568436656 3597 25635 29232
13590439668 1116 954  2070
13630577991 6960 690  7650
13682846555 1938 2910 4848
13729199489 240  0    240
13736230513 2481 24681 27162
13768778790 120  120  240
13846544121 264  0    264
13956435636 132  1512 1644
13966251146 240  0    240
13975057813 11058 48243 59301
13992314666 3008 3720 6728
15043685818 3659 3538 7197
15910133277 3156 2936 6092
15959002129 1938 180  2118
18271575951 1527 2106 3633
18390173782 9531 2412 11943
84188413    4116 1432 5548
```

2）期望输出数据。

期望输出数据如下：

13509468723	7335	110349	117684	
13736230513	2481	24681	27162	
13956435636	132	1512	1644	
13846544121	264	0	264	

...

2. 实现思路分析

输入数据是手机号码与流量构成的文本文件，输出数据需要按照 FlowBean 对象中的总流量进行降序排序。将 FlowBean 对象作为 map 阶段的输出键，然后使 FlowBean 继承 WritableComparable 接口，在重写 compareTo()方法时，编写按照总流量倒序排列的代码逻辑。

在 map()方法中，将 FlowBean 对象作为键、将手机号码作为值输出，即可按照 FlowBean 对象中的总流量进行排序。

在 reduce()方法中，输入值是以迭代器的形式给出的，所以我们遍历这个迭代器，与输入键重新组合并交换位置输出，使最后输出的数据依然是手机号码在前、流量在后，与期望输出数据相同。

3. 代码实现

（1）使 FlowBean 继承 WritableComparable 接口，重写 compareTo()方法。

```java
package com.atguigu.mapreduce.writablecompable;

import org.apache.hadoop.io.WritableComparable;
import java.io.DataInput;
import java.io.DataOutput;
import java.io.IOException;

public class FlowBean implements WritableComparable<FlowBean> {

    private long upFlow;      //上行流量
    private long downFlow;    //下行流量
    private long sumFlow;     //总流量

    //提供无参构造方法
    public FlowBean() {
    }

    //生成3个属性的getter()方法和setter()方法
    public long getUpFlow() {
        return upFlow;
    }

    public void setUpFlow(long upFlow) {
        this.upFlow = upFlow;
    }

    public long getDownFlow() {
        return downFlow;
    }

    public void setDownFlow(long downFlow) {
        this.downFlow = downFlow;
    }

    public long getSumFlow() {
        return sumFlow;
    }
```

```java
    public void setSumFlow(long sumFlow) {
        this.sumFlow = sumFlow;
    }

    public void setSumFlow() {
        this.sumFlow = this.upFlow + this.downFlow;
    }

    //实现序列化和反序列化方法，注意顺序必须保持一致
    @Override
    public void write(DataOutput out) throws IOException {
        out.writeLong(this.upFlow);
        out.writeLong(this.downFlow);
        out.writeLong(this.sumFlow);

    }

    @Override
    public void readFields(DataInput in) throws IOException {
        this.upFlow = in.readLong();
        this.downFlow = in.readLong();
        this.sumFlow = in.readLong();
    }

    //重写toString()方法，最后要输出FlowBean
    @Override
    public String toString() {
        return upFlow + "\t" + downFlow + "\t" + sumFlow;
    }

    @Override
    public int compareTo(FlowBean o) {

        //按照总流量进行比较，并且进行倒序排序
        if(this.sumFlow > o.sumFlow){
            return -1;
        }else if(this.sumFlow < o.sumFlow){
            return 1;
        }else {
            return 0;
        }
    }
}
```

（2）编写 Mapper 类，在 map()方法中将 FlowBean 作为键、将手机号码作为值输出。

```java
package com.atguigu.mapreduce.writablecompable;

import org.apache.hadoop.io.LongWritable;
import org.apache.hadoop.io.Text;
import org.apache.hadoop.mapreduce.Mapper;
import java.io.IOException;
```

```java
public class FlowMapper extends Mapper<LongWritable, Text, FlowBean, Text> {
    private FlowBean outK = new FlowBean();
    private Text outV = new Text();

    @Override
    protected void map(LongWritable key, Text value, Context context) throws IOException, InterruptedException {

        //1 获取一行数据
        String line = value.toString();

        //2 按照"\t"符号切割数据
        String[] split = line.split("\t");

        //3 封装outK、outV
        outK.setUpFlow(Long.parseLong(split[1]));
        outK.setDownFlow(Long.parseLong(split[2]));
        outK.setSumFlow();
        outV.set(split[0]);

        //4 写出outK、outV
        context.write(outK,outV);
    }
}
```

(3) 编写 Reducer 类，输入数据的泛型需要与 Mapper 类输出数据的泛型保持一致，输出键的泛型为 FlowBean，输出值的泛型为 Text。输出数据的泛型与之相反，输出键的泛型为 Text，输出值的泛型为 FlowBean。

```java
package com.atguigu.mapreduce.writablecompable;

import org.apache.hadoop.io.Text;
import org.apache.hadoop.mapreduce.Reducer;
import java.io.IOException;

public class FlowReducer extends Reducer<FlowBean, Text, Text, FlowBean> {
    @Override
    protected void reduce(FlowBean key, Iterable<Text> values, Context context) throws IOException, InterruptedException {

        //遍历 values 迭代器
        for (Text value : values) {
            //调换 KV 位置，反向写出
            context.write(value,key);
        }
    }
}
```

(4) 编写 Driver 类，进行必要配置。需要注意的是，输入数据是 5.3.3 节中案例的输出数据。

```java
package com.atguigu.mapreduce.writablecompable;

import org.apache.hadoop.conf.Configuration;
import org.apache.hadoop.fs.Path;
import org.apache.hadoop.io.Text;
import org.apache.hadoop.mapreduce.Job;
```

```java
import org.apache.hadoop.mapreduce.lib.input.FileInputFormat;
import org.apache.hadoop.mapreduce.lib.output.FileOutputFormat;
import java.io.IOException;

public class FlowDriver {

    public static void main(String[] args) throws IOException, ClassNotFoundException, InterruptedException {

        //1 获取 Job 对象 job
        Configuration conf = new Configuration();
        Job job = Job.getInstance(conf);

        //2 关联当前 Driver 类
        job.setJarByClass(FlowDriver.class);

        //3 关联 Mapper 组件和 Reducer 组件
        job.setMapperClass(FlowMapper.class);
        job.setReducerClass(FlowReducer.class);

        //4 设置 map 端输出 KV 的数据类型
        job.setMapOutputKeyClass(FlowBean.class);
        job.setMapOutputValueClass(Text.class);

        //5 设置程序最终输出 KV 的数据类型
        job.setOutputKeyClass(Text.class);
        job.setOutputValueClass(FlowBean.class);

        //6 设置输入路径和输出路径
        FileInputFormat.setInputPaths(job, new Path("D:\\inputflow2"));
        FileOutputFormat.setOutputPath(job, new Path("D:\\comparout"));

        //7 提交 Job
        boolean b = job.waitForCompletion(true);
        System.exit(b ? 0 : 1);
    }
}
```

（5）重新运行程序，查看输出结果。

```
13509468723	7335	110349	117684
13975057813	11058	48243	59301
13568436656	3597	25635	29232
13736230513	2481	24681	27162
18390173782	9531	2412	11943
13630577991	6960	690	7650
15043685818	3659	3538	7197
13992314666	3008	3720	6728
15910133277	3156	2936	6092
13560439638	918	4938	5856
84188413	4116	1432	5548
13682846555	1938	2910	4848
18271575951	1527	2106	3633
```

15959002129	1938	180	2118
13590439668	1116	954	2070
13956435636	132	1512	1644
13470253144	180	180	360
13846544121	264	0	264
13729199489	240	0	240
13768778790	120	120	240
13966251146	240	0	240

5.5.6 WritableComparable 排序案例实操（区内排序）

1. 需求分析

在 5.3.3 节统计每个手机号码耗费的总上行流量、总下行流量、总流量的案例基础上，将统计结果按照手机归属地省份输出到不同文件，并且每个文件中都按照总流量内部有序。

输入数据与 5.5.5 节中案例的输入数据相同，是 5.3.3 节中案例的输出结果。

2. 实现思路分析

基于 5.5.5 节的实现思路，增加自定义分区器，按照省份手机号码设置分区，如图 5-16 所示。

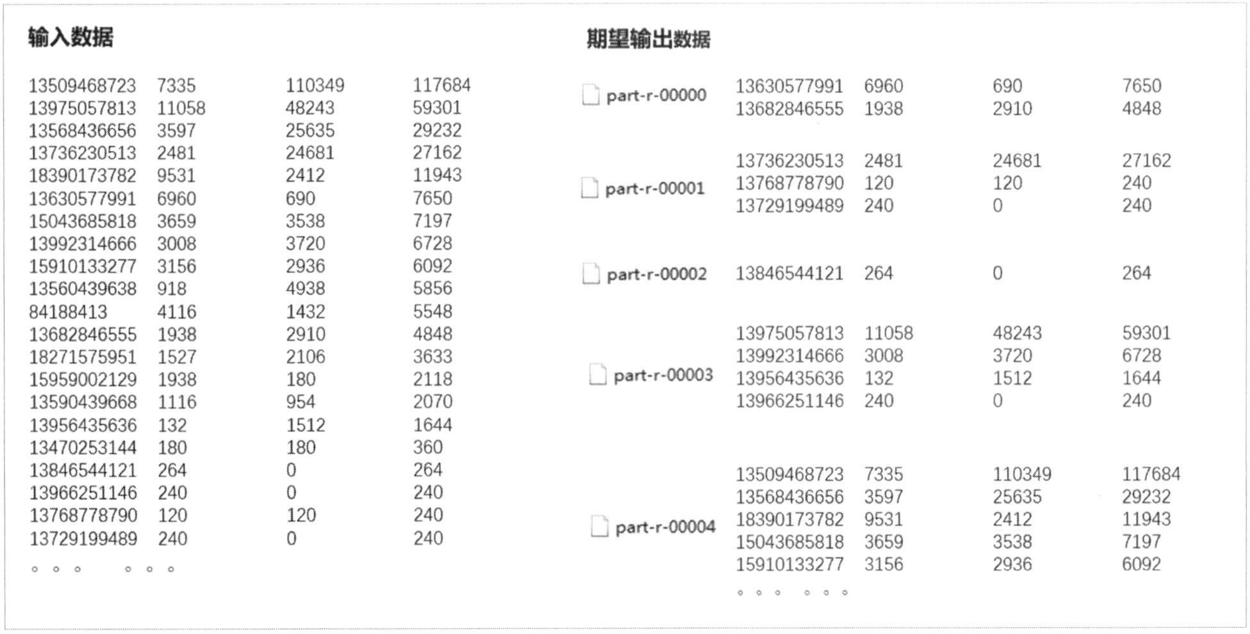

图 5-16 实现思路分析

3. 案例实操

（1）自定义分区器 ProvincePartitioner2，指定泛型需要与 Mapper 类输出数据的泛型保持一致，分别是 FlowBean 和 Text，其中 FlowBean 需要实现 WritableComparable 接口。

```
package com.atguigu.mapreduce.partitionercompable;

import org.apache.hadoop.io.Text;
import org.apache.hadoop.mapreduce.Partitioner;

public class ProvincePartitioner2 extends Partitioner<FlowBean, Text> {

    @Override
    public int getPartition(FlowBean flowBean, Text text, int numPartitions) {
```

```java
        //获取手机号码前3位
        String phone = text.toString();
        String prePhone = phone.substring(0, 3);

        //定义一个分区号变量partition, 根据prePhone的值设置分区号
        int partition;
        if("136".equals(prePhone)){
            partition = 0;
        }else if("137".equals(prePhone)){
            partition = 1;
        }else if("138".equals(prePhone)){
            partition = 2;
        }else if("139".equals(prePhone)){
            partition = 3;
        }else {
            partition = 4;
        }

        //最后返回分区号partition
        return partition;
    }
}
```

（2）编写驱动类FlowDriver，配置自定义分区器ProvincePartitioner2，代码如下：

```java
package com.atguigu.mapreduce.writablecompable;

import org.apache.hadoop.conf.Configuration;
import org.apache.hadoop.fs.Path;
import org.apache.hadoop.io.Text;
import org.apache.hadoop.mapreduce.Job;
import org.apache.hadoop.mapreduce.lib.input.FileInputFormat;
import org.apache.hadoop.mapreduce.lib.output.FileOutputFormat;
import java.io.IOException;

public class FlowDriver {

    public static void main(String[] args) throws IOException, ClassNotFoundException, InterruptedException {

        //1 获取Job对象job
        Configuration conf = new Configuration();
        Job job = Job.getInstance(conf);

        //2 关联当前Driver类
        job.setJarByClass(FlowDriver.class);

        //3 关联Mapper组件和Reducer组件
        job.setMapperClass(FlowMapper.class);
        job.setReducerClass(FlowReducer.class);

        //4 设置map端输出KV的数据类型
        job.setMapOutputKeyClass(FlowBean.class);
```

```java
            job.setMapOutputValueClass(Text.class);

            //5 设置程序最终输出 KV 的数据类型
            job.setOutputKeyClass(Text.class);
            job.setOutputValueClass(FlowBean.class);

            //6 设置自定义分区器
job.setPartitionerClass(com.atguigu.mapreduce.partitionercompable.ProvincePartitioner2.class);

            //7 设置对应的 ReduceTask 的数量
            job.setNumReduceTasks(5);

            //8 设置输入路径和输出路径
            FileInputFormat.setInputPaths(job, new Path("flowoutput"));
            FileOutputFormat.setOutputPath(job, new Path("output3"));

            //9 提交 Job
            boolean b = job.waitForCompletion(true);
            System.exit(b ? 0 : 1);
    }
}
```

（3）重新运行程序，查看输出结果，生成了 5 个结果文件。分别查看 5 个结果文件中的内容，可以看到每个文件都是在分区内有序的，具体如下。

- part-r-00000 文件。

```
13630577991  6960    690    7650
13682846555  1938    2910   4848
```

- part-r-00001 文件。

```
13736230513  2481    24681  27162
13729199489  240     0      240
13768778790  120     120    240
```

- part-r-00002 文件。

```
13846544121  264     0      264
```

- part-r-00003 文件。

```
13975057813  11058   48243  59301
13992314666  3008    3720   6728
13956435636  132     1512   1644
13966251146  240     0      240
```

- part-r-00004 文件。

```
13509468723  7335    110349 117684
13568436656  3597    25635  29232
18390173782  9531    2412   11943
15043685818  3659    3538   7197
15910133277  3156    2936   6092
13560439638  918     4938   5856
84188413     4116    1432   5548
18271575951  1527    2106   3633
15959002129  1938    180    2118
```

```
13590439668    1116    954    2070
13470253144    180     180    360
```

5.5.7 Combiner 合并

Combiner 组件是 MapReduce 程序中除 Mapper 组件和 Reducer 组件外的一种组件。在介绍 shuffle 机制时，我们曾经提到过，通过配置 Combiner 组件，可以减少 map 阶段与 reduce 阶段之间的传输数据量。Combiner 组件的实现方法与 Reducer 组件的实现方法类似，父类是 Reducer 类。Combiner 组件与 Reducer 组件之间的区别在于运行的位置。Combiner 组件会在每个 MapTask 所在的节点上运行，而 Reducer 组件会接收所有 MapTask 的输出结果，运行在 ReduceTask 上。

配置 Combiner 组件的意义在于对每个 MapTask 的输出数据进行局部汇总，从而减少网络传输数据量。

需要注意的是，应用 Combiner 组件的前提是不能影响最终的业务逻辑，并且 Combiner 组件的输出键和输出值的数据类型需要与 Reducer 组件的输入键与输入值的数据类型一一对应。

假设第一个 MapTask 的输出数据如下：
(a,3)
(a,5)
(a,7)

第二个 MapTask 的输出数据如下：
(a,1)
(a,4)

使用 reduce()方法可以针对每个键求平均值，所以输入数据和输出数据如下：
(a,[3,5,7,2,6])→(a,(3+5+7+1+4)/5) 计算结果为(a,4)

如果使用 Combiner 组件对 MapTask 的输出数据进行合并，那么对第一个 MapTask 的输出数据进行合并后的结果如下：
(a,(3+5+7)/3) 计算结果为(a,5)

对第二个 MapTask 的输出数据进行合并后的结果如下：
(a,(1+4)/2) 计算结果为(a,2.5)

reduce()方法的输入数据和输出数据如下：
(a,[5,2.5])→(a,(5+2.5)/2) 计算结果为(a,3.75)

显然，计算结果是不正确的。

通过上面的案例可知，在配置 Combiner 组件时应该慎重，在合并数据的同时不能影响最终的计算逻辑。

用户自定义 Combiner 组件的实现步骤如下。

（1）同样以 5.3.3 节中的案例为例，自定义一个 Combiner 组件，即 WordCountCombiner 类，使其继承 Reducer 类，重写 reduce()方法，代码如下：

```java
public class WordCountCombiner extends Reducer<Text, IntWritable, Text, IntWritable> {

    private IntWritable outV = new IntWritable();

    @Override
    protected void reduce(Text key, Iterable<IntWritable> values, Context context) throws IOException, InterruptedException {

        int sum = 0;
        for (IntWritable value : values) {
            sum += value.get();
        }

        outV.set(sum);
```

```
        context.write(key,outV);
    }
}
```

（2）在 Job 驱动类 Driver 中增加以下代码，配置自定义的 Combiner 组件。

```
job.setCombinerClass(WordCountCombiner.class);
```

5.5.8 Combiner 合并案例实操

1. 需求分析

对输入文件进行词频统计，在统计过程中，使用 Combiner 组件对每个 MapTask 的输出数据进行局部汇总，从而减少网络传输数据量。

1）输入数据。

输入数据如下：

```
banzhang ni hao
xihuan hadoop banzhang
banzhang ni hao
xihuan hadoop banzhang
```

2）期望效果。

使用 Combiner 组件对 map 的输出数据进行合并，从而减少网络传输数据量。

2. 实现思路分析

方案一：单独编写 Combiner 组件，在 Combiner 组件中对统计单词进行初步汇总。

方案二：直接将 Reducer 组件作为 Combiner 组件，在 Driver 驱动类中进行指定。

3. 案例实操——方案一

（1）以 5.2.3 节中的 WordCount 程序为基础，添加一个 WordCountCombiner 类，使其继承 Reducer 类，重写 reduce()方法，代码如下：

```java
package com.atguigu.mapreduce.combiner;

import org.apache.hadoop.io.IntWritable;
import org.apache.hadoop.io.Text;
import org.apache.hadoop.mapreduce.Reducer;
import java.io.IOException;

public class WordCountCombiner extends Reducer<Text, IntWritable, Text, IntWritable> {

private IntWritable outV = new IntWritable();

    @Override
    protected void reduce(Text key, Iterable<IntWritable> values, Context context) throws
IOException, InterruptedException {

        int sum = 0;
        for (IntWritable value : values) {
            sum += value.get();
        }

        //封装 outV
        outV.set(sum);
```

```
        //写出 key 和 outV
        context.write(key,outV);
    }
}
```

（2）在 WordCountDriver 驱动类中配置 Combiner 组件。

```
// 指定需要使用Combiner 组件，以及用哪个类作为Combiner 组件的逻辑
job.setCombinerClass(WordCountCombiner.class);
```

4．案例实操——方案二

以 5.2.3 节中的 WordCount 程序为基础，在 WordCountDriver 驱动类中指定 WordCountReducer 类作为 Combiner 组件。

```
// 指定需要使用Combiner 组件，以及用哪个类作为Combiner 组件的逻辑
job.setCombinerClass(WordCountReducer.class);
```

对于以上两种解决方案，直接运行程序，观察控制台中打印的日志，如图 5-17 所示，可以发现在使用 Combiner 组件后，网络传输数据量明显减少。

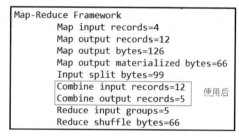

图 5-17 使用 Combiner 组件前后的对比

5.6 MapReduce 框架原理之 OutputFormat 数据输出

在讲解完数据的处理过程后，下面讲解 MapReduce 的最后一个阶段——数据输出。

5.6.1 OutputFormat 接口的实现类

OutputFormat 接口是 MapReduce 输出类的基类，所有的 MapReduce 输出类都实现了 OutputFormat 接口。OutputFormat 接口的实现类如图 5-18 所示。

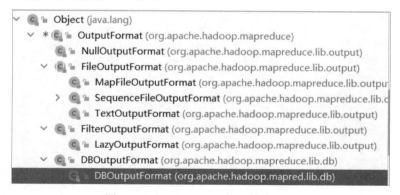

图 5-18 OutputFormat 接口的实现类

下面介绍 OutputFormat 接口的几种常见的实现类。

1．TextOutputFormat 类

TextOutputFormat 类是默认的 MapReduce 输出类，它将每条记录都写成一个文本行，键和值可以是

任意数据类型，TextOutputFormat 类使用 toString()方法将其转换成字符串并写入文本，键与值之间默认使用"\t"进行分隔。

2. SequenceFileOutputFormat 类

SequenceFileOutputFormat 类将 MapReduce 的输出数据写入一个 SequenceFile 文件，SequenceFile 文件是 Hadoop 的一种顺序文件存储格式，格式紧凑、易于压缩。如果输出数据后续将作为其他 MapReduce 程序的输入数据，那么将数据输出为 SequenceFile 文件是一个很好的选择。

3. 自定义 OutputFormat 类

当用户需要将数据输出至 MySQL、HBase、Elasticsearch 等存储框架中时，Hadoop 就不能提供对应的 OutputFormat 类了，需要用户自定义 OutputFormat 类。

自定义 OutputFormat 类的步骤如下。

（1）自定义一个类，使其继承 RecordWriter 类，在该类中创建文件的输出流及文件的输出方法。

（2）自定义一个类，使其继承 FileOutputFormat 类，重写 getRecordWriter()方法，在 getRecordWriter()方法中创建自定义的 RecordWriter 类并返回。

5.6.2 自定义 OutputFormat 类的案例实操

1. 需求分析

过滤输入的日志信息，将包含 atguigu 字段的网址数据输出到 e:/atguigu.log 文件中，将不包含 atguigu 字段的网址数据输出到 e:/other.log 文件中。

1）输入数据。

输入数据如下：

```
http://www.baidu.com
http://www.google.com
http://cn.bing.com
http://www.atguigu.com
http://www.sohu.com
http://www.sina.com
http://www.sin2a.com
http://www.sin2desa.com
http://www.sindsafa.com
```

2）期望输出数据。

期望输出数据如下：

```
atguigu.log
    http://www.atguigu.com
other.log
    http://cn.bing.com
    http://www.baidu.com
    http://www.google.com
    http://www.sin2a.com
    http://www.sin2desa.com
    http://www.sina.com
    http://www.sindsafa.com
    http://www.sohu.com
```

2. 案例实操

（1）编写 LogMapper 类，在 map()方法中不对原数据进行任何处理，直接将一行数据写出。

```
package com.atguigu.mapreduce.outputformat;
```

```java
import org.apache.hadoop.io.LongWritable;
import org.apache.hadoop.io.NullWritable;
import org.apache.hadoop.io.Text;
import org.apache.hadoop.mapreduce.Mapper;

import java.io.IOException;

public class LogMapper extends Mapper<LongWritable, Text,Text, NullWritable> {
    @Override
    protected void map(LongWritable key, Text value, Context context) throws IOException, InterruptedException {
        //不进行任何处理,直接写出一行Log数据
        context.write(value,NullWritable.get());
    }
}
```

(2)编写 LogReducer 类,在 reduce()方法中不进行任何特殊处理,只将数据迭代写出即可。

```java
package com.atguigu.mapreduce.outputformat;

import org.apache.hadoop.io.NullWritable;
import org.apache.hadoop.io.Text;
import org.apache.hadoop.mapreduce.Reducer;

import java.io.IOException;

public class LogReducer extends Reducer<Text, NullWritable,Text, NullWritable> {
    @Override
    protected void reduce(Text key, Iterable<NullWritable> values, Context context) throws IOException, InterruptedException {
        // 防止有相同的数据,迭代写出
        for (NullWritable value : values) {
            context.write(key,NullWritable.get());
        }
    }
}
```

(3)自定义一个 LogOutputFormat 类,使其继承 FileOutputFormat 类,重写 getRecordWriter()方法,代码如下。getRecordWriter()方法主要用于创建一个 RecordWriter 对象并返回。RecordWriter 类需要用户自定义。

```java
package com.atguigu.mapreduce.outputformat;

import org.apache.hadoop.io.NullWritable;
import org.apache.hadoop.io.Text;
import org.apache.hadoop.mapreduce.RecordWriter;
import org.apache.hadoop.mapreduce.TaskAttemptContext;
import org.apache.hadoop.mapreduce.lib.output.FileOutputFormat;

import java.io.IOException;

public class LogOutputFormat extends FileOutputFormat<Text, NullWritable> {
    @Override
    public RecordWriter<Text, NullWritable> getRecordWriter(TaskAttemptContext job) throws IOException, InterruptedException {
```

```
        //创建一个自定义的RecordWriter对象并返回
        LogRecordWriter logRecordWriter = new LogRecordWriter(job);
        return logRecordWriter;
    }
}
```

（4）编写 LogRecordWriter 类，使其继承 RecordWriter 类，在构造方法中创建文件输出流，在 write()方法中编写对文件内容进行判断并分流写出的逻辑代码。

```
package com.atguigu.mapreduce.outputformat;

import org.apache.hadoop.fs.FSDataOutputStream;
import org.apache.hadoop.fs.FileSystem;
import org.apache.hadoop.fs.Path;
import org.apache.hadoop.io.IOUtils;
import org.apache.hadoop.io.NullWritable;
import org.apache.hadoop.io.Text;
import org.apache.hadoop.mapreduce.RecordWriter;
import org.apache.hadoop.mapreduce.TaskAttemptContext;

import java.io.IOException;

public class LogRecordWriter extends RecordWriter<Text, NullWritable> {

    private FSDataOutputStream atguiguOut;
    private FSDataOutputStream otherOut;

    public LogRecordWriter(TaskAttemptContext job) {
        try {
            //获取文件系统对象
            FileSystem fs = FileSystem.get(job.getConfiguration());
            //用文件系统对象创建两个输出流，对应不同的目录
            atguiguOut = fs.create(new Path("d:/hadoop/atguigu.log"));
            otherOut = fs.create(new Path("d:/hadoop/other.log"));
        } catch (IOException e) {
            e.printStackTrace();
        }
    }

    @Override
    public void write(Text key, NullWritable value) throws IOException, InterruptedException {
        String log = key.toString();
        //根据一行Log数据是否包含atguigu字段，判断两条输出流输出的内容
        if (log.contains("atguigu")) {
            atguiguOut.writeBytes(log + "\n");
        } else {
            otherOut.writeBytes(log + "\n");
        }
    }

    @Override
    public void close(TaskAttemptContext context) throws IOException, InterruptedException {
        //关闭流
```

```
        IOUtils.closeStream(atguiguOut);
        IOUtils.closeStream(otherOut);
    }
}
```

（5）编写 LogDriver 类，将 OutputFormat 类设置为自定义的 LogOutputFormat 类。

```
package com.atguigu.mapreduce.outputformat;

import org.apache.hadoop.conf.Configuration;
import org.apache.hadoop.fs.Path;
import org.apache.hadoop.io.NullWritable;
import org.apache.hadoop.io.Text;
import org.apache.hadoop.mapreduce.Job;
import org.apache.hadoop.mapreduce.lib.input.FileInputFormat;
import org.apache.hadoop.mapreduce.lib.output.FileOutputFormat;

import java.io.IOException;

public class LogDriver {
    public static void main(String[] args) throws IOException, ClassNotFoundException, InterruptedException {

        Configuration conf = new Configuration();
        Job job = Job.getInstance(conf);

        job.setJarByClass(LogDriver.class);
        job.setMapperClass(LogMapper.class);
        job.setReducerClass(LogReducer.class);

        job.setMapOutputKeyClass(Text.class);
        job.setMapOutputValueClass(NullWritable.class);

        job.setOutputKeyClass(Text.class);
        job.setOutputValueClass(NullWritable.class);

        //设置自定义的 OutputFormat 类
        job.setOutputFormatClass(LogOutputFormat.class);

        FileInputFormat.setInputPaths(job, new Path("D:\\input"));
        //虽然我们自定义了 OutputFormat 类，但是因为自定义的 OutputFormat 类继承自 FileoutputFormat 类，
        //而 FileoutputFormat 类要输出一个_SUCCESS 文件，所以还需要指定一个输出目录
        FileOutputFormat.setOutputPath(job, new Path("D:\\logoutput"));

        boolean b = job.waitForCompletion(true);
        System.exit(b ? 0 : 1);
    }
}
```

（6）运行程序，观察输出结果，发现生成了两个结果文件，分别为 atguigu.log 文件和 other.log 文件。
- atguigu.log 文件中的内容如下：

http://www.atguigu.com
- other.log 文件中的内容如下：

http://cn.bing.com

```
http://www.baidu.com
http://www.google.com
http://www.sin2a.com
http://www.sin2desa.com
http://www.sina.com
http://www.sindsafa.com
http://www.sohu.com
```

5.7 MapReduce 工作流程

MapReduce 在 map 阶段的详细工作流程如图 5-19 所示，在 reduce 阶段的详细工作流程如图 5-20 所示。

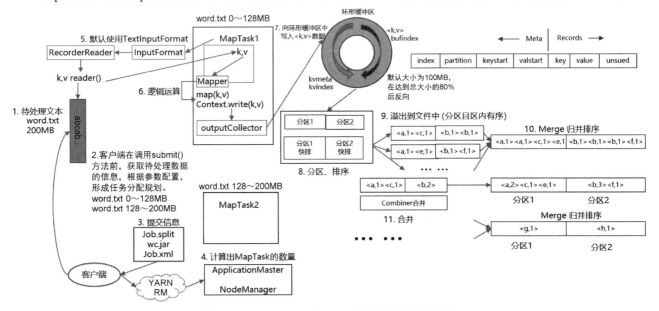

图 5-19　MapReduce 在 map 阶段的详细工作流程

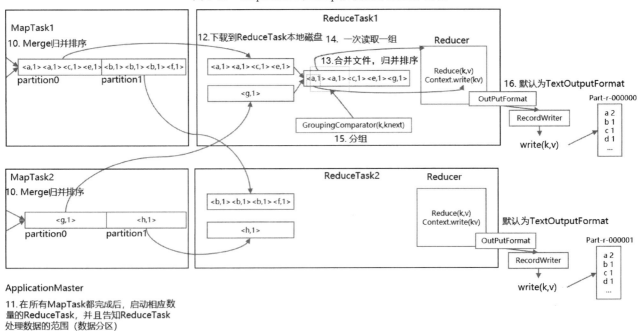

图 5-20　MapReduce 在 reduce 阶段的详细工作流程

在图 5-19 和图 5-20 中，步骤 7～16 是 shuffle 过程，具体的 shuffle 过程如下。

（1）MapTask 收集 map()方法输出的键/值对，并且将其存储于内存缓冲区中。

（2）从内存缓冲区中不断溢出本地磁盘文件，可能会溢出多个文件。

（3）多个溢出文件会被合并成大的溢出文件。

（4）在数据溢出及合并的过程中，都要调用 Partitioner 对数据进行分区和针对 key 进行排序。

（5）ReduceTask 根据自己的分区号，在各个 MapTask 服务器中获取相应的结果分区数据。

（6）ReduceTask 会抓取同一个分区中来自不同 MapTask 的结果文件，ReduceTask 会再次对这些文件进行合并（归并排序）。

（7）在合并成大文件后，shuffle 过程结束，进入 ReduceTask 的逻辑运算过程（从文件中取出一个一个的键/值对 Group，调用用户自定义的 reduce()方法进行处理）。

注意：
- shuffle 过程中的缓冲区大小会影响 MapReduce 程序的执行效率，原则上，缓冲区越大，磁盘 I/O 的次数越少，执行速度就越快。
- 缓冲区的大小可以通过参数进行调整，参数 mapreduce.task.io.sort.mb 的默认值为 100MB。

5.8 Join

学习过数据库的读者对数据集之间的连接（Join）操作不会陌生，MapReduce 也可以进行数据集之间的连接操作。

下面举例说明什么是数据集之间的连接操作。假设一个数据集中存储的是某电商网站中的所有用户信息，另一个数据集中存储的是该网站中的所有订单信息，需要分析计算各年龄段的消费额分布情况。很明显，年龄信息需要从用户数据集中获取，而消费额信息需要从订单数据集中获取，因此需要将两个数据集通过共同的用户 ID 字段进行连接操作。

如果连接操作发生在 map 阶段，则称为 Map Join；如果连接操作发生在 reduce 阶段，则称为 Reduce Join。下面详细讲解两种连接操作的原理并进行案例实操。

5.8.1 Reduce Join

Reduce Join 是比较常见的连接操作，并且对数据集没有特定要求。

map 阶段的主要工作：首先为来自不同表或文件的 key/value 对打标签，用于区别不同来源的记录；然后将连接字段作为 key，将其他部分和新加的标志作为 value；最后进行输出。

reduce 阶段的主要工作：在 shuffle 过程中的 reduce 端，将连接字段作为 key 的分组已经完成，在 reduce()方法中将每个分组中来自不同文件的记录（在 map 阶段已经打标签）分开并进行合并。

5.8.2 Reduce Join 案例实操

1. 需求分析

有两个数据集，分别是订单数据表和商品信息表。订单数据表 order.txt 中的数据如下，3 个字段分别表示订单编号、商品编号、数量。

```
1001    01    1
1002    02    2
1003    03    3
1004    01    4
1005    02    5
1006    03    6
```

商品信息表 pd.txt 中的数据如下，两个字段分别表示商品编号和商品名称。
```
01  小米
02  华为
03  格力
```
根据商品编号，将商品信息表中的数据合并到订单数据表中，结果如表 5-3 所示。

表 5-3 合并结果

订 单 编 号	商 品 名 称	数 量
1001	小米	1
1004	小米	4
1002	华为	2
1005	华为	5
1003	格力	3
1006	格力	6

2．实现思路分析

通过将关联条件作为 map 阶段输出的 key，将两个表中满足 Join 条件的数据及携带数据来源的文件信息发送给同一个 ReduceTask，在 Reduce 组件中进行数据的串联操作，如图 5-21 所示。

图 5-21 Reduce Join 的实现思路

3．代码实现

（1）创建订单数据表和商品信息表合并后的 TableBean 类，代码如下：

```
package com.atguigu.mapreduce.reducejoin;

import org.apache.hadoop.io.Writable;

import java.io.DataInput;
import java.io.DataOutput;
```

```java
import java.io.IOException;

public class TableBean implements Writable {

    private String id;          //订单编号
    private String pid;         //商品编号
    private int amount;         //数量
    private String pname;       //商品名称
    private String flag;        //标志字段,用于判断是订单数据表,还是商品信息表

    public TableBean() {
    }

    public String getId() {
        return id;
    }

    public void setId(String id) {
        this.id = id;
    }

    public String getPid() {
        return pid;
    }

    public void setPid(String pid) {
        this.pid = pid;
    }

    public int getAmount() {
        return amount;
    }

    public void setAmount(int amount) {
        this.amount = amount;
    }

    public String getPname() {
        return pname;
    }

    public void setPname(String pname) {
        this.pname = pname;
    }

    public String getFlag() {
        return flag;
    }

    public void setFlag(String flag) {
        this.flag = flag;
```

```java
    }

    @Override
    public String toString() {
        return id + "\t" + pname + "\t" + amount;
    }

    @Override
    public void write(DataOutput out) throws IOException {
        out.writeUTF(id);
        out.writeUTF(pid);
        out.writeInt(amount);
        out.writeUTF(pname);
        out.writeUTF(flag);
    }

    @Override
    public void readFields(DataInput in) throws IOException {
        this.id = in.readUTF();
        this.pid = in.readUTF();
        this.amount = in.readInt();
        this.pname = in.readUTF();
        this.flag = in.readUTF();
    }
}
```

（2）编写 TableMapper 类，需要重写 setup()方法。setup()方法在 MapTask 被初始化时调用。在 setup() 方法中，通过 Context 对象获取文件名称。重写 map()方法，针对不同的文件，对数据进行适当的切分操作，并且将切分结果写入 TableBean 对象作为 value、将商品编号作为 key 发送出去。

```java
package com.atguigu.mapreduce.reducejoin;

import org.apache.hadoop.io.LongWritable;
import org.apache.hadoop.io.Text;
import org.apache.hadoop.mapreduce.InputSplit;
import org.apache.hadoop.mapreduce.Mapper;
import org.apache.hadoop.mapreduce.lib.input.FileSplit;

import java.io.IOException;

public class TableMapper extends Mapper<LongWritable,Text,Text,TableBean> {

    private String filename;
    private Text outK = new Text();
    private TableBean outV = new TableBean();

    @Override
    protected void setup(Context context) throws IOException, InterruptedException {
        //获取对应文件名称
        InputSplit split = context.getInputSplit();
        FileSplit fileSplit = (FileSplit) split;
        filename = fileSplit.getPath().getName();
    }
```

```java
@Override
protected void map(LongWritable key, Text value, Context context) throws IOException,
InterruptedException {

    //获取一行
    String line = value.toString();

    //判断是哪个文件,然后针对不同的文件进行不同的操作
    if(filename.contains("order")){    //对订单数据表的处理
        String[] split = line.split("\t");
        //封装outK
        outK.set(split[1]);
        //封装outV
        outV.setId(split[0]);
        outV.setPid(split[1]);
        outV.setAmount(Integer.parseInt(split[2]));
        outV.setPname("");
        outV.setFlag("order");
    }else {                            //对商品信息表的处理
        String[] split = line.split("\t");
        //封装outK
        outK.set(split[0]);
        //封装outV
        outV.setId("");
        outV.setPid(split[0]);
        outV.setAmount(0);
        outV.setPname(split[1]);
        outV.setFlag("pd");
    }

    //写出 outK、outV
    context.write(outK,outV);
}
}
```

（3）编写 TableReducer 类，重写 reduce()方法。在 reduce()方法中编辑数据连接操作的逻辑代码，先对本组数据按照来源进行分类，分成订单信息和商品信息，再将商品信息中的商品名称写入订单信息。

```java
package com.atguigu.mapreduce.reducejoin;

import org.apache.commons.beanutils.BeanUtils;
import org.apache.hadoop.io.NullWritable;
import org.apache.hadoop.io.Text;
import org.apache.hadoop.mapreduce.Reducer;

import java.io.IOException;
import java.lang.reflect.InvocationTargetException;
import java.util.ArrayList;

public class TableReducer extends Reducer<Text,TableBean,TableBean, NullWritable> {

    @Override
```

```java
protected void reduce(Text key, Iterable<TableBean> values, Context context) throws
IOException, InterruptedException {

    ArrayList<TableBean> orderBeans = new ArrayList<>();
    TableBean pdBean = new TableBean();

    for (TableBean value : values) {

        //判断数据来自哪个表
        if("order".equals(value.getFlag())){    //订单数据表

            //创建一个临时的TableBean对象，用于接收value
            TableBean tmpOrderBean = new TableBean();

            try {
                BeanUtils.copyProperties(tmpOrderBean,value);
            } catch (IllegalAccessException e) {
                e.printStackTrace();
            } catch (InvocationTargetException e) {
                e.printStackTrace();
            }

            //将临时的TableBean对象添加到集合orderBeans中
            orderBeans.add(tmpOrderBean);
        }else {                                  //商品信息表
            try {
                BeanUtils.copyProperties(pdBean,value);
            } catch (IllegalAccessException e) {
                e.printStackTrace();
            } catch (InvocationTargetException e) {
                e.printStackTrace();
            }
        }
    }

    //遍历集合orderBeans，将每个orderBean对象的商品编号替换为商品名称，然后写出
    for (TableBean orderBean : orderBeans) {

        orderBean.setPname(pdBean.getPname());

        //写出修改后的orderBean对象
        context.write(orderBean,NullWritable.get());
    }
}
}
```

（4）编写 TableDriver 类，进行必要的配置，代码如下：

```java
package com.atguigu.mapreduce.reducejoin;

import org.apache.hadoop.conf.Configuration;
import org.apache.hadoop.fs.Path;
import org.apache.hadoop.io.NullWritable;
```

```java
import org.apache.hadoop.io.Text;
import org.apache.hadoop.mapreduce.Job;
import org.apache.hadoop.mapreduce.lib.input.FileInputFormat;
import org.apache.hadoop.mapreduce.lib.output.FileOutputFormat;

import java.io.IOException;

public class TableDriver {
    public static void main(String[] args) throws IOException, ClassNotFoundException, InterruptedException {
        Job job = Job.getInstance(new Configuration());

        job.setJarByClass(TableDriver.class);
        job.setMapperClass(TableMapper.class);
        job.setReducerClass(TableReducer.class);

        job.setMapOutputKeyClass(Text.class);
        job.setMapOutputValueClass(TableBean.class);

        job.setOutputKeyClass(TableBean.class);
        job.setOutputValueClass(NullWritable.class);

        FileInputFormat.setInputPaths(job, new Path("D:\\input"));
        FileOutputFormat.setOutputPath(job, new Path("D:\\output"));

        boolean b = job.waitForCompletion(true);
        System.exit(b ? 0 : 1);
    }
}
```

4. 测试

运行程序，查看输出结果，具体如下：

```
1004    小米 4
1001    小米 1
1005    华为 5
1002    华为 2
1006    格力 6
1003    格力 3
```

5. 缺点

在 Reduce Join 中，合并操作是在 reduce 阶段完成的，reduce 端的数据处理压力很大，map 端的运算负载则很低，资源利用率不高，并且在 reduce 阶段极易产生数据倾斜。数据倾斜是指，如果某个商品编号的订单数据远远超过其余商品编号的订单数据，那么处理该商品编号的 ReduceTask 的处理时间会大大延长。如何解决数据倾斜问题，将在后续章节中进行详细介绍。

5.8.3　Map Join

1. 适用场景

Map Join 适用于一个表很小（以至于可以分发至集群中的每个节点上）、另一个表很大的场景。

2. 优点

在 map 端缓存多个表，提前处理业务逻辑，从而增加 map 端业务，减轻 reduce 端的数据处理压力，可以尽可能地减少数据倾斜。

3. 具体方法

（1）在 Driver 驱动类中加载缓存。

```
//缓存普通文件到Task运行节点中
job.addCacheFile(new URI("file:///e:/cache/pd.txt"));
//如果是集群运行的，则需要设置HDFS路径
job.addCacheFile(new URI("hdfs://hadoop102:8020/cache/pd.txt"));
```

（2）在 Mapper 组件的 setup()方法中，将文件读取到缓存集合中；在 map()方法中，实现数据连接操作。

5.8.4 Map Join 案例实操

1. 需求分析

与 5.8.2 节中的案例需求相同，但是要求使用 Map Join 实现。

2. 代码实现

（1）在 MapJoinDriver 驱动类中添加缓存文件。因为所有的数据连接操作都在 Mapper 组件中完成，不需要 Reducer 组件，所以将 ReduceTask 的数量设置为 0 个。

```java
package com.atguigu.mapreduce.mapjoin;

import org.apache.hadoop.conf.Configuration;
import org.apache.hadoop.fs.Path;
import org.apache.hadoop.io.NullWritable;
import org.apache.hadoop.io.Text;
import org.apache.hadoop.mapreduce.Job;
import org.apache.hadoop.mapreduce.lib.input.FileInputFormat;
import org.apache.hadoop.mapreduce.lib.output.FileOutputFormat;

import java.io.IOException;
import java.net.URI;
import java.net.URISyntaxException;

public class MapJoinDriver {

    public static void main(String[] args) throws IOException, URISyntaxException, ClassNotFoundException, InterruptedException {

        // 1 获取Job对象job
        Configuration conf = new Configuration();
        Job job = Job.getInstance(conf);

        // 2 关联Driver类
        job.setJarByClass(MapJoinDriver.class);

        // 3 关联Mapper组件
        job.setMapperClass(MapJoinMapper.class);

        // 4 设置map端输出KV的数据类型
        job.setMapOutputKeyClass(Text.class);
```

```
        job.setMapOutputValueClass(NullWritable.class);

        // 5 设置程序最终输出 KV 的数据类型
        job.setOutputKeyClass(Text.class);
        job.setOutputValueClass(NullWritable.class);

        // 6 加载缓存数据
        job.addCacheFile(new URI("file:///D:/input/tablecache/pd.txt"));
        // Map Join 的逻辑不需要 reduce 阶段,设置 ReduceTask 的数量为 0 个
        job.setNumReduceTasks(0);

        // 7 设置输入路径和输出路径
        FileInputFormat.setInputPaths(job, new Path("D:\\input"));
        FileOutputFormat.setOutputPath(job, new Path("D:\\output"));

        // 8 提交 Job
        boolean b = job.waitForCompletion(true);
        System.exit(b ? 0 : 1);
    }
}
```

(2) 在 MapJoinMapper 类中的 setup()方法中读取缓存文件,将缓存文件中的数据写入 HashMap,在 map()方法中编写数据连接的逻辑代码。

```
package com.atguigu.mapreduce.mapjoin;

import org.apache.commons.lang.StringUtils;
import org.apache.hadoop.fs.FSDataInputStream;
import org.apache.hadoop.fs.FileSystem;
import org.apache.hadoop.fs.Path;
import org.apache.hadoop.io.IOUtils;
import org.apache.hadoop.io.LongWritable;
import org.apache.hadoop.io.NullWritable;
import org.apache.hadoop.io.Text;
import org.apache.hadoop.mapreduce.Mapper;

import java.io.BufferedReader;
import java.io.IOException;
import java.io.InputStreamReader;
import java.net.URI;
import java.util.HashMap;
import java.util.Map;

public class MapJoinMapper extends Mapper<LongWritable, Text, Text, NullWritable> {

    private Map<String, String> pdMap = new HashMap<>();
    private Text text = new Text();

    //在任务开始前,将商品信息表 pd.txt 中的数据缓存进 pdMap
    @Override
    protected void setup(Context context) throws IOException, InterruptedException {

        //通过缓存文件得到商品信息表 pd.txt 中的数据
```

```java
        URI[] cacheFiles = context.getCacheFiles();
        Path path = new Path(cacheFiles[0]);

        //获取文件系统对象并开流
        FileSystem fs = FileSystem.get(context.getConfiguration());
        FSDataInputStream fis = fs.open(path);

        //将字节流转换为字符流,方便按行读取
        BufferedReader reader = new BufferedReader(new InputStreamReader(fis, "UTF-8"));

        //逐行读取,按行处理
        String line;
        while (StringUtils.isNotEmpty(line = reader.readLine())) {
            //切割一行
            //01 小米
            String[] split = line.split("\t");
            pdMap.put(split[0], split[1]);
        }

        //关闭流
        IOUtils.closeStream(reader);
    }

    @Override
    protected void map(LongWritable key, Text value, Context context) throws IOException, InterruptedException {

        //获取pid
        String pid = fields[1];
        //读取订单数据表order.txt中的数据
        //1001    01    1
        String[] fields = value.toString().split("\t");

        //根据订单数据表order.txt中每行数据的pid(商品编号),从pdMap中取出pname(商品名称)
        String pname = pdMap.get(fields[1]);

        //将订单数据表order.txt中每行数据的pid替换为pname
        text.set(fields[0] + "\t" + pname + "\t" + fields[2]);

        //写出
        context.write(text,NullWritable.get());
    }
}
```

3. 测试

运行程序,查看输出结果,具体如下:

```
1004    小米 4
1001    小米 1
1005    华为 5
1002    华为 2
1006    格力 6
1003    格力 3
```

5.9 数据清洗

ETL（Extract-Transform-Load）主要用于描述将数据从来源端经过抽取（Extract）、转换（Transform）、加载（Load）至目的端的过程。ETL 通常应用于数据仓库，但其使用场景并不限于数据仓库。

在运行核心业务 MapReduce 程序前，通常需要先对数据进行清洗，清除不符合用户要求的数据。数据清洗过程通常只需运行 Mapper 程序，无须运行 Reducer 程序。

1. 需求分析

有一份日志数据，要求过滤掉日志中字段数量不超过 11 个的数据。

输入数据如下：

```
194.237.142.21 - - [18/Sep/2013:06:49:18 +0000] "GET /wp-content/uploads/2013/07/rstudio-git3.png HTTP/1.1" 304 0 "-" "Mozilla/4.0 (compatible;)"
183.49.46.228 - - [18/Sep/2013:06:49:23 +0000] "-" 400 0 "-" "-"
163.177.71.12 - - [18/Sep/2013:06:49:33 +0000] "HEAD / HTTP/1.1" 200 20 "-" "DNSPod-Monitor/1.0"
163.177.71.12 - - [18/Sep/2013:06:49:36 +0000] "HEAD / HTTP/1.1" 200 20 "-" "DNSPod-Monitor/1.0"
101.226.68.137 - - [18/Sep/2013:06:49:42 +0000] "HEAD / HTTP/1.1" 200 20 "-" "DNSPod-Monitor/1.0"
101.226.68.137 - - [18/Sep/2013:06:49:45 +0000] "HEAD / HTTP/1.1" 200 20 "-" "DNSPod-Monitor/1.0"
60.208.6.156 - - [18/Sep/2013:06:49:48 +0000] "GET /wp-content/uploads/2013/07/rcassandra.png HTTP/1.0" 200 185524 "http://cos.name/category/software/packages/" "Mozilla/5.0 (Windows NT 6.1) AppleWebKit/537.36 (KHTML, like Gecko) Chrome/29.0.1547.66 Safari/537.36"
222.68.172.190 - - [18/Sep/2013:06:49:57 +0000] "GET /images/my.jpg HTTP/1.1" 200 19939 "http://www.angularjs.cn/A00n" "Mozilla/5.0 (Windows NT 6.1) AppleWebKit/537.36 (KHTML, like Gecko) Chrome/29.0.1547.66 Safari/537.36"
222.68.172.190 - - [18/Sep/2013:06:50:08 +0000] "-" 400 0 "-" "-"
...
...
...
```

2. 代码实现

（1）编写 WebLogMapper 类，在 map()方法中实现对日志数据的过滤，代码如下：

```java
package com.atguigu.mapreduce.weblog;
import java.io.IOException;
import org.apache.hadoop.io.LongWritable;
import org.apache.hadoop.io.NullWritable;
import org.apache.hadoop.io.Text;
import org.apache.hadoop.mapreduce.Mapper;

public class WebLogMapper extends Mapper<LongWritable, Text, Text, NullWritable>{

    @Override
    protected void map(LongWritable key, Text value, Context context) throws IOException, InterruptedException {

        // 1 获取1行数据
        String line = value.toString();
```

```
        // 2 解析日志
        boolean result = parseLog(line,context);

        // 3 日志不合法，退出
        if (!result) {
            return;
        }

        // 4 日志合法，直接写出
        context.write(value, NullWritable.get());
    }

    // 封装解析日志的方法
    private boolean parseLog(String line, Context context) {

        // 1 截取
        String[] fields = line.split(" ");

        // 2 日志长度大于11为合法
        if (fields.length > 11) {
            return true;
        }else {
            return false;
        }
    }
}
```

（2）编写 WebLogDriver 类，进行必要的配置，代码如下：

```
package com.atguigu.mapreduce.weblog;
import org.apache.hadoop.conf.Configuration;
import org.apache.hadoop.fs.Path;
import org.apache.hadoop.io.NullWritable;
import org.apache.hadoop.io.Text;
import org.apache.hadoop.mapreduce.Job;
import org.apache.hadoop.mapreduce.lib.input.FileInputFormat;
import org.apache.hadoop.mapreduce.lib.output.FileOutputFormat;

public class WebLogDriver {
    public static void main(String[] args) throws Exception {

        // 输入路径和输出路径，根据实际的输入路径和输出路径进行设置
        args = new String[] { "D:/input/inputlog", "D:/output1" };

        // 1 获取 Job 对象 job
        Configuration conf = new Configuration();
        Job job = Job.getInstance(conf);

        // 2 关联 Driver 类
        job.setJarByClass(WebLogDriver.class);

        // 3 关联 Mapper 组件
        job.setMapperClass(WebLogMapper.class);
```

```
        // 4 设置程序最终输出 KV 的数据类型
        job.setOutputKeyClass(Text.class);
        job.setOutputValueClass(NullWritable.class);

        // 5 设置 ReduceTask 的数量为 0 个
        job.setNumReduceTasks(0);

        // 6 设置输入路径和输出路径
        FileInputFormat.setInputPaths(job, new Path(args[0]));
        FileOutputFormat.setOutputPath(job, new Path(args[1]));

        // 7 提交 Job
        boolean b = job.waitForCompletion(true);
        System.exit(b ? 0 : 1);
    }
}
```

(3) 运行程序,查看输出结果。

5.10 Hadoop 中的数据压缩

在文件存储过程中,适当使用压缩可以节省存储空间。下面讲解 Hadoop 中的数据压缩。

5.10.1 数据压缩概述

数据压缩可以节省磁盘存储空间,提高数据在网络和磁盘间的传输速度。但是,对数据进行压缩会增加 CPU 的开销。因此,在 MapReduce 中,应用数据压缩的原则如下。
- 对于运算密集型的任务,少用数据压缩。
- 对于 IO 密集型的任务,多用数据压缩。

MapReduce 支持的压缩编码如表 5-4 所示。

表 5-4 MapReduce 支持的压缩编码

压缩格式	Hadoop 是否自带	算 法	文件扩展名	是否可切片	在换成压缩格式后,原来的程序是否需要修改
DEFLATE	是,直接使用	DEFLATE	.deflate	否	和文本处理一样,不需要修改
GZIP	是,直接使用	DEFLATE	.gz	否	和文本处理一样,不需要修改
bzip2	是,直接使用	bzip2	.bz2	是	和文本处理一样,不需要修改
LZO	否,需要安装	LZO	.lzo	是	需要创建索引,还需要指定输入格式
Snappy	是,直接使用	Snappy	.snappy	否	和文本处理一样,不需要修改

以上压缩格式的压缩性能经过测试后的结果如表 5-5 所示。

表 5-5 压缩性能对比

压缩格式	原始文件大小	压缩文件大小	压缩速度	解压缩速度
GZIP	8.3GB	1.8GB	17.5MB/s	58MB/s
bzip2	8.3GB	1.1GB	2.4MB/s	9.5MB/s
LZO	8.3GB	2.9GB	49.3MB/s	74.6MB/s

1. 压缩格式的选择

在选择压缩格式时需要重点考虑压缩/解压缩速度、压缩率、压缩后是否支持切片等因素。

1）GZIP。
- 优点：压缩率较高。
- 缺点：不支持切片，压缩/解压缩速度一般。

2）bzip2。
- 优点：压缩率高，支持切片。
- 缺点：压缩/解压缩速度慢。

3）LZO。
- 优点：压缩/解压缩速度较快，支持切片。
- 缺点：压缩率一般，支持切片需要额外创建索引。

4）Snappy。
- 优点：压缩/解压缩速度快。
- 缺点：不支持切片，压缩率一般。

2．压缩位置的选择

可以在 MapReduce 的任意阶段压缩数据，如图 5-22 所示。

图 5-22　压缩数据的位置

1）输入端采用压缩。

无须显示指定使用的编解码方式。Hadoop 自动检查文件扩展名，如果扩展名能够匹配，就会用恰当的编码/解码方式对文件进行压缩/解压缩。

在企业开发中，需要考虑以下两点。
- 如果数据量小于数据块大小，那么重点考虑采用压缩/解压缩速度较快的 LZO 和 Snappy。
- 如果数据量非常大，那么重点考虑采用支持切片的 bzip2 和 LZO。

2）Mapper 输出采用压缩。

在企业开发中，为了减少 MapTask 和 ReduceTask 之间的网络 I/O，有时需要对 Mapper 输出采用压缩。重点考虑压缩/解压缩速度较快的 LZO、Snappy。

3）Reducer 输出采用压缩。

在企业开发中，需要考虑以下两点。
- 如果数据永久保存，那么考虑采用压缩率较高的 bzip2 和 GZIP。
- 如果作为下一个 MapReduce 输入，那么需要考虑数据量及是否支持切片。

5.10.2　压缩参数配置

为了支持多种压缩/解压缩格式，Hadoop 引入了编解码器，压缩格式及与之对应的编解码器如表 5-6 所示。

表 5-6　压缩格式及与之对应的编解码器

压 缩 格 式	对应的编解码器
DEFLATE	org.apache.hadoop.io.compress.DefaultCodec
GZIP	org.apache.hadoop.io.compress.GzipCodec

续表

压 缩 格 式	对应的编解码器
bzip2	org.apache.hadoop.io.compress.BZip2Codec
LZO	com.hadoop.compression.lzo.LzopCodec
Snappy	org.apache.hadoop.io.compress.SnappyCodec

Hadoop 中数据压缩的参数配置如表 5-7 所示。

表 5-7 Hadoop 中数据压缩的参数配置

参 数	默 认 值	阶 段	建 议
io.compression.codecs （在 core-site.xml 文件中配置）	无，需要在命令行中执行 hadoop checknative 命令查看	输入压缩	Hadoop 使用文件扩展名判断是否支持某种编解码器
mapreduce.map.output.compress （在 mapred-site.xml 文件中配置）	false	mapper 输出	将该参数的值设置为 true，表示启用压缩功能
mapreduce.map.output.compress.codec （在 mapred-site.xml 文件中配置）	org.apache.hadoop.io.compress.DefaultCodec	mapper 输出	企业通常使用 LZO 或 Snappy 编解码器在此阶段压缩数据
mapreduce.output.fileoutputformat.compress （在 mapred-site.xml 文件中配置）	false	reducer 输出	将该参数的值设置为 true，表示启用压缩功能
mapreduce.output.fileoutputformat.compress.codec （在 mapred-site.xml 文件中配置）	org.apache.hadoop.io.compress.DefaultCodec	reducer 输出	使用标准工具或编解码器，如 GZIP 和 bzip2

5.10.3 压缩案例实操

1．Mapper 输出采用压缩

即使 MapReduce 的输入文件和输出文件都是未压缩的文件，也可以对 MapTask 的中间输出结果进行数据压缩，因为需要将其写入硬盘并通过网络将其传输到 ReduceTask 所在的节点中，对其进行压缩可以提高很多性能，这些工作只要设置两个属性即可，具体实现如下。

（1）以 5.2.3 节中的 WordCount 程序为基础，在 Driver 驱动类中启用 Mapper 输出压缩功能，并且设置压缩格式为 bzip2，代码如下：

```java
package com.atguigu.mapreduce.compress;
import java.io.IOException;
import org.apache.hadoop.conf.Configuration;
import org.apache.hadoop.fs.Path;
import org.apache.hadoop.io.IntWritable;
import org.apache.hadoop.io.Text;
import org.apache.hadoop.io.compress.BZip2Codec;
import org.apache.hadoop.io.compress.CompressionCodec;
import org.apache.hadoop.io.compress.GzipCodec;
import org.apache.hadoop.mapreduce.Job;
import org.apache.hadoop.mapreduce.lib.input.FileInputFormat;
import org.apache.hadoop.mapreduce.lib.output.FileOutputFormat;

public class WordCountDriver {

    public static void main(String[] args) throws IOException, ClassNotFoundException, InterruptedException {

        Configuration conf = new Configuration();
```

```java
        // 启用 Mapper 输出压缩功能
        conf.setBoolean("mapreduce.map.output.compress", true);

        // 设置 Mapper 输出压缩方式
        conf.setClass("mapreduce.map.output.compress.codec", BZip2Codec.class,CompressionCodec.class);

        Job job = Job.getInstance(conf);

        job.setJarByClass(WordCountDriver.class);

        job.setMapperClass(WordCountMapper.class);
        job.setReducerClass(WordCountReducer.class);

        job.setMapOutputKeyClass(Text.class);
        job.setMapOutputValueClass(IntWritable.class);

        job.setOutputKeyClass(Text.class);
        job.setOutputValueClass(IntWritable.class);

        FileInputFormat.setInputPaths(job, new Path(args[0]));
        FileOutputFormat.setOutputPath(job, new Path(args[1]));

        boolean result = job.waitForCompletion(true);

        System.exit(result ? 0 : 1);
    }
}
```

(2) Mapper 组件保持不变,代码如下:

```java
package com.atguigu.mapreduce.compress;
import java.io.IOException;
import org.apache.hadoop.io.IntWritable;
import org.apache.hadoop.io.LongWritable;
import org.apache.hadoop.io.Text;
import org.apache.hadoop.mapreduce.Mapper;

public class WordCountMapper extends Mapper<LongWritable, Text, Text, IntWritable>{

    Text k = new Text();
    IntWritable v = new IntWritable(1);

    @Override
    protected void map(LongWritable key, Text value, Context context)throws IOException, InterruptedException {

        // 1 获取一行
        String line = value.toString();

        // 2 切割
        String[] words = line.split(" ");
```

```java
        // 3 循环写出
        for(String word:words){
            k.set(word);
            context.write(k, v);
        }
    }
}
```

（3）Reducer 组件保持不变，代码如下：

```java
package com.atguigu.mapreduce.compress;
import java.io.IOException;
import org.apache.hadoop.io.IntWritable;
import org.apache.hadoop.io.Text;
import org.apache.hadoop.mapreduce.Reducer;

public class WordCountReducer extends Reducer<Text, IntWritable, Text, IntWritable>{

    IntWritable v = new IntWritable();

    @Override
    protected void reduce(Text key, Iterable<IntWritable> values,
            Context context) throws IOException, InterruptedException {

        int sum = 0;

        // 1 汇总
        for(IntWritable value:values){
            sum += value.get();
        }

        v.set(sum);

        // 2 输出
        context.write(key, v);
    }
}
```

（4）运行程序，发现对输出结果没有影响，数据压缩只发生在 map 阶段。

2. Reducer 输出采用压缩

（1）以 5.2.3 节中的 WordCount 程序为基础，在 Driver 驱动类中启用 reducer 输出压缩功能，并且设置压缩格式为 bzip2。

```java
package com.atguigu.mapreduce.compress;
import java.io.IOException;
import org.apache.hadoop.conf.Configuration;
import org.apache.hadoop.fs.Path;
import org.apache.hadoop.io.IntWritable;
import org.apache.hadoop.io.Text;
import org.apache.hadoop.io.compress.BZip2Codec;
import org.apache.hadoop.io.compress.DefaultCodec;
import org.apache.hadoop.io.compress.GzipCodec;
import org.apache.hadoop.io.compress.Lz4Codec;
```

```java
import org.apache.hadoop.io.compress.SnappyCodec;
import org.apache.hadoop.mapreduce.Job;
import org.apache.hadoop.mapreduce.lib.input.FileInputFormat;
import org.apache.hadoop.mapreduce.lib.output.FileOutputFormat;

public class WordCountDriver {

    public static void main(String[] args) throws IOException, ClassNotFoundException, InterruptedException {

        Configuration conf = new Configuration();

        Job job = Job.getInstance(conf);

        job.setJarByClass(WordCountDriver.class);

        job.setMapperClass(WordCountMapper.class);
        job.setReducerClass(WordCountReducer.class);

        job.setMapOutputKeyClass(Text.class);
        job.setMapOutputValueClass(IntWritable.class);

        job.setOutputKeyClass(Text.class);
        job.setOutputValueClass(IntWritable.class);

        FileInputFormat.setInputPaths(job, new Path(args[0]));
        FileOutputFormat.setOutputPath(job, new Path(args[1]));

        // 启用 Reducer 输出压缩功能
        FileOutputFormat.setCompressOutput(job, true);

        // 设置 Reducer 输出压缩方式
        FileOutputFormat.setOutputCompressorClass(job, BZip2Codec.class);
        // FileOutputFormat.setOutputCompressorClass(job, GzipCodec.class);
        // FileOutputFormat.setOutputCompressorClass(job, DefaultCodec.class);

        boolean result = job.waitForCompletion(true);

        System.exit(result?0:1);
    }
}
```

（2）Mapper 组件和 Reducer 组件均保持不变，运行程序，分别设置使用 3 种压缩格式，并且查看输出结果。

使用 bzip2 压缩格式的输出结果如图 5-23 所示。

名称	修改日期	类型	大小
._SUCCESS.crc	2021/11/11 16:23	CRC 文件	1 KB
.part-r-00000.bz2.crc	2021/11/11 16:23	CRC 文件	1 KB
_SUCCESS	2021/11/11 16:23		0 KB
part-r-00000.bz2	2021/11/11 16:23	WinRAR 压缩文件	1 KB

图 5-23　使用 bzip2 压缩格式的输出结果

使用 GZIP 压缩格式的输出结果如图 5-24 所示。

名称	修改日期	类型	大小
_SUCCESS.crc	2021/11/11 16:22	CRC 文件	1 KB
.part-r-00000.gz.crc	2021/11/11 16:22	CRC 文件	1 KB
_SUCCESS	2021/11/11 16:22		0 KB
part-r-00000.gz	2021/11/11 16:22	WinRAR 压缩文件	1 KB

图 5-24 使用 GZIP 压缩格式的数据结果

使用 DEFAULT 压缩格式的输出结果如图 5-25 所示。

名称	修改日期	类型	大小
_SUCCESS.crc	2021/11/11 16:22	CRC 文件	1 KB
.part-r-00000.deflate.crc	2021/11/11 16:22	CRC 文件	1 KB
_SUCCESS	2021/11/11 16:22	.	0 KB
part-r-00000.deflate	2021/11/11 16:22	DEFLATE 文件	1 KB

图 5-25 使用 DEFAULT 压缩格式的输出结果

5.11 本章总结

本章主要讲解了 Hadoop 中的重要计算框架 MapReduce，MapReduce 程序的整体设计都是为了海量数据集的计算考虑的，并且为用户提供了尽可能简洁的 API。在本章内容的学习过程中，读者可能会感觉 MapReduce 的整体运行是繁杂的，但是在实际的使用过程中，通常只需要考虑 Mapper 组件和 Reducer 组件的代码逻辑。此外，了解 MapReduce 的底层运行机制，可以在开发过程中更快地定位问题并提出有针对性的解决方案。MapReduce 设计过程中用到的各种理念和数据处理思路，有助于读者学习其他大数据技术框架。

第6章

资源调度器 YARN

在 Hadoop 设计之初，没有一套成熟的资源管理系统，MapReduce 程序运行在 Hadoop 集群上，遇到了各种资源调度问题，如资源利用率低、可靠性差等。在 Hadoop 2.0 中，将资源管理系统独立出来，使 Hadoop 具有了更强的可扩展性，这个资源管理系统就是 YARN（Yet Another Resource Negotiator，另一种资源调取器）。本章将对 YARN 进行介绍，讲解 Hadoop 是如何通过 YARN 飞升为大数据领域稳固基石的。

6.1 YARN 概述

根据不同的业务需求，Hadoop 可以搭建不同规模的集群，几台、几十台、几百台甚至上千台，那么如何进行资源的管理和分配呢？如何决定给不同的任务分配多少计算资源呢？在 Hadoop 的初代版本中，已经对资源管理系统有了一定的考虑，我们将这个阶段称为 MapReduce 1，区别在于 Hadoop 2.0 及更高版本中使用 YARN。

在 MapReduce 1 中，通过两种进程调度 MapReduce 任务，分别是一个 JobTracker 和多个 TaskTracker。其中，JobTracker 主要负责协调管理任务，记录任务进度；TaskTracker 主要负责运行任务并向 JobTracker 提交任务进度。总体来看，MapReduce 1 已经具备了一个资源管理系统的雏形，但是也存在一些问题。

- JobTracker 同时负责管理调度资源、跟踪整个集群的任务进度、重启失败任务和延迟任务、维护计数器等工作，负载过大，一旦发生故障，就会造成严重后果。
- MapReduce 1 使用基于槽位（slot）的资源分配模型，槽位是资源分配单位，分为 map slot 和 reduce slot，彼此之间不能共享资源，造成了很大的资源浪费。
- MapReduce 1 的资源管理体系只能服务于 MapReduce，其他类型的应用程序不能分享 Hadoop 集群的计算资源。

YARN 的推出很好地解决了 MapReduce 1 的以上局限性问题。YARN 中存在一个全局的资源管理器（ResourceManager）及每个应用都会单独开启的作业管理器（ApplicationMaster）。使用 NodeManager 管理每台节点服务器中的资源，可以更灵活地运行 MapTask 或 ReduceTask。此外，通过 YARN 提供的容器资源调度策略，可以使 Hadoop 的计算资源向除 MapReduce 外的其他类型的分布式应用程序开放。

与 MapReduce 1 相比，YARN 具有以下优点。

- 可扩展性。由于每个应用程序都由其对应的 ApplicationMaster 进行进度管理，因此相较于 MapReduce 1 使用 JobTracker 兼顾资源调度和具体任务的进度管理，YARN 更具扩展性。
- 可用性。YARN 将资源管理器和任务管理器分开，降低了主节点（master）的负担，并且分别实现了高可用（HA，High Availability）。资源管理器可以通过 Zookeeper 实现高可用，针对每个应用程序的任务管理器都设计了失败重启机制。
- 资源利用率更高。MapReduce 1 将计算资源简单地分成 map slot 和 reduce slot，只能供对应类型的 Task 运行；而 YARN 为每个节点服务器都提供了资源管理角色 NodeManager，不是只提供固定大小的计算资源，而是根据应用程序的需求提供资源，资源管理更加灵活、精细。

- 支持非 MapReduce 应用程序的计算需求。

总而言之，YARN 是一个资源调度平台，负责为运算程序提供服务器运算资源，相当于一个分布式的操作系统，而 MapReduce 等运算程序相当于运行在操作系统上的应用程序。

6.1.1 基本架构

YARN 的基本架构如图 6-1 所示。YARN 采用了常见的 Master-Slaver 架构，其中，资源管理器 ResourceManager 担任 Master 角色，负责整个框架的资源统一管理和调度；NodeManager 担任 Slave 角色，负责任务的执行及当前节点的资源管理。

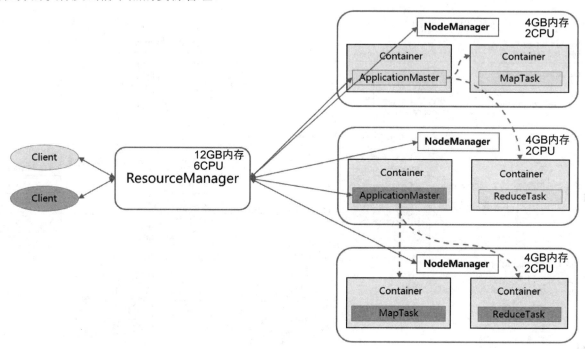

图 6-1　YARN 的基本架构

YARN 主要由 ResourceManager、NodeManager、ApplicationMaster 和 Container 等组件构成。

1. 资源管理器 ResourceManager

ResourceManager（RM）是一个全局资源管理器，负责管理整个集群的资源，主要作用如下。
- 处理客户端（Client）请求。
- 监控 NodeManager。
- 启动或监控 ApplicationMaster。
- 分配与调度资源。

2. 节点管理器 NodeManager

NodeManager（NM）是每个节点上的资源和任务的管理器，主要作用如下。
- 管理单个节点上的资源和运行任务。
- 处理来自 ResourceManager 的命令。
- 定时汇报本节点的资源使用情况及各个 Container 的运行状态。
- 处理来自 ApplicationMaster 的命令。

3. ApplicationMaster

用户提交的每个应用程序中均包含一个 ApplicationMaster（AM），其主要作用如下。

- 为应用程序申请资源并分配给内部的任务。
- 与 NodeManager 通信，以便启动或停止任务。
- 监控任务的运行状态，并且在任务运行失败时重新申请资源，以便重启任务。

4．容器 Container

Container（容器）是 YARN 中的资源抽象，它封装了某个节点服务器的多维度资源，如内存、CPU、磁盘、网络等。

6.1.2 工作机制

YARN 的工作机制如图 6-2 所示。

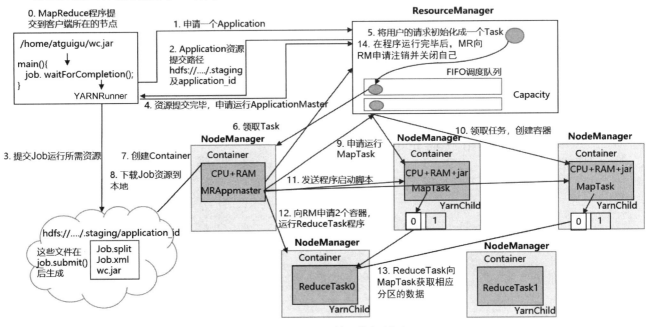

图 6-2　YARN 的工作机制

（1）客户端通过调用 job.waitForCompletion()方法，将 MapReduce 任务提交给整个集群，并且在客户端创建一个 YARNRunner 对象。

（2）YARNRunner 对象向资源管理器 ResourceManager 申请一个 Application。

（3）ResourceManager 将该 Application 的资源提交路径返回给 YARNRunner 对象。

（4）YARNRunner 对象根据 ResourceManager 给出的资源提交路径，将该程序所需的资源（如文件分片信息、运行参数信息、运行 jar 包等）提交给 HDFS。

（5）在程序将资源提交完毕后，客户端向 ResourceManager 正式申请运行 ApplicationMaster。

（6）ResourceManager 将客户端的请求初始化成一个 Task，并且为该 Task 调度分配资源。

（7）集群中的一个空闲 NodeManager 领取 Task。

（8）该 NodeManager 会创建一个 Container，并且在该 Container 中启动 Task 的 ApplicationMaster，ApplicationMaster 首先向 ResourceManager 注册，以便用户通过 ResourceManager 查看任务的运行状态。

（9）ApplicationMaster 从 HDFS 中将程序运行资源复制到本地节点中。

（10）ApplicationMaster 根据复制的程序运行资源决定需要运行几个 MapTask，并且向 ResourceManager 申请运行 MapTask 所需的资源（此处假设需要运行 2 个 MapTask）。

（11）ResourceManager 将 MapTask 任务分配给另外 2 个 NodeManager，这 2 个 NodeManager 分别领取 MapTask 任务并创建 Container。

（12）ApplicationMaster 向 2 个接收到任务的 NodeManager 发送程序启动脚本，这 2 个 NodeManager

分别启动 MapTask，MapTask 对数据进行计算，最终生成对应 ReduceTask 数量的分区文件。MapTask 在运行过程中，不断向 ApplicationMaster 汇报各任务的运行状态和进度，以便让 ApplicationMaster 掌握各任务的运行状态和进展，并且在任务失败时尝试重启。

（13）ApplicationMaster 在等待 MapTask 进行到一定程度（完成的 MapTask 数量占总 MapTask 数量的 5%以上，在这个案例中，至少要一个 MapTask 运行完成）后，会向 ResourceManager 申请运行 ReduceTask 所需的资源（此处假设需要运行 2 个 ReduceTask）。

（14）ResourceManager 为 ReduceTask 分配 NodeManager，NodeManager 领取任务并创建 2 个 Container。ApplicationMaster 向这 2 个 Container 发送任务启动脚本，启动 ReduceTask。ReduceTask 向 MapTask 获取相应分区的数据，完成 ReduceTask 的计算任务，将数据输出至指定路径下。

（15）在所有程序运行完毕后，ApplicationMaster 会向 ResourceManager 申请注销并关闭自己。

以上就是一个 MapReduce 任务通过 YARN 实现资源调度的全流程。此外，客户端每隔 5 秒都会使用 waitForCompletion()方法检查作业是否完成。在作业完成后，ApplicationMaster 和 Container 会清理工作状态。作业的信息会存储于作业历史服务器中，以备用户核查。

对以上流程进行详细解析，可以发现，整个流程大体可以分为两个阶段。一是客户端向 ResourceManager 请求启动 ApplicationMaster，ResourceManager 为任务分配第一个 Container，并且要求在这个 Container 中启动 ApplicationMaster。二是由 Application 获取任务运行资源，并且开始申请资源，监控任务运行全过程，直到任务完成。

一旦 ApplicationMaster 开始运行，关于此任务的运行状态监控和管理就全部转交了过来，大大减轻了 ResourceManager 的运行负担。ResourceManager 主要负责处理资源申请，YARN 为此提供了多种资源调度器和可配置策略供我们选择。

HDFS、YARN 和 MapReduce 之间的关系如图 6-3 所示。在客户端向 ResourceManager 提交任务后，ResourceManager 为 MapReduce 程序分配资源，并且在任意节点上启动 ApplicationMaster。ApplicationMaster 为各任务申请资源。其中，MapTask 的计算资源会被优先分配到数据块所在的节点中。在计算完成后，将输出数据上传至 HDFS 集群中。在这个过程中，HDFS 集群主要负责存储输入数据和输出数据。在图 6-3 中，NameNode 主要负责处理存储数据的读/写请求、管理元数据信息等工作，DataNode 主要负责文件的具体存储工作，SecondaryNameNode 主要负责与 NameNode 协调完成定期合并 EditLog 与 FsImage 文件的工作。

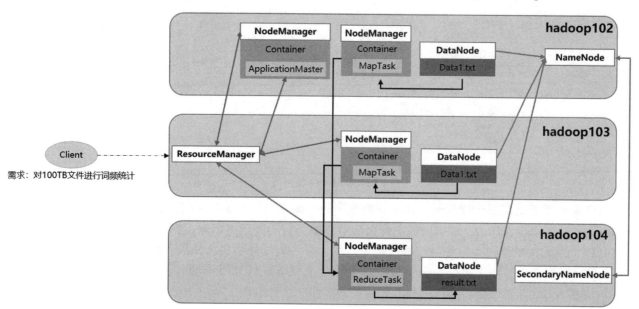

图 6-3　HDFS、YARN 和 MapReduce 之间的关系

6.2 YARN 的资源调度器和调度算法

在一个庞大又忙碌的服务器集群上,用户在不断地提交运行任务,任务从属于不同的工程、部门,需要的资源有多有少,那么我们应该遵循怎样的规则进行资源的分配,才是效率最高的呢?

资源调度器是 YARN 的核心组件之一,负责整个集群的资源调度工作,主要解决如何根据管理人员的既定策略分配资源的问题。

目前,YARN 的资源调度器主要有 3 种:FIFO 调度器(FIFO Scheduler)、容量调度器(Capacity Scheduler)和公平调度器(Fair Scheduler)。Apache Hadoop 3.1.3 默认使用的资源调度器是容量调度器。CDH 框架默认使用的资源调度器是公平调度器。

具体使用哪一种调度器在 yarn-default.xml 文件中进行配置,具体配置参数如下:

```xml
<property>
    <description>The class to use as the resource scheduler.</description>
    <name>yarn.resourcemanager.scheduler.class</name>
    <value>org.apache.hadoop.yarn.server.resourcemanager.scheduler.capacity.CapacityScheduler
</value>
</property>
```

6.2.1 FIFO 调度器

在 Hadoop 的最初版本中,MapReduce 1 采用的是简单的 FIFO(First In First Out,先进先出)调度器。FIFO 调度器是简单的单队列调度器,如图 6-4 所示,它将任务按照其到达的时间进行排序,队列中先到达的任务先获得资源。

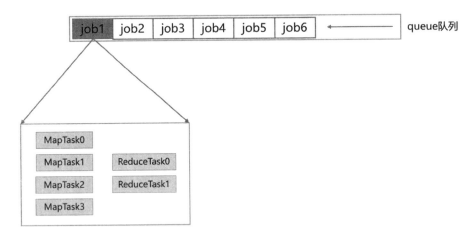

图 6-4 FIFO 调度器

FIFO 调度器的优点在于简单易懂,不需要进行任何配置。在图 6-4 中,job1 位于队列前方,包含 4 个 MapTask 和 2 个 ReduceTask,每当集群中有新的节点服务器资源空闲出来,都会优先分配给 job1。

FIFO 调度器的缺点也十分明显,不支持多队列,会造成任务堵塞。如果 job1 占用了集群的所有资源且任务运行需要很长时间,那么队列后方的任务将无法获取任何计算资源。所以在生产环境中很少使用 FIFO 调度器。

6.2.2 容量调度器

容量调度器是 Yahoo 开发的多用户调度器,如图 6-5 所示。

图 6-5 容量调度器

容量调度器具有以下特点。
- 多队列：管理员可以事先配置多个队列，可以为每个队列配置一定的资源量。例如，在图 6-5 中，为 queueA 配置 20%的资源量，为 queueB 配置 50%的资源量，为 queueC 配置 30%的资源量，每个队列内部都使用 FIFO 调度器。
- 灵活性：如果一个队列中的资源有剩余，那么可以暂时将其共享给其他需要资源的队列；如果有新的任务被提交到该队列中，那么其他队列借调的资源会归还给该队列，这种队列称为弹性队列（Queue Elasticity）。
- 容量保证：一个队列可能会出现初期运行任务少，资源使用少，其他队列任务多，资源需求大，过多占用本队列的资源，导致在后期需要资源时无资源可用的情况。针对这种情况，管理员可以为每个队列设置资源最低保证和资源使用上限，避免队列在需要资源时无资源可用和过度侵占其他队列资源。
- 多用户：支持多用户共享集群和多应用程序同时运行。为了防止同一个用户的作业独占队列中的资源，该调度器会对同一个用户提交的作业所占的资源量进行限定。

容量调度器的资源分配算法如图 6-6 所示。

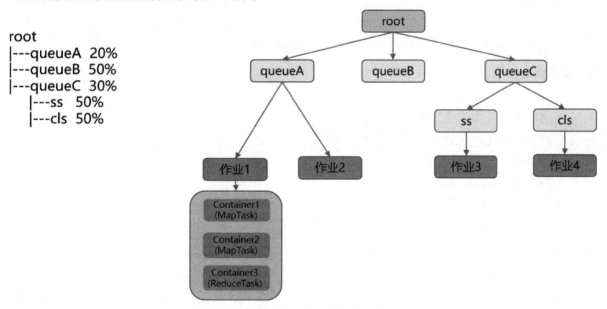

图 6-6 容量调度器的资源分配算法

（1）队列资源分配。从根队列 root 开始，使用深度优先算法，优先给资源占用率最低的队列分配资源。

（2）作业资源分配。在选中队列后，默认按照提交作业的优先级和提交时间顺序分配资源。

（3）容器资源分配。在选中作业后，同一个作业请求的容器可能是多样化的，按照容器的优先级分配资源。如果优先级相同，那么按照数据本地性原则分配资源，优先级顺序如下。

① 任务和数据在同一个节点中。
② 任务和数据在同一个机架中。
③ 任务和数据不在同一个节点中，也不在同一个机架中。

6.2.3 公平调度器

公平调度器是 Facebook 开发的多用户调度器，如图 6-7 所示。

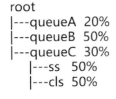

图 6-7 公平调度器

公平调度器与容量调度器相同，都支持进行多队列配置，以队列为单位划分资源。对公平调度器与容量调度器进行对比，二者的相同点和不同点如下。

1．公平调度器与容量调度器的相同点

- 多队列：支持多队列多作业。
- 灵活性：如果一个队列中的资源有剩余，那么可以暂时将其共享给其他需要资源的队列；如果有新的任务被提交到该队列中，那么其他队列借调的资源会归还给该队列。
- 容量保证：管理员可以为每个队列设置资源最低保证和资源使用上限。
- 多用户：支持多用户共享集群和多应用程序同时运行。为了防止同一个用户的作业独占队列中的资源，该调度器会对同一个用户提交的作业所占的资源量进行限定。

2．公平调度器与容量调度器的不同点

1）核心调度策略不同。
- 容量调度器：优先选择资源利用率低的队列。
- 公平调度器：优先选择对资源的缺额比例大的队列。

2）每个队列都可以设置不同的资源调度策略。
- 容量调度器：FIFO、DRF。
- 公平调度器：FIFO、FAIR、DRF。

3．公平调度器的特点

根据图 6-7 可知，公平调度器可以为所有运行的应用程序公平地分配资源。在配置文件中，管理员可以配置多个队列，并且为各个队列配置资源权重，用于公平共享计算资源。在图 6-7 中，当集群资源按照 2∶5∶3 的比例给 queueA、queueB 和 queueC 分配资源时，集群分配被认为是公平的。queueC 中有两个子队列，当这两个子队列的资源分配比例是 1∶1 时，被认为是公平的。

公平调度器的设计目标是，在时间尺度上，所有作业都可以获得公平的资源。在某个时刻，一个作业应获取资源和实际获取资源的差距称为缺额。公平调度器的缺额如图 6-8 所示。根据图 6-8 可知，队列中有 5 个任务，在理想情况下，5 个任务应该共享资源，但是在实际情况下，job15 只占用了很少的资源。调度器会优先为缺额大的作业分配资源。

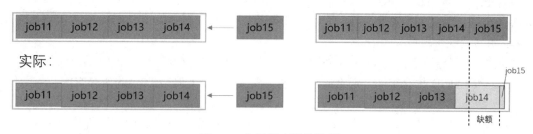

图 6-8 公平调度器的缺额

4. 公平调度器的资源调度策略

公平调度器可以为每个子队列配置不同的资源调度策略,包括 FIFO 策略、Fair 策略和 DRF 策略。

1）FIFO 策略。

如果公平调度器采用 FIFO 策略为每个队列分配资源,那么公平调度器相当于容量调度器。

2）Fair 策略。

Fair 策略是一种基于最大最小公平算法实现的资源多路复用方式。在默认情况下,每个队列内部都采用 Fair 策略分配资源。这意味着,如果一个队列中有两个任务在同时运行,那么每个任务可以得到 1/2 的资源；如果一个队列中有 3 个任务同时运行,那么每个任务可以得到 1/3 的资源。

如果公平调度器采用 Fair 策略为每个队列分配资源,那么其资源分配流程与容量调度器的资源分配流程相同,都是首先选择队列,然后选择作业,最后选择容器。以上 3 步,每一步都按照 Fair 策略分配资源。Fair 策略的队列资源分配方法如图 6-9 所示。

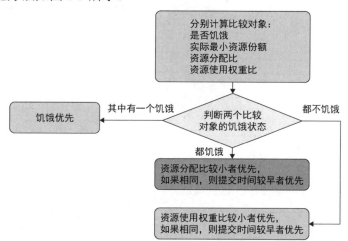

图 6-9 Fair 策略的队列资源分配方法

分别计算比较对象（队列、作业或容器）的实际最小资源份额、是否饥饿、资源分配比和资源使用权重比,这 4 个参数的计算公式如下。

- 实际最小资源份额：mindshare=Min（资源需求量,配置的最小资源）。
- 是否饥饿：isNeedy=资源使用量<mindshare。
- 资源分配比：minshareRatio=资源使用量/Max(mindshare,1)。
- 资源使用权重比：useToWeightRatio=资源使用量/权重。

首先判断两个比较对象的饥饿状态,如果其中有一个饥饿,那么饥饿对象优先；如果二者都不饥饿,那么资源使用权重比较小者优先,如果资源使用权重比相同,那么提交时间较早者优先；如果二者都饥饿,那么资源分配比较小者优先,如果资源分配比相同,那么提交时间较早者优先。

3）DRF 策略。

DRF（Dominant Resource Fairness,主资源公平调度）策略扩展了最大最小公平策略,使其能够支持多维资源的调度。我们之前说的资源，都是单一的，如只考虑内存（也是 YARN 默认的情况）。但是资源有很多种，如内存空间、CPU、网络带宽等，因此很难衡量两个应用程序应该分配的资源比例。在 YARN 中，使用 DRF 策略如何调度资源呢？

假设集群中有 100 个 CPU 和 10TB 内存空间，应用程序 A 需要 2 个 CPU 和 300GB 内存空间，应用程序 B 需要 6 个 CPU 和 100GB 内存空间，那么应用程序 A 需要 2%的 CPU 和 3%的内存空间，应用程序 B 需要 6%的 CPU 和 1%的内存空间。也就是说，应用程序 A 是内存主导的，应用程序 B 是 CPU 主导的。针对这种情况，我们可以使用 DRF 策略对不同应用程序针对不同类型的资源（CPU 和内存）进行单独的比例限制。

DRF 策略可以解决多资源和复杂需求的资源分配问题，应用十分广泛。

下面通过一些案例讲解公平调度器的 DRF 策略。

① 队列资源分配。

DRF 策略的队列资源分配方法如图 6-10 所示，使用公平调度器配置了 3 个队列，并且给出了 3 个队列及其子队列的资源分配权重。

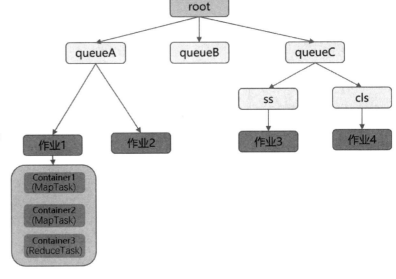

图 6-10 DRF 策略的队列资源分配方法

假设集群总资源为 100（这里模糊单位），那么 3 个队列对资源的需求如下：

```
queueA: 20
queueB: 50
queueC: 30
```

针对以上要求，YARN 会进行第一次计算，三者平分资源，每个队列分得的资源与实际需求资源的差额分别如下：

```
平分资源值：100/3=33.33
queueA: 分得 33.33 → 与需求相差+13.33
queueB: 分得 33.33 → 与需求相差-16.67
queueC: 分得 33.33 → 与需求相差+3.33
```

第二次计算，将第一次计算中多余的资源分配给资源不足的队列，具体如下：

```
多余的资源 13.33+3.33=16.66
queueA: 分得 20
queueB: 分得 33.33 + 16.66 = 50
queueC: 分得 30
```

通过两次计算，不再有多余资源，计算结束，按照以上资源分配结果进行分配。

如果通过计算仍有多余资源,那么继续计算,直到没有多余资源。

② 作业资源分配。

作业资源分配有两种情况,分别是作业不配置权重的情况和作业配置权重的情况。

a. 作业不配置权重。

假设有一条队列的总资源为 12(这里模糊单位),共有 4 个作业需要资源,并且这 4 个作业对资源的需求分别如下:

```
job1: 1
job2: 2
job3: 6
job4: 5
```

针对以上要求,YARN 会进行第一次计算,4 个作业平分资源,每个作业分得的资源与实际需求资源的差额分别如下:

```
平分资源值: 12 / 4 = 3
job1: 分得3 → 与需求相差+2
job2: 分得3 → 与需求相差+1
job3: 分得3 → 与需求相差-3
job4: 分得3 → 与需求相差-2
```

第二次计算,将第一次计算中多余的资源分配给资源不足的作业,具体如下:

```
多余的资源: (2 + 1)/2 = 1.5
job1: 分得1
job2: 分得2
job3: 分得3 + 1.5 = 4.5→ 与需求相差-1.5
job4: 分得3 + 1.5 = 4.5→ 与需求相差-0.5
```

通过两次计算,不再有多余资源,计算结束,按照以上资源分配结果进行分配。

如果通过计算仍有多余资源,那么继续计算,直至没有多余资源。

b. 作业配置权重。

假设有一条队列的总资源为 16(这里模糊单位),共有 4 个作业需要资源,并且这 4 个作业对资源的需求分别如下:

```
job1: 4
job2: 2
job3: 10
job4: 4
```

每个作业的权重分别如下:

```
job1: 5
job2: 8
job3: 1
job4: 2
```

针对以上要求,YARN 会进行第一次计算,4 个作业按照权重分配资源,每个作业分得的资源与实际需求资源的差额分别如下:

```
平分资源值: 16 / (5+8+1+2) = 1
job1: 分得5 → 与需求相差+1
job2: 分得8 → 与需求相差+6
job3: 分得1 → 与需求相差-9
job4: 分得2 → 与需求相差-2
```

第二次计算,将第一次计算中多余的资源按照权重分配给资源不足的作业,具体如下:

```
多余的资源: (1 + 6)/(1 + 2) = 7/3
job1: 分得4
```

```
job2：分得 2
job3：分得 1 + 7/3 = 3.33 → 与需求相差-6.67
job4：分得 2 + 7/3 × 2 = 6.67 → 与需求相差+2.67
```

第三次计算，将多余的资源按照权重分给资源不足的作业，具体如下：

```
多余的资源：2.66
job1：分得 4
job2：分得 2
job3：分得 3.33 + 2.67 = 6 → 与需求相差-4
job4：分得 4
```

通过三次计算，不再有多余资源，计算结束，按照以上资源分配结果进行分配。

如果通过计算仍有多余资源，那么继续计算，直到没有多余资源。

6.3 YARN 实操

本节主要讲解 YARN 的常用命令行命令及关键配置。

6.3.1 常用的命令行命令

对于 YARN 的状态，除了可以在 hadoop103:8088 页面查看，还可以通过命令行操作查看。

执行 Hadoop 的官方示例程序 WordCount，然后使用 YARN 的命令行命令查看任务执行情况。

```
[atguigu@hadoop102 hadoop-3.1.3]$ hadoop jar share/hadoop/mapreduce/hadoop-mapreduce-examples-3.1.3.jar wordcount /input /output
```

1. 查看任务命令 yarn application

1）yarn application -list。

yarn application -list 命令主要用于列出所有的 Application。

```
[atguigu@hadoop102 hadoop-3.1.3]$ yarn application -list
2021-02-06 10:21:19,238 INFO client.RMProxy: Connecting to ResourceManager at hadoop103/192.168.10.103:8032
Total number of applications (application-types: [], states: [SUBMITTED, ACCEPTED, RUNNING] and tags: []):0
            Application-Id    Application-Name    Application-Type          User
Queue           State         Final-State         Progress
Tracking-URL
```

2）yarn application -list -appStates。

yarn application -list -appStates 命令主要用于根据 Application 的状态筛选 Application，Application 的状态类型包括 ALL、NEW、NEW_SAVING、SUBMITTED、ACCEPTED、RUNNING、FINISHED、FAILED、KILLED。

执行以下命令，筛选所有处于 FINISHED 状态的 Application。

```
[atguigu@hadoop102 hadoop-3.1.3]$ yarn application -list -appStates FINISHED
2021-02-06 10:22:20,029 INFO client.RMProxy: Connecting to ResourceManager at hadoop103/192.168.10.103:8032
Total number of applications (application-types: [], states: [FINISHED] and tags: []):1
            Application-Id    Application-Name    Application-Type          User
Queue           State         Final-State         Progress
Tracking-URL
application_1612577921195_0001      word count          MAPREDUCE      atguigu
default         FINISHED        SUCCEEDED           100%
    http://hadoop102:19888/jobhistory/job/job_1612577921195_0001
```

3）yarn Application -kill <ApplicationId>。

yarn Application -kill <ApplicationId>命令主要用于杀死 Application，该命令需要指定 ApplicationId。

```
[atguigu@hadoop102 hadoop-3.1.3]$ yarn application -kill application_1612577921195_0001
2021-02-06 10:23:48,530 INFO client.RMProxy: Connecting to ResourceManager at hadoop103/
192.168.10.103:8032
Application application_1612577921195_0001 has already finished
```

2. 查看日志命令 yarn logs

1）yarn logs -applicationId <ApplicationId>。

yarn logs -applicationId <ApplicationId>命令主要用于查询指定 ApplicationId 的 Application 日志。

```
[atguigu@hadoop102 hadoop-3.1.3]$ yarn logs -applicationId application_1612577921195_0001
```

2）yarn logs -applicationId <ApplicationId> -containerId <ContainerId>。

yarn logs -applicationId <ApplicationId> -containerId <ContainerId>命令主要用于查询指定 ApplicationId 的 ContainerId 日志。

```
[atguigu@hadoop102 hadoop-3.1.3]$ yarn logs -applicationId application_1612577921195_0001
-containerId container_1612577921195_0001_01_000001
```

3. 查看尝试运行的任务命令 yarn applicationattempt

1）yarn applicationattempt -list <ApplicationId>。

yarn applicationattempt -list <ApplicationId>命令主要用于列出所有尝试运行的 Application。

```
[atguigu@hadoop102 hadoop-3.1.3]$ yarn applicationattempt -list application_1612577921195_0001
2021-02-06 10:26:54,195 INFO client.RMProxy: Connecting to ResourceManager at hadoop103/
192.168.10.103:8032
Total number of application attempts :1
    ApplicationAttempt-Id    State    AM-Container-Id    Tracking-URL
appattempt_1612577921195_0001_000001          FINISHED
    container_1612577921195_0001_01_000001
    http://hadoop103:8088/proxy/application_1612577921195_0001/
```

2）yarn applicationattempt -status <ApplicationAttemptId>。

yarn applicationattempt -status <ApplicationAttemptId>命令主要用于打印指定 ApplicationAttemptId 的 Application 的 ApplicationAttemp 状态。

```
[atguigu@hadoop102 hadoop-3.1.3]$ yarn applicationattempt -status appattempt_1612577921195_
0001_000001
2021-02-06 10:27:55,896 INFO client.RMProxy: Connecting to ResourceManager at hadoop103/
192.168.10.103:8032
Application Attempt Report :
    ApplicationAttempt-Id : appattempt_1612577921195_0001_000001
    State : FINISHED
    AMContainer : container_1612577921195_0001_01_000001
    Tracking-URL : http://hadoop103:8088/proxy/application_1612577921195_0001/
    RPC Port : 34756
    AM Host : hadoop104
    Diagnostics :
```

4. 查看容器命令 yarn container

1）yarn container -list <ApplicationAttemptId>。

yarn container -list <ApplicationAttemptId>命令主要用于列出所有的 Container。

```
[atguigu@hadoop102 hadoop-3.1.3]$ yarn container -list appattempt_1612577921195_0001_000001
```

```
2021-02-06 10:28:41,396 INFO client.RMProxy: Connecting to ResourceManager at hadoop103/
192.168.10.103:8032
Total number of containers :0
  Container-Id    Start Time    Finish Time    State    Host    Node Http Address
```

2）yarn container -status <ContainerId>。

yarn container -status <ContainerId>命令主要用于打印指定 ContainerId 的 Container 状态。

```
[atguigu@hadoop102 hadoop-3.1.3]$ yarn container -status container_1612577921195_0001_
01_000001
2021-02-06 10:29:58,554 INFO client.RMProxy: Connecting to ResourceManager at hadoop103/
192.168.10.103:8032
Container with id 'container_1612577921195_0001_01_000001' doesn't exist in RM or Timeline
Server.
```

注意：只有在任务运行的过程中才能看到 Container 的状态。

5. 查看节点状态命令 yarn node

yarn node -list -all 命令主要用于列出所有 NodeManager。

```
[atguigu@hadoop102 hadoop-3.1.3]$ yarn node -list -all
2021-02-06 10:31:36,962 INFO client.RMProxy: Connecting to ResourceManager at hadoop103/
192.168.10.103:8032
Total Nodes:3
       Node-Id    Node-State  Node-Http-Address   Number-of-Running-Containers
  hadoop103:38168      RUNNING      hadoop103:8042                           0
  hadoop102:42012      RUNNING      hadoop102:8042                           0
  hadoop104:39702      RUNNING      hadoop104:8042                           0
```

6. 更新配置命令 yarn rmadmin

yarn rmadmin -refreshQueues 命令主要用于重新加载队列配置。

```
[atguigu@hadoop102 hadoop-3.1.3]$ yarn rmadmin -refreshQueues
2021-02-06 10:32:03,331 INFO client.RMProxy: Connecting to ResourceManager at hadoop103/
192.168.10.103:8033
```

7. 查看队列命令 yarn queue

yarn queue -status <QueueName>命令主要用于打印指定名称的队列的状态信息。

```
[atguigu@hadoop102 hadoop-3.1.3]$ yarn queue -status default
2021-02-06 10:32:33,403 INFO client.RMProxy: Connecting to ResourceManager at hadoop103/
192.168.10.103:8032
Queue Information :
Queue Name : default
   State : RUNNING
   Capacity : 100.0%
   Current Capacity : .0%
   Maximum Capacity : 100.0%
   Default Node Label expression : <DEFAULT_PARTITION>
   Accessible Node Labels : *
   Preemption : disabled
   Intra-queue Preemption : disabled
```

6.3.2 核心参数

在 YARN 的使用过程中，用户可以通过调节一些核心参数实现所需的资源调度效果。下面分类展示重要的核心参数。

ResourceManager 的相关参数如表 6-1 所示。

表 6-1 ResourceManager 的相关参数

参　　数	含　　义	默 认 值
yarn.resourcemanager.scheduler.class	配置调度器	容量调度器
yarn.resourcemanager.scheduler.client.thread-count	ResourceManager 处理调度器请求的线程数量	50

NodeManager 的相关参数如表 6-2 所示。

表 6-2 NodeManager 的相关参数

参　　数	含　　义	默 认 值
yarn.nodemanager.resource.detect-hardware-capabilities	是否根据系统内存空间自动计算	false
yarn.nodemanager.resource.count-logical-processors-as-cores	是否将虚拟核数当作 CPU 核数	false
yarn.nodemanager.resource.pcores-vcores-multiplier	虚拟核数和物理核数的乘数。例如,对于 4 核 8 线程的 CPU,该参数的值应该为 2	1.0
yarn.nodemanager.resource.memory-mb	NodeManager 使用的内存空间	8GB
yarn.nodemanager.resource.system-reserved-memory-mb	NodeManager 为系统保留了多少内存空间。此参数与 yarn.nodemanager.resource.memory-mb 参数配置一个即可	—
yarn.nodemanager.resource.cpu-vcores	NodeManager 使用的 CPU 核数	8 个
yarn.nodemanager.pmem-check-enabled	是否开启物理内存空间检查限制 Container	true（开启）
yarn.nodemanager.vmem-check-enabled	是否开启虚拟内存空间检查限制 Container	true（开启）
yarn.nodemanager.vmem-pmem-ratio	虚拟内存空间与物理内存空间的比值	2.1

Container 的相关参数如表 6-3 所示。

表 6-3 Container 的相关参数

参　　数	含　　义	默 认 值
yarn.scheduler.minimum-allocation-mb	容器最小内存空间	1GB
yarn.scheduler.maximum-allocation-mb	容器最大内存空间	8GB
yarn.scheduler.minimum-allocation-vcores	容器最小 CPU 核数	1 个
yarn.scheduler.maximum-allocation-vcores	容器最大 CPU 核数	4 个

6.3.3 核心参数配置案例

1．需求分析

1）需求。

使用词频统计程序,统计 1GB 数据中每个单词出现的次数。

2）服务器现状。

现有服务器 3 台,每台都有 4GB 内存、4 核 CPU、4 线程。

3）具体需求分析。

假设 1GB 的数据不存在小文件问题,将数据块大小设置为 128MB,则会分为 8 个数据块进行存储。在 MapReduce 程序中会产生 8 个 MapTask,需要运行 1 个 ReduceTask,还需要运行 MRAppMaster 进程,即 ApplicationMaster 角色,共计需要运行 10 个进程。为了尽量平衡各节点的负载,按照（4,3,3）的形式分配进程。

2．代码实现

（1）修改 yarn-site.xml 文件中的配置参数,具体如下:

```xml
<!-- 选择调度器，默认容量 -->
<property>
    <description>The class to use as the resource scheduler.</description>
    <name>yarn.resourcemanager.scheduler.class</name>
    <value>org.apache.hadoop.yarn.server.resourcemanager.scheduler.capacity.CapacityScheduler
</value>
</property>

<!-- ResourceManager 处理调度器请求的线程数量，默认值为 50；如果提交的任务数大于 50，则可以增加该值，但
是不能超过 12（3 台 × 4 线程 = 12 线程），去除其他应用程序，实际不能超过 8 -->
<property>
    <description>Number of threads to handle scheduler interface.</description>
    <name>yarn.resourcemanager.scheduler.client.thread-count</name>
    <value>8</value>
</property>

<!-- 是否根据系统内存空间自动计算，默认值为 false。如果该节点中有很多其他应用程序，那么建议手动配置；如果
该节点中没有其他应用程序，那么可以采用自动配置 -->
<property>
    <description>Enable auto-detection of node capabilities such as
    memory and CPU.
    </description>
    <name>yarn.nodemanager.resource.detect-hardware-capabilities</name>
    <value>false</value>
</property>

<!-- 是否将虚拟核数当作 CPU 核数，默认值为 false，采用物理 CPU 核数 -->
<property>
    <description>Flag to determine if logical processors(such as
    hyperthreads) should be counted as cores. Only applicable on Linux
    when yarn.nodemanager.resource.cpu-vcores is set to -1 and
    yarn.nodemanager.resource.detect-hardware-capabilities is true.
    </description>
    <name>yarn.nodemanager.resource.count-logical-processors-as-cores</name>
    <value>false</value>
</property>

<!-- 虚拟核数和物理核数的乘数，默认值为 1.0 -->
<property>
    <description>Multiplier to determine how to convert phyiscal cores to
    vcores. This value is used if yarn.nodemanager.resource.cpu-vcores
    is set to -1(which implies auto-calculate vcores) and
    yarn.nodemanager.resource.detect-hardware-capabilities is set to true. The number of
vcores will be calculated as number of CPUs * multiplier.
    </description>
    <name>yarn.nodemanager.resource.pcores-vcores-multiplier</name>
    <value>1.0</value>
</property>

<!-- NodeManager 使用的内存空间，默认为 8GB，将其修改为 4GB -->
<property>
```

```xml
    <description>Amount of physical memory, in MB, that can be allocated
    for containers. If set to -1 and
    yarn.nodemanager.resource.detect-hardware-capabilities is true, it is
    automatically calculated(in case of Windows and Linux).
    In other cases, the default is 8192MB.
    </description>
    <name>yarn.nodemanager.resource.memory-mb</name>
    <value>4096</value>
</property>

<!-- NodeManager 使用的 CPU 核数，当不按照硬件环境自动设定时，默认是 8 个，将其修改为 4 个 -->
<property>
    <description>Number of vcores that can be allocated
    for containers. This is used by the RM scheduler when allocating
    resources for containers. This is not used to limit the number of
    CPUs used by YARN containers. If it is set to -1 and
    yarn.nodemanager.resource.detect-hardware-capabilities is true, it is
    automatically determined from the hardware in case of Windows and Linux.
    In other cases, number of vcores is 8 by default.</description>
    <name>yarn.nodemanager.resource.cpu-vcores</name>
    <value>4</value>
</property>

<!-- 容器最小内存空间，默认为 1GB -->
<property>
    <description>The minimum allocation for every container request at the RM in MBs.
    Memory requests lower than this will be set to the value of this property. Additionally,
    a node manager that is configured to have less memory than this value will be shut down
    by the resource manager.
    </description>
    <name>yarn.scheduler.minimum-allocation-mb</name>
    <value>1024</value>
</property>

<!-- 容器最大内存空间，默认为 8GB，将其修改为 2GB -->
<property>
    <description>The maximum allocation for every container request at the RM in MBs.
    Memory requests higher than this will throw an InvalidResourceRequestException.
    </description>
    <name>yarn.scheduler.maximum-allocation-mb</name>
    <value>2048</value>
</property>

<!-- 容器最小 CPU 核数，默认为 1 个 -->
<property>
    <description>The minimum allocation for every container request at the RM in terms of
    virtual CPU cores. Requests lower than this will be set to the value of this property.
    Additionally, a node manager that is configured to have fewer virtual cores than this
    value will be shut down by the resource manager.
    </description>
    <name>yarn.scheduler.minimum-allocation-vcores</name>
```

```xml
        <value>1</value>
</property>

<!-- 容器最大CPU核数，默认为4个，将其修改为2个 -->
<property>
    <description>The maximum allocation for every container request at the RM in terms of virtual CPU cores. Requests higher than this will throw an
    InvalidResourceRequestException.</description>
    <name>yarn.scheduler.maximum-allocation-vcores</name>
    <value>2</value>
</property>

<!-- 虚拟内存空间检查限制，默认为开启，将其修改为关闭 -->
<property>
    <description>Whether virtual memory limits will be enforced for
    containers.</description>
    <name>yarn.nodemanager.vmem-check-enabled</name>
    <value>false</value>
</property>

<!-- 设置虚拟内存空间与物理内存空间的比值，默认值为2.1 -->
<property>
    <description>Ratio between virtual memory to physical memory when setting memory limits
for containers. Container allocations are expressed in terms of physical memory, and
virtual memory usage is allowed to exceed this allocation by this ratio.
    </description>
    <name>yarn.nodemanager.vmem-pmem-ratio</name>
    <value>2.1</value>
</property>
```

（2）在执行任务时，可能会出现以下报错。

```
container [pid=26086,containerID=container_1482373104195_0001_02_000001] is running beyond virtual memory limits. Current usage: 161.4 MB of 200 MB physical memory used; 879.8 MB of 420.0 MB virtual memory used. Killing container.
```

（3）出现以上报错的原因是YARN对物理内存空间和虚拟内存空间的检查机制。要避免报错，需要注意yarn.xml文件中的两项关于内存空间的配置参数，分别是yarn.nodemanager.vmem-check-enabled和yarn.nodemanager. vmem-pmem-ratio，具体如下：

```xml
<!-- 虚拟内存空间检查限制，默认为开启，将其修改为关闭 -->
<property>
    <description>Whether virtual memory limits will be enforced for
    containers.</description>
    <name>yarn.nodemanager.vmem-check-enabled</name>
    <value>false</value>
</property>

<!-- 设置虚拟内存空间与物理内存空间的比值中，默认值为2.1，可适当调大 -->
<property>
    <description>Ratio between virtual memory to physical memory when setting memory limits
for containers. Container allocations are expressed in terms of physical memory, and
virtual memory usage is allowed to exceed this allocation by this ratio.
    </description>
    <name>yarn.nodemanager.vmem-pmem-ratio</name>
```

```
        <value>2.1</value>
</property>
```
（4）分发配置文件，代码如下：

```
[atguigu@hadoop102 hadoop-3.1.3]$ xsync /opt/module/hadoop-3.1.3/etc/hadoop/yarn-site.xml
```

注意：如果集群中每台节点服务器的硬件资源都不一致，那么对每个 NodeManager 单独进行配置。

（5）重启 YARN 集群，代码如下：

```
[atguigu@hadoop102 hadoop-3.1.3]$ sbin/stop-yarn.sh
[atguigu@hadoop103 hadoop-3.1.3]$ sbin/start-yarn.sh
```

（6）执行 WordCount 程序，代码如下：

```
[atguigu@hadoop102 hadoop-3.1.3]$ hadoop jar share/hadoop/mapreduce/hadoop-mapreduce-examples-3.1.3.jar wordcount /input /output
```

（7）查看 ResourceManager 的 Web 端页面中的任务执行结果，如图 6-11 所示。

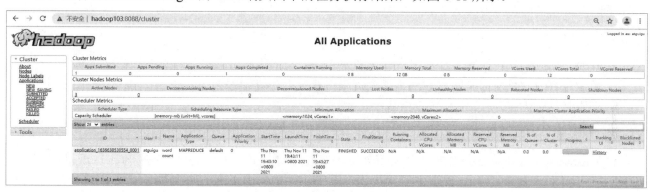

图 6-11　任务执行结果

单击具体任务，可以看到该任务的详细资源配置信息，如图 6-12 所示。

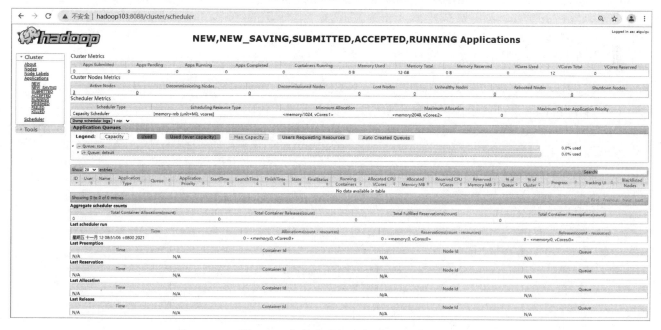

图 6-12　任务的详细资源配置信息

6.3.4　容量调度器配置案例

容量调度器是一个支持多队列的调度器，在使用容量调度器时，常见的两个问题如下。

问题一：容量调度器默认只有 1 个 default 队列，不能满足实际需求，那么在生产环境中以什么标准划分队列？

- 按照计算框架。按照不同的计算框架（如 Hive、Spark、Flink 等）划分队列。
- 按照业务模块。按照不同的业务模块（如登录注册业务、购物车业务、下单业务、业务部门 A、业务部门 B 等）划分队列。

问题二：创建多队列有什么好处？

- 可以避免因为某个用户的代码失误，在代码中使用递归死循环等不当逻辑将所有资源耗尽。
- 实现不同业务的任务降级使用，在特殊时期，可以保证重要任务的队列资源充足。例如，在电商大型促销活动期间，要给促销活动分配更多的计算资源，对于不太重要的任务，可以降低其权重，但是依然能获取一定的资源。

下面使用容量调度器实现一个配置案例。

1. 需求分析

- default 队列的资源额定容量占总内存空间的 40%、最大资源占有量为总资源的 60%，hive 队列的资源额定容量占总内存空间的 60%、最大资源占有量为总资源的 80%。
- 配置队列优先级。

2. 配置多队列的容量调度器

（1）capacity-scheduler.xml 文件中的配置如下，在根队列下配置 2 个队列，分别为 default 队列和 hive 队列。

```xml
<!-- 指定多队列，增加hive队列 -->
<property>
    <name>yarn.scheduler.capacity.root.queues</name>
    <value>default,hive</value>
    <description>
      The queues at the this level (root is the root queue).
    </description>
</property>
```

（2）配置 default 队列的资源额定容量和最大资源占有量，代码如下：

```xml
<!-- 将default队列的资源额定容量减少为40%，默认值为100% -->
<property>
    <name>yarn.scheduler.capacity.root.default.capacity</name>
    <value>40</value>
</property>

<!-- 将default队列的最大资源占有量减少为60%，默认值为100% -->
<property>
    <name>yarn.scheduler.capacity.root.default.maximum-capacity</name>
    <value>60</value>
</property>
```

（3）配置 hive 队列的资源额定容量和最大资源占有量，代码如下：

```xml
<!-- 指定hive队列的资源额定容量 -->
<property>
    <name>yarn.scheduler.capacity.root.hive.capacity</name>
    <value>60</value>
```

```xml
</property>

<!-- 允许单个用户最多可获取的队列资源的倍数，默认值为1，确保无论集群有多空闲，单个用户都不会占用超过队列
配置的资源 -->
<property>
    <name>yarn.scheduler.capacity.root.hive.user-limit-factor</name>
    <value>1</value>
</property>

<!-- 指定hive队列的最大资源占有量 -->
<property>
    <name>yarn.scheduler.capacity.root.hive.maximum-capacity</name>
    <value>80</value>
</property>
```

（4）对hive队列进行必要的配置，代码如下：

```xml
<!-- 启动hive队列 -->
<property>
    <name>yarn.scheduler.capacity.root.hive.state</name>
    <value>RUNNING</value>
</property>

<!-- 哪些用户有权向队列提交作业 -->
<property>
    <name>yarn.scheduler.capacity.root.hive.acl_submit_applications</name>
    <value>*</value>
</property>

<!-- 哪些用户有权操作队列，管理员权限（查看/杀死） -->
<property>
    <name>yarn.scheduler.capacity.root.hive.acl_administer_queue</name>
    <value>*</value>
</property>

<!-- 哪些用户有权配置所提交任务的优先级 -->
<property>
    <name>yarn.scheduler.capacity.root.hive.acl_application_max_priority</name>
    <value>*</value>
</property>

<!-- 提交任务时的超时时间设置方式: yarn application -appId appId -updateLifetime Timeout-->

<!-- 如果在提交任务时指定了超时时间，那么指定的超时时间以此参数的值为上限，默认值为-1，表示无上限 -->
<property>
    <name>yarn.scheduler.capacity.root.hive.maximum-application-lifetime</name>
    <value>-1</value>
</property>

<!-- 如果在提交任务时没指定超时时间，则使用此参数的值作为默认值，默认值为-1，表示无上限 -->
```

```xml
<property>
    <name>yarn.scheduler.capacity.root.hive.default-application-lifetime</name>
    <value>-1</value>
</property>
```

（5）分发配置文件，代码如下：

```
[atguigu@hadoop102 hadoop-3.1.3]$ xsync /opt/module/hadoop-3.1.3/etc/hadoop/capacity-scheduler.xml
```

（6）重启 YARN，或者执行以下命令刷新队列。

```
[atguigu@hadoop102 hadoop-3.1.3]$ yarn rmadmin -refreshQueues
```

（7）观察 YARN 的 Web 端页面，可以看到两个队列的配置情况，如图 6-13 所示。

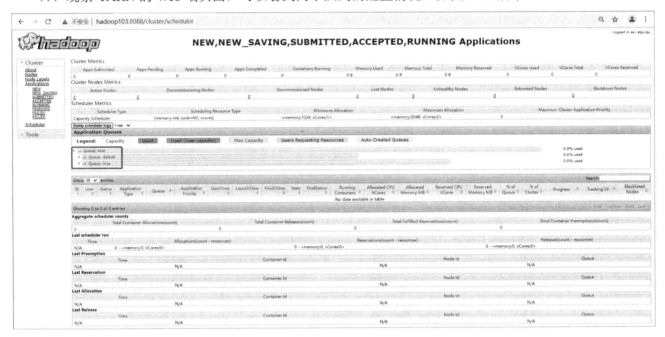

图 6-13　两个队列的配置情况

3．向 hive 队列提交任务

（1）使用 hadoop jar 命令运行 Hadoop 提供的 WordCount 程序，通过 mapreduce.job.queuename 参数向 hive 队列提交任务，代码如下：

```
[atguigu@hadoop102 hadoop-3.1.3]$ hadoop jar share/hadoop/mapreduce/hadoop-mapreduce-examples-3.1.3.jar wordcount -D mapreduce.job.queuename=hive /input /output
```

注意：-D 表示在运行时改变参数值。

（2）在代码中对队列进行配置。

在默认情况下，将任务提交到 default 队列中，如果希望向 hive 队列提交任务，则需要在 Driver 驱动类中配置 mapreduce.job.queuename 参数，代码如下：

```java
public class WcDrvier {

    public static void main(String[] args) throws IOException, ClassNotFoundException, InterruptedException {

        Configuration conf = new Configuration();

        conf.set("mapreduce.job.queuename","hive");
```

```
    //1. 获取一个 Job 对象 job
    Job job = Job.getInstance(conf);

    ...

    //6. 提交 Job
    boolean b = job.waitForCompletion(true);
    System.exit(b ? 0 : 1);
    }
}
```

在配置完成后，重新打包 jar 包，即可将任务提交到 hive 队列中，如图 6-14 所示。

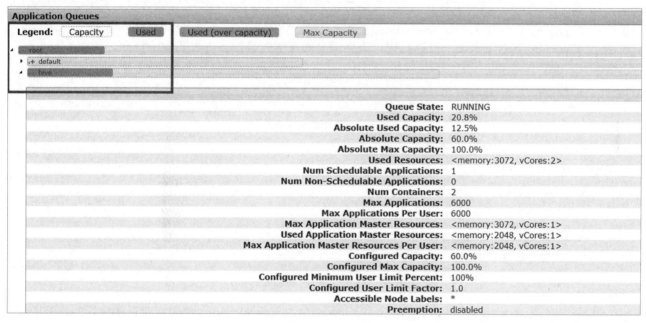

图 6-14　任务提交情况

4．任务优先级

容量调度器支持任务优先级的配置，在资源紧张时，优先级高的任务可以优先获取资源。在默认情况下，YARN 将所有任务的优先级限制为 0，如果要使用任务的优先级功能，则需要解除该限制。

（1）修改 yarn-site.xml 文件，添加以下参数，将任务的最高优先级设置为 5。

```
<property>
    <name>yarn.cluster.max-application-priority</name>
    <value>5</value>
</property>
```

（2）分发配置文件，并且重启 YARN，代码如下：

```
[atguigu@hadoop102 hadoop]$ xsync yarn-site.xml
[atguigu@hadoop103 hadoop-3.1.3]$ sbin/stop-yarn.sh
[atguigu@hadoop103 hadoop-3.1.3]$ sbin/start-yarn.sh
```

（3）模拟资源紧张环境，连续提交以下任务，直到新提交的任务申请不到资源为止，如图 6-15 所示。

```
[atguigu@hadoop102 hadoop-3.1.3]$ hadoop jar /opt/module/hadoop-3.1.3/share/hadoop/mapreduce/
hadoop-mapreduce-examples-3.1.3.jar pi 5 2000000
```

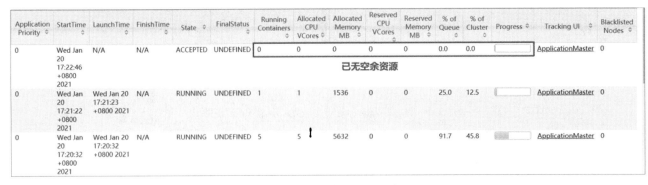

图 6-15　模拟资源紧张环境

（4）再次提交任务，将任务优先级设置为最高级 5，代码如下：

```
[atguigu@hadoop102 hadoop-3.1.3]$ hadoop jar /opt/module/hadoop-3.1.3/share/hadoop/mapreduce/hadoop-mapreduce-examples-3.1.3.jar pi -D mapreduce.job.priority=5 5 2000000
```

查看 YARN 的 Web 端页面，可以看到，最后提交的高优先级任务优先获取了资源，如图 6-16 所示。

图 6-16　高优先级任务优先获取资源

（5）可以使用 yarn application -appID <ApplicationId> -updatePriority 命令修改正在执行的任务的优先级，示例代码如下：

```
[atguigu@hadoop102 hadoop-3.1.3]$ yarn application -appID application_1611133087930_0009 -updatePriority 5
```

6.3.5　公平调度器配置案例

使用公平调度器实现一个配置案例。

1．需求分析

创建两个队列，分别是 test 和 atguigu（以用户所属组命名），期望实现以下效果：如果用户在提交任务时指定队列，那么将任务提交到指定队列中运行；如果未指定队列，那么将 test 用户的任务到提交 root.group.test 队列中运行，将 atguigu 用户的任务提交到 root.group.atguigu 队列中运行（group 为用户所属组）。

公平调度器的配置涉及两个文件，一个是配置文件 yarn-site.xml，另一个是公平调度器队列分配文件 fair-scheduler.xml（文件名可自定义）。

2．配置多队列的公平调度器

（1）修改 yarn-site.xml 文件，添加以下参数。

```
<property>
```

```xml
        <name>yarn.resourcemanager.scheduler.class</name>
<value>org.apache.hadoop.yarn.server.resourcemanager.scheduler.fair.FairScheduler</value>
        <description>配置使用公平调度器</description>
</property>

<property>
        <name>yarn.scheduler.fair.allocation.file</name>
        <value>/opt/module/hadoop-3.1.3/etc/hadoop/fair-scheduler.xml</value>
        <description>指明公平调度器队列分配配置文件</description>
</property>

<property>
        <name>yarn.scheduler.fair.preemption</name>
        <value>false</value>
        <description>禁止队列间资源抢占</description>
</property>
```

（2）创建 fair-scheduler.xml 文件，并且在该文件中添加以下内容。

```xml
<?xml version="1.0"?>
<allocations>
  <!-- 单个队列中 ApplicationMaster 占用资源的最大比例，取值范围为 0~1，企业一般配置 0.1 -->
  <queueMaxAMShareDefault>0.5</queueMaxAMShareDefault>
  <!-- 单个队列最大资源的默认值 test atguigu default -->
  <queueMaxResourcesDefault>4096mb,4vcores</queueMaxResourcesDefault>

  <!-- 增加一个队列 test -->
  <queue name="test">
    <!-- 队列最小资源 -->
    <minResources>2048mb,2vcores</minResources>
    <!-- 队列最大资源 -->
    <maxResources>4096mb,4vcores</maxResources>
    <!-- 队列中最多同时运行的应用程序数量，默认为 50 个，根据线程数量进行配置 -->
    <maxRunningApps>4</maxRunningApps>
    <!-- 队列中 ApplicationMaster 占用资源的最大比例 -->
    <maxAMShare>0.5</maxAMShare>
    <!-- 该队列的资源权重，默认值为 1.0 -->
    <weight>1.0</weight>
    <!-- 队列内部的资源分配策略 -->
    <schedulingPolicy>fair</schedulingPolicy>
  </queue>
  <!-- 增加一个队列 atguigu -->
  <queue name="atguigu" type="parent">
    <!-- 队列最小资源 -->
    <minResources>2048mb,2vcores</minResources>
    <!-- 队列最大资源 -->
    <maxResources>4096mb,4vcores</maxResources>
    <!-- 队列中最多同时运行的应用程序数量，默认为 50 个，根据线程数量进行配置 -->
    <maxRunningApps>4</maxRunningApps>
    <!-- 队列中 ApplicationMaster 占用资源的最大比例 -->
    <maxAMShare>0.5</maxAMShare>
```

```xml
<!-- 该队列的资源权重，默认值为 1.0 -->
<weight>1.0</weight>
<!-- 队列内部的资源分配策略 -->
<schedulingPolicy>fair</schedulingPolicy>
</queue>

<!-- 任务队列分配策略，可以配置多层规则，从第一个规则开始匹配，直到匹配成功 -->
<queuePlacementPolicy>
  <!-- 在提交任务时指定队列，如果未指定队列，则继续匹配下一个规则；如果值为 false，则表示指定的队列不存在，不允许自动创建-->
  <rule name="specified" create="false"/>
  <!-- 提交到 root.group.username 队列中，如果 root.group 不存在，则不允许自动创建；如果 root.group.username 不存在，则允许自动创建 -->
  <rule name="nestedUserQueue" create="true">
      <rule name="primaryGroup" create="false"/>
  </rule>
  <!-- 最后一个规则必须为 reject 或 default。Reject 表示拒绝提交任务,default 表示将任务提交到 default 队列中 -->
  <rule name="reject" />
</queuePlacementPolicy>
</allocations>
```

（3）分发配置文件并重启 YARN，代码如下：

```
[atguigu@hadoop102 hadoop]$ xsync yarn-site.xml
[atguigu@hadoop102 hadoop]$ xsync fair-scheduler.xml

[atguigu@hadoop103 hadoop-3.1.3]$ sbin/stop-yarn.sh
[atguigu@hadoop103 hadoop-3.1.3]$ sbin/start-yarn.sh
```

3．测试提交任务

（1）通过 atguigu 用户在提交任务时指定队列 test，代码如下：

```
[atguigu@hadoop102 hadoop-3.1.3]$ hadoop jar /opt/module/hadoop-3.1.3/share/hadoop/mapreduce/hadoop-mapreduce-examples-3.1.3.jar pi -Dmapreduce.job.queuename=root.test 1 1
```

按照配置规则，任务会被提交到指定的 root.test 队列中，如图 6-17 所示。

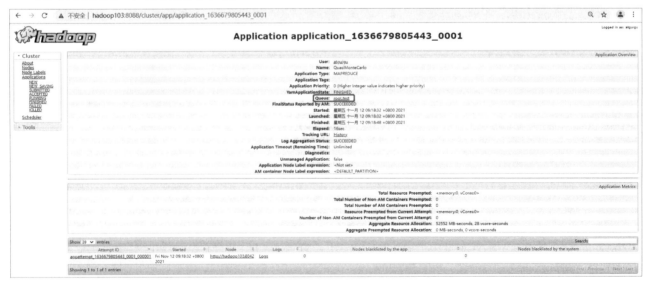

图 6-17　任务所属队列 root.test

（2）通过 atguigu 用户在提交任务时不指定队列，代码如下：
[atguigu@hadoop102 hadoop-3.1.3]$ hadoop jar /opt/module/hadoop-3.1.3/share/hadoop/mapreduce/hadoop-mapreduce-examples-3.1.3.jar pi 1 1

按照配置规则，任务会被提交到 root.atguigu.atguigu 队列中，如图 6-18 所示。

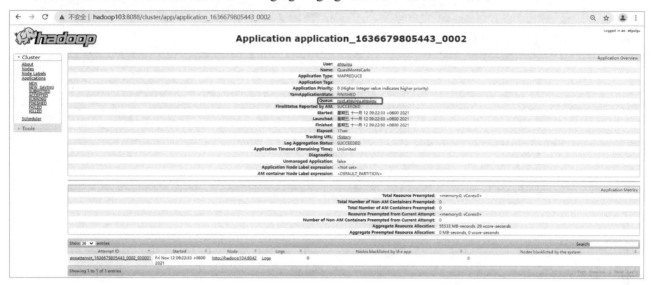

图 6-18　任务所属队列 root.atguigu.atguigu

6.3.6　Tool 接口案例

回顾一下我们在通过自己编写的 WordCount 程序向集群提交任务时执行的命令，具体如下：
[atguigu@hadoop102 hadoop-3.1.3]$ hadoop jar wc.jar com.atguigu.mapreduce.wordcount2.WordCountDriver /input /output1

使用-D 参数动态修改程序运行时的重要参数会报错。这是因为，程序认为"-Dmapreduce.job.queuename=root.test"是第一个输入参数，即编写代码时传入的数据输入路径。
[atguigu@hadoop102 hadoop-3.1.3]$ hadoop jar wc.jar com.atguigu.mapreduce.wordcount2.WordCountDriver -Dmapreduce.job.queuename=root.test /input /output1

但是在通过 Hadoop 官方提供的示例程序 WordCount 提交程序时，可以使用-D 参数动态修改参数。那么如何编写代码，才能实现动态修改参数的功能呢？答案是使用 Tool 接口。具体步骤如下：

（1）新建 Maven 工程 YarnDemo，在 pom.xml 文件中添加以下依赖。
```xml
<dependencies>
    <dependency>
        <groupId>org.apache.hadoop</groupId>
        <artifactId>hadoop-client</artifactId>
        <version>3.1.3</version>
    </dependency>
</dependencies>
```

（2）创建 com.atguigu.yarn 包。
（3）创建 WordCount 类，使其实现 Tool 接口，代码如下：
```java
package com.atguigu.yarn;

import org.apache.hadoop.conf.Configuration;
import org.apache.hadoop.fs.Path;
import org.apache.hadoop.io.IntWritable;
import org.apache.hadoop.io.LongWritable;
```

```java
import org.apache.hadoop.io.Text;
import org.apache.hadoop.mapreduce.Job;
import org.apache.hadoop.mapreduce.Mapper;
import org.apache.hadoop.mapreduce.Reducer;
import org.apache.hadoop.mapreduce.lib.input.FileInputFormat;
import org.apache.hadoop.mapreduce.lib.output.FileOutputFormat;
import org.apache.hadoop.util.Tool;

import java.io.IOException;

public class WordCount implements Tool {

    private Configuration conf;

    @Override
    public int run(String[] args) throws Exception {

        Job job = Job.getInstance(conf);

        job.setJarByClass(WordCountDriver.class);

        job.setMapperClass(WordCountMapper.class);
        job.setReducerClass(WordCountReducer.class);

        job.setMapOutputKeyClass(Text.class);
        job.setMapOutputValueClass(IntWritable.class);
        job.setOutputKeyClass(Text.class);
        job.setOutputValueClass(IntWritable.class);

        FileInputFormat.setInputPaths(job, new Path(args[0]));
        FileOutputFormat.setOutputPath(job, new Path(args[1]));

        return job.waitForCompletion(true) ? 0 : 1;
    }

    @Override
    public void setConf(Configuration conf) {
        this.conf = conf;
    }

    @Override
    public Configuration getConf() {
        return conf;
    }

    public static class WordCountMapper extends Mapper<LongWritable, Text, Text, IntWritable> {

        private Text outK = new Text();
        private IntWritable outV = new IntWritable(1);

        @Override
```

```java
        protected void map(LongWritable key, Text value, Context context) throws IOException, InterruptedException {

            String line = value.toString();
            String[] words = line.split(" ");

            for (String word : words) {
                outK.set(word);

                context.write(outK, outV);
            }
        }
    }

    public static class WordCountReducer extends Reducer<Text, IntWritable, Text, IntWritable> {
        private IntWritable outV = new IntWritable();

        @Override
        protected void reduce(Text key, Iterable<IntWritable> values, Context context) throws IOException, InterruptedException {

            int sum = 0;

            for (IntWritable value : values) {
                sum += value.get();
            }
            outV.set(sum);

            context.write(key, outV);
        }
    }
}
```

（4）创建 WordCountDriver 类，代码如下：

```java
package com.atguigu.yarn;

import org.apache.hadoop.conf.Configuration;
import org.apache.hadoop.util.Tool;
import org.apache.hadoop.util.ToolRunner;
import java.util.Arrays;

public class WordCountDriver {

    private static Tool tool;

    public static void main(String[] args) throws Exception {
        // 1 创建配置文件
        Configuration conf = new Configuration();

        // 2 判断是否有 Tool 接口
        switch (args[0]){
```

```
            case "wordcount":
                tool = new WordCount();
                break;
            default:
                throw new RuntimeException(" No such tool: "+ args[0] );
        }
        // 3 使用ToolRunner执行程序
        // Arrays.copyOfRange将原数组中的元素放到新数组中
        int run = ToolRunner.run(conf, tool, Arrays.copyOfRange(args, 1, args.length));

        System.exit(run);
    }
}
```

（5）首先将程序打包，然后将 jar 包重命名为 YarnDemo.jar，最后将 jar 包上传至 Hadoop 的安装目录下。

（6）在 HDFS 中准备输入文件/input，执行以下命令，向集群提交该 jar 包，验证程序编写是否成功。

```
[atguigu@hadoop102 hadoop-3.1.3]$ yarn jar YarnDemo.jar com.atguigu.yarn.WordCountDriver wordcount /input /output
```

需要注意的是，此时提交的 3 个参数，第 1 个参数主要用于指定特定的 Tool 接口实现类，第 2 个和第 3 个参数分别为数据的输入路径和输出路径。如果用户希望加入其他配置参数，则可以在 wordcount 后面使用-D 参数添加，代码如下；如果需要多个参数，那么继续使用-D 参数添加其他参数。

```
[atguigu@hadoop102 hadoop-3.1.3]$ yarn jar YarnDemo.jar com.atguigu.yarn.WordCountDriver wordcount -Dmapreduce.job.queuename=root.test /input /output1
```

6.4 本章总结

本章主要讲解了 Hadoop 的资源调度框架——YARN。在引入 YARN 后，Hadoop 集群就可以将资源开放给更多计算引擎，将更多的大数据框架纳入其中了。用户在使用 Hadoop 时，通常对运行在幕后的资源调度器是无知无觉的，但是其作用不能忽视。了解 YARN 的底层运行机制，对实际开发大有裨益，对资源调度器参数进行合理的调整，可以显著提高程序的运行效率。

第 7 章 高可用 HA

经过前面的讲解，相信读者对 Hadoop 是如何运行的已经有了全面的了解。Hadoop 的两大重要组件 HDFS 和 YARN，都使用 Master-Slave 部署模式（分布式架构最常采用的模式之一），但是这种架构都存在一个严重的问题——单点故障，如果充当 Master 角色的节点发生故障，那么整个集群将不能工作。针对这种情况，开发人员需要增加容错机制。本章将详细讲解 Hadoop 是如何解决这个问题的。

7.1 ZooKeeper 详解

在讲解 Hadoop 如何实现高可用前，首先需要讲解 ZooKeeper。

7.1.1 ZooKeeper 入门

ZooKeeper 是 Apache 公司为分布式应用程序提供协调服务的开源项目。

从设计模式的角度来看，ZooKeeper 是一个基于观察者模式设计的分布式服务管理框架，它负责存储和管理大家都关心的数据，然后接受观察者的注册，一旦这些数据的状态发生了变化，ZooKeeper 就会通知已经在 ZooKeeper 上注册的观察者，令其做出相应的反应。事实上，我们可以认为 ZooKeeper 就是文件系统和通知机制的汇总。

ZooKeeper 的基本架构如图 7-1 所示。

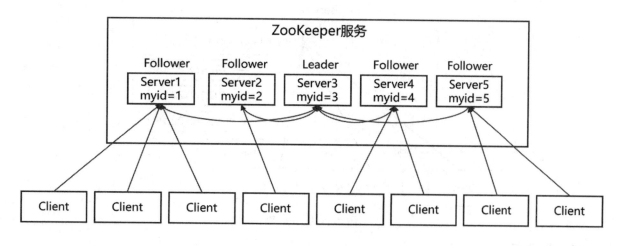

图 7-1 ZooKeeper 的基本架构

根据图 7-1 可知，ZooKeeper 具有以下特点。

- ZooKeeper 是由一个领导者（Leader）和多个跟随者（Follower）组成的集群。
- ZooKeeper 集群中只要有超过一半的节点存活，就能正常提供服务。

- 具有全局数据一致性。每个 Server 中都存储一份相同的数据副本，Client 无论连接到哪个 Server，数据都是一致的。
- 来自同一个 Client 的更新请求会按其发送顺序依次执行。
- 数据更新具有原子性。一次数据更新要么成功，要么失败。
- 具有实时性。在一定的时间范围内，Client 能读取最新数据。

ZooKeeper 的数据模型如图 7-2 所示，其结构与 UNIX 文件系统的结构类似，在整体上可以将其看作一棵树，节点称为 ZNode。ZNode 既可以存储数据，又可以存储其他节点。一个 ZNode 默认可以存储 1MB 数据，每个 ZNode 都可以通过其路径唯一标识。

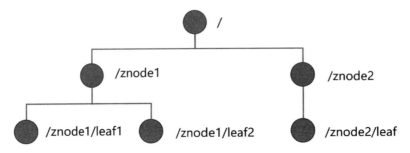

图 7-2　ZooKeeper 的数据模型

ZooKeeper 可以提供的服务包括统一命名、统一配置管理、统一集群管理、动态感知服务器节点上下线、软负载均衡等。

1．统一命名

在分布式环境中，经常需要对应用程序或服务进行统一命名，以便识别。这是因为服务器的 IP 地址不容易记住，但是域名容易记住。如图 7-3 所示，在统一命名后，3 个客户端可以通过访问域名"www.baidu.com"访问正确的服务器地址。

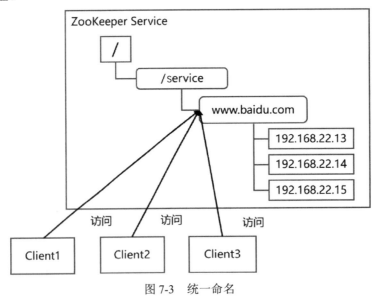

图 7-3　统一命名

2．统一配置管理

在分布式环境中，配置文件同步非常常见。一般要求一个集群中所有节点的配置信息保持一致，如 Kafka 集群、Hadoop 集群等。在对配置文件进行修改后，希望能够将其快速同步到各个节点上。但是集群的数量可能非常庞大，同步配置的过程会十分烦琐。

统一配置管理可以交由 ZooKeeper 实现。将配置信息写入 ZooKeeper 中的一个 ZNode，各个客户端都会监听这个 ZNode，一旦这个 ZNode 中的配置信息被修改，ZooKeeper 就会通知各个客户端，如图 7-4 所示。

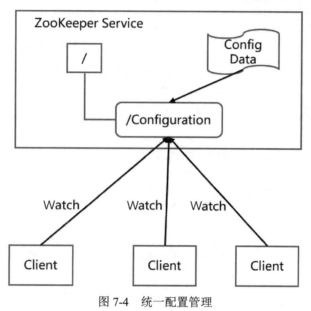

图 7-4　统一配置管理

3．统一集群管理

在分布式环境中，实时掌握每个节点的状态是必要的，可以根据节点的实时状态做出适当的调整。

ZooKeeper 可以实时监控节点的状态变化情况，如图 7-5 所示，将节点信息写入 ZooKeeper 中的一个 ZNode，监听这个 ZNode，可以获取节点的实时状态变化情况。

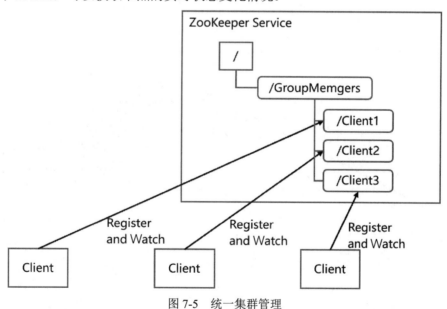

图 7-5　统一集群管理

4．动态感知服务器节点的上下线

服务器节点在启动时，会到 ZooKeeper 集群上注册节点信息，客户端会从 ZooKeeper 处获取当前在线的服务器列表，并且注册监听。当服务器节点下线时，ZooKeeper 会向注册监听的客户端发送节点下线通知，客户端会重新获取在线的服务器列表，并且注册监听，如图 7-6 所示。

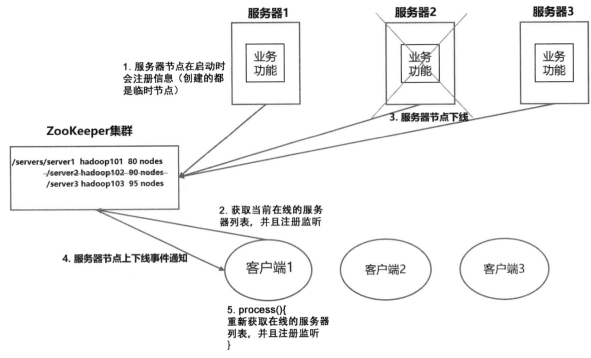

图 7-6 动态感知服务器节点的上下线

5. 软负载均衡

在 ZooKeeper 中记录每台服务器的访问数，让访问负载最小的服务器处理最新的客户端请求，实现软负载均衡，如图 7-7 所示。

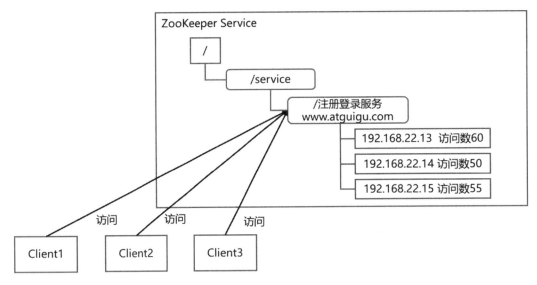

图 7-7 软负载均衡

7.1.2 ZooKeeper 安装

ZooKeeper 服务有两种不同的运行模式，一种是独立模式，另一种是集群模式。独立模式是指只安装一个 ZooKeeper 服务器，模式运行简单，通常只应用于测试环境中，在开发环境中不会使用，因为这种模式不能保证高可用性。集群模式是指安装多台 ZooKeeper 服务器，形成一个集群，在这种模式下，只要有超过半数的节点存活，就能对外提供服务。本节只对集群模式的安装进行讲解。

1. 集群规划

在 hadoop102、hadoop103 和 hadoop104 节点服务器上部署 ZooKeeper。

2. 解压缩并安装

（1）将 ZooKeeper 安装包解压缩到/opt/module/目录下。

```
[atguigu@hadoop102 software]$ tar -zxvf apache-zookeeper-3.5.7-bin.tar.gz -C /opt/module/
```

（2）将/opt/module/apache-zookeeper-3.5.7-bin 重命名为 zookeeper-3.5.7。

```
[atguigu@hadoop102 module]$ mv apache-zookeeper-3.5.7-bin/ zookeeper-3.5.7
```

（3）将/opt/module/zookeeper-3.5.7 目录下的内容同步到 hadoop103、hadoop104 节点服务器中。

```
[atguigu@hadoop102 module]$ xsync zookeeper-3.5.7/
```

3. 配置服务器编号

（1）在/opt/module/zookeeper-3.5.7 目录下创建一个 zkData 文件夹。

```
[atguigu@hadoop102 zookeeper-3.5.7]$ mkdir zkData
```

（2）在/opt/module/zookeeper-3.5.7/zkData 目录下创建一个 myid 文件。

```
[atguigu@hadoop102 zkData]$ vi myid
```

在 myid 文件中，根据 zoo.cfg 文件中配置的 Server ID 与节点服务器 IP 地址的对应关系，添加与 Server 对应的编号，如在 hadoop102 节点服务器中添加 2。

```
2
```

注意：在创建 myid 文件时，一定要在 Linux 操作系统中创建，在文本编辑工具中创建有可能出现乱码。

（3）将配置好的 myid 文件复制到其他节点服务器中，并且在 hadoop103、hadoop104 节点服务器中将 myid 文件中的内容分别修改为 3、4。

```
[atguigu@hadoop102 zookeeper-3.5.7]$ xsync zkData
```

4. 配置 zoo.cfg 文件

（1）将/opt/module/zookeeper-3.5.7/conf 目录下的配置文件 zoo_sample.cfg 重命名为 zoo.cfg。

```
[atguigu@hadoop102 conf]$ mv zoo_sample.cfg zoo.cfg
```

（2）打开 zoo.cfg 文件。

```
[atguigu@hadoop102 conf]$ vim zoo.cfg
```

在 zoo.cfg 文件中找到以下配置信息，将数据存储目录 dataDir 设置为前面自行创建的 zkData 文件夹。

```
dataDir=/opt/module/zookeeper-3.5.7/zkData
```

增加以下配置，指出 ZooKeeper 集群中 3 台节点服务器的相关信息。

```
#######################cluster##########################
server.2=hadoop102:2888:3888
server.3=hadoop103:2888:3888
server.4=hadoop104:2888:3888
```

（3）解读配置参数。

```
Server.A=B:C:D。
```

- A 是一个数字，表示第几台节点服务器。
- B 是这台节点服务器的 IP 地址。
- C 是这台节点服务器与集群中的 Leader 交换信息的端口。
- 当集群中的 Leader 无法正常运行时，需要选举出一个新的 Leader。D 表示选举时服务器之间互相通信的端口。

在集群模式下配置 myid 文件，该文件在 dataDir 目录下，其中有一个数据是 A 的值，ZooKeeper 在启动时读取该文件，并且将该文件中的数据与 zoo.cfg 文件中的配置信息进行比较，从而判断到底是哪台服务器。

在 ZooKeeper 的配置文件 zoo.cfg 中，对其他参数的相关说明如下。

```
tickTime =2000
```

tickTime 参数主要用于配置 ZooKeeper 的服务器与客户端之间的心跳时间，单位为毫秒。

心跳时间是 ZooKeeper 使用的基本时间，是服务器之间或客户端与服务器之间维持心跳的时间间隔，也就是每隔 tickTime 时间，都会发送一个心跳，时间单位为毫秒。可以设置最小的会话超时时间为两倍心跳时间。

```
initLimit =10
```

initLimit 参数主要用于配置集群中的 Follower 与 Leader 在初始化同步数据时能容忍的最大心跳数量（tickTime 的数量），主要用于限定集群中的 ZooKeeper 服务器连接 Leader 的时限。

```
syncLimit =5
```

syncLimit 参数是在运行过程中，集群中 Follower 与 Leader 之间进行通信时能容忍的最大心跳数量。

```
dataDir
```

dataDir 是数据文件目录及数据持久化目录，主要用于存储 ZooKeeper 中的数据。

```
clientPort =2181
```

clientPort 参数主要用于配置监听客户端连接的端口，默认为 2181。

（4）分发配置文件 zoo.cfg，代码如下：

```
[atguigu@hadoop102 conf]$ xsync zoo.cfg
```

5．集群操作

（1）在 3 台节点服务器中分别启动 ZooKeeper，代码如下：

```
[atguigu@hadoop102 zookeeper-3.5.7]# bin/zkServer.sh start
[atguigu@hadoop103 zookeeper-3.5.7]# bin/zkServer.sh start
[atguigu@hadoop104 zookeeper-3.5.7]# bin/zkServer.sh start
```

（2）执行以下命令，在 3 台节点服务器中查看 ZooKeeper 的服务状态。

```
[atguigu@hadoop102 zookeeper-3.5.7]# bin/zkServer.sh status
JMX enabled by default
Using config: /opt/module/zookeeper-3.5.7/bin/../conf/zoo.cfg
Mode: follower
[atguigu@hadoop103 zookeeper-3.5.7]# bin/zkServer.sh status
JMX enabled by default
Using config: /opt/module/zookeeper-3.5.7/bin/../conf/zoo.cfg
Mode: leader
[atguigu@hadoop104 zookeeper-3.5.7]# bin/zkServer.sh status
JMX enabled by default
Using config: /opt/module/zookeeper-3.5.7/bin/../conf/zoo.cfg
Mode: follower
```

7.1.3　ZooKeeper 的内部原理

ZooKeeper 能对外提供可靠的分布式管理服务，依赖于其高可用的设计。ZooKeeper 运行在一组节点服务器上，客户端无论连接哪一台节点服务器，都可以获取同样的服务。此外，只要有半数以上的节点服务器存活，ZooKeeper 就可以正常提供服务。ZooKeeper 中每台节点服务器的角色都是相同的，存储的数据也是相同的，哪台节点服务器会成为 Leader 是由选举机制决定的。

ZooKeeper 的选举机制使用的是半数机制，也就是说，只要集群中有半数以上的节点服务器存活，集群就可用。因此，ZooKeeper 的集群适合部署奇数台节点服务器。

ZooKeeper 虽然在配置文件中并没有指定 Leader 和 Follower。但是，ZooKeeper 在工作时，有一台节点服务器为 Leader，其他节点服务器均为 Follower，Leader 是通过内部的选举机制临时产生的。

下面使用一个简单的案例说明整个选举过程。假设有 5 台节点服务器组成的 ZooKeeper 集群，这 5 台节点服务器的 id 为 1~5，如图 7-8 所示。这 5 台节点服务器都是最新启动的，没有历史数据。依序启动这 5 台节点服务器，观察会发生什么。

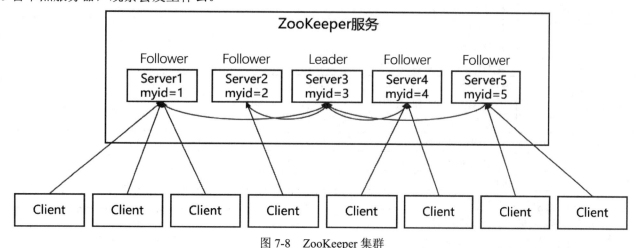

图 7-8　ZooKeeper 集群

（1）启动 Server1，发起一次选举。Server1 投自己一票。此时 Server1 的票数为 1 票，少于半数（3 票），选举无法完成，Server1 的状态保持为 LOOKING。

（2）启动 Server2，发起一次选举。Server1 和 Server2 分别投自己一票并交换选票信息。此时，Server1 发现 Server2 的 id 比自己目前投票推举的服务器（Server1）id 大，将选票改投给 Server2。此时，Server1 的票数为 0 票，Server2 的票数为 2 票，少于半数，选举无法完成，Server1 和 Server2 的状态保持为 LOOKING。

（3）启动 Server3，发起一次选举。此时，Server1 和 Server2 都将选票改投给 Server3。此次投票结果为，Server1 的票数为 0 票，Server2 的票数为 0 票，服务器 3 的票数为 3 票。此时，Server3 的票数已经超过半数，Server3 当选 Leader。将 Server1、Server2 的状态修改为 FOLLOWING，将 Server3 的状态修改为 LEADING。

（4）启动 Server4，发起一次选举。此时，Server1、Server2、Server3 的状态已经不是 LOOKING 状态了，因此不会更改选票信息。交换选票信息的结果为，Server3 的票数为 3 票，Server4 的票数为 1 票。此时，Server4 服从多数服务器，将选票改投给 Server3，并且将状态修改为 FOLLOWING。

（5）启动 Server5，与 Server4 一样，将 Server5 的状态修改为 FOLLOWING。

在正常运行过程中，如果 Leader 宕机了，则会再进行一次选举，只要有超过半数的节点服务器存活，就可以产生新的 Leader。所以，如果只有不到半数的节点服务器存活，那么 ZooKeeper 不会对外提供服务。

7.1.4　ZooKeeper 的命令操作

操作 ZooKeeper 的方式有很多，最常用的方式是客户端命令、Java API 操作。

1．客户端命令

除了四字命令，ZooKeeper 还可以开启客户端，执行客户端命令，查看节点和数据情况。常用的客户端命令如表 7-1 所示。

表 7-1 常用的客户端命令

命令基本语法	功 能 描 述
help	显示所有操作命令
ls path	使用 ls 命令查看当前 ZNode 的子节点 [可监听]。 -w: 监听子节点的变化情况。 -s: 附加次级信息
create	普通创建。 -s: 含有序列。 -e: 临时（重启或超时消失）
get path	获取节点的值 [可监听]。 -w: 监听节点内容的变化情况。 -s: 附加次级信息
set	设置节点的具体值
stat	查看节点状态
delete	删除节点
deleteall	递归删除节点

下面介绍以上命令的相关应用。

1）在任意节点服务器上启动客户端。

```
[atguigu@hadoop103 zookeeper-3.5.7]$ bin/zkCli.sh
```

2）显示所有操作命令。

```
[zk: localhost:2181(CONNECTED) 1] help
```

3）查看当前 ZNode 中包含的内容。

```
[zk: localhost:2181(CONNECTED) 0] ls /
[zookeeper]
```

4）查看当前节点中的详细数据。

```
[zk: localhost:2181(CONNECTED) 1] ls -s /
[zookeeper]
cZxid = 0x0
ctime = Thu Jan 01 08:00:00 CST 1970
mZxid = 0x0
mtime = Thu Jan 01 08:00:00 CST 1970
pZxid = 0x0
cversion = -1
dataVersion = 0
aclVersion = 0
ephemeralOwner = 0x0
dataLength = 0
numChildren = 1
```

在上述节点详细数据中，各参数的含义如下。

- cZxid：创建节点事务的 Zxid。每次修改 ZooKeeper 状态，都会收到一个 Zxid 形式的时间戳，即 ZooKeeper 的事务 ID。事务 ID 是 ZooKeeper 中所有修改操作的总次序。每个修改操作都有唯一的 Zxid，如果 Zxid1 小于 Zxid2，那么 Zxid1 在 Zxid2 之前发生。
- ctime：ZNode 被创建的毫秒数（从 1970 年开始）。
- mZxid：ZNode 最后更新的事务 Zxid。
- mtime：ZNode 最后修改的毫秒数（从 1970 年开始）。
- pZxid：ZNode 最后更新的子节点 Zxid。

- cversion：ZNode 子节点的修改次数。
- dataVersion：ZNode 数据的修改次数。
- aclVersion：ZNode 访问控制列表的修改次数。
- ephemeralOwner：如果是临时节点，那么该参数的值是 ZNode 拥有者的 session id；如果不是临时节点，那么该参数的值是 0。
- dataLength：ZNode 的数据长度。
- numChildren：ZNode 的子节点数量。

5）创建节点测试。

ZooKeeper 中的节点可以分为以下两类。
- 持久化（Persistent）节点：在客户端和服务器端断开连接后，创建的节点依然存在。
- 临时（Ephemeral）节点：在客户端和服务器端断开连接后，创建的节点被自动删除。

在创建 ZNode 时设置顺序标识，会在 ZNode 名称后附加一个值，顺序号是一个单调递增的计数器，由父节点维护。在分布式系统中，使用顺序号可以对所有事件进行全局排序，客户端可以通过顺序号推断事件的顺序。

根据以上的内容，可以将节点细分为以下 4 种。
- 持久化节点：在客户端与 ZooKeeper 断开连接后，该节点依然存在。
- 持久化顺序编号节点：在客户端与 ZooKeeper 断开连接后，该节点依然存在，只是 ZooKeeper 会对该类节点按照名称进行顺序编号。
- 临时节点：在客户端与 ZooKeeper 断开连接后，该节点被删除。
- 临时顺序编号节点：在客户端与 ZooKeeper 断开连接后，该节点被删除，但 ZooKeeper 会对该类节点按照名称进行顺序编号。

下面进行以下节点创建测试。

① 在根节点下分别创建两个持久化节点。

```
[zk: localhost:2181(CONNECTED) 3] create /sanguo "jinlian"
Created /sanguo
[zk: localhost:2181(CONNECTED) 4] create /sanguo/shuguo "liubei"
Created /sanguo/shuguo
```

获取节点的值。

```
[zk: localhost:2181(CONNECTED) 5] get /sanguo
jinlian
cZxid = 0x100000003
ctime = Wed Aug 29 00:03:23 CST 2018
mZxid = 0x100000003
mtime = Wed Aug 29 00:03:23 CST 2018
pZxid = 0x100000004
cversion = 1
dataVersion = 0
aclVersion = 0
ephemeralOwner = 0x0
dataLength = 7
numChildren = 1
[zk: localhost:2181(CONNECTED) 6]
[zk: localhost:2181(CONNECTED) 6] get /sanguo/shuguo
liubei
cZxid = 0x100000004
ctime = Wed Aug 29 00:04:35 CST 2018
```

```
mZxid = 0x100000004
mtime = Wed Aug 29 00:04:35 CST 2018
pZxid = 0x100000004
cversion = 0
dataVersion = 0
aclVersion = 0
ephemeralOwner = 0x0
dataLength = 6
numChildren = 0
```

② 创建临时节点。

```
[zk: localhost:2181(CONNECTED) 7] create -e /sanguo/wuguo "zhouyu"
Created /sanguo/wuguo
```

在当前客户端查看节点。

```
[zk: localhost:2181(CONNECTED) 3] ls /sanguo
[wuguo, shuguo]
```

退出当前客户端，然后重启客户端。

```
[zk: localhost:2181(CONNECTED) 12] quit
[atguigu@hadoop104 zookeeper-3.5.7]$ bin/zkCli.sh
```

再次查看根目录，临时节点/sanguo/wuguo 已被删除。

```
[zk: localhost:2181(CONNECTED) 0] ls /sanguo
[shuguo]
```

③ 创建持久化顺序编号节点。

创建一个普通的根节点/sanguo/weiguo。

```
[zk: localhost:2181(CONNECTED) 1] create /sanguo/weiguo "caocao"
Created /sanguo/weiguo
```

创建 3 个持久化顺序编号节点。

```
[zk: localhost:2181(CONNECTED) 2] create -s /sanguo/weiguo/xiaoqiao "jinlian"
Created /sanguo/weiguo/xiaoqiao0000000000
[zk: localhost:2181(CONNECTED) 3] create -s /sanguo/weiguo/daqiao "jinlian"
Created /sanguo/weiguo/daqiao0000000001
[zk: localhost:2181(CONNECTED) 4] create -s /sanguo/weiguo/diaocan "jinlian"
Created /sanguo/weiguo/diaocan0000000002
```

6）节点值的修改及监听测试。

客户端会注册监听它关心的目录节点，当目录节点发生变化（数据改变、节点删除、子目录节点增加、子目录节点删除）时，ZooKeeper 会通知客户端。监听机制可以保证 ZooKeeper 中存储的任何数据的任何改变都能快速地响应到监听该节点的应用程序上。

创建监听器的原理如图 7-9 所示。

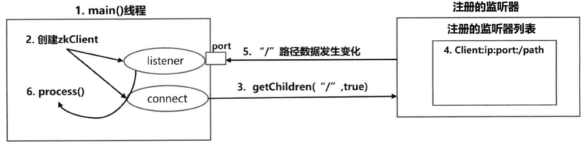

图 7-9　创建监听器的原理

(1) 创建 main() 线程。
(2) 在 main 线程中创建 ZooKeeper 客户端，这时会创建两个线程，一个负责网络连接通信（connet），一个负责监听（listener）。
(3) 通过 connect 线程将注册的监听事件发送给 ZooKeeper。
(4) 将注册的监听事件添加到 ZooKeeper 的注册监听器列表中。
(5) 如果 ZooKeeper 监听到有数据或路径发生变化，就会将这个消息发送给 listener 线程。
(6) listener 线程内部执行 process() 方法。
常见的监听事件包括监听节点数据的变化和监听子节点增减的变化。
① 修改并监听节点数据的变化。
修改节点数据值，代码如下：

```
[zk: localhost:2181(CONNECTED) 6] set /sanguo/weiguo "simayi"
```

在 hadoop104 节点服务器上注册监听 /sanguo 节点数据的变化，代码如下：

```
[zk: localhost:2181(CONNECTED) 26] [zk: localhost:2181(CONNECTED) 8] get -w /sanguo
```

在 hadoop103 节点服务器上修改 /sanguo 节点的数据，代码如下：

```
[zk: localhost:2181(CONNECTED) 1] set /sanguo "xisi"
```

可以观察到，hadoop104 节点服务器监听到了节点数据的变化，代码如下：

```
WATCHER::
WatchedEvent state:SyncConnected type:NodeDataChanged path:/sanguo
```

② 修改并监听子节点增减的变化。
在 hadoop104 节点服务器上注册监听 /sanguo 节点的子节点增减变化，代码如下：

```
[zk: localhost:2181(CONNECTED) 1] ls -w /sanguo
[aa0000000001, server101]
```

在 hadoop103 节点服务器的 /sanguo 节点下创建子节点，代码如下：

```
[zk: localhost:2181(CONNECTED) 2] create /sanguo/jin "simayi"
Created /sanguo/jin
```

可以观察到，hadoop104 节点服务器监听到了子节点增减的变化，代码如下：

```
WATCHER::
WatchedEvent state:SyncConnected type:NodeChildrenChanged path:/sanguo
```

7) 删除节点，代码如下：

```
[zk: localhost:2181(CONNECTED) 4] delete /sanguo/jin
```

8) 递归删除节点，代码如下：

```
[zk: localhost:2181(CONNECTED) 15] deleteall /sanguo/shuguo
```

9) 查看节点状态，代码如下：

```
[zk: localhost:2181(CONNECTED) 17] stat /sanguo
cZxid = 0x100000003
ctime = Wed Aug 29 00:03:23 CST 2018
mZxid = 0x100000011
mtime = Wed Aug 29 00:21:23 CST 2018
pZxid = 0x100000014
cversion = 9
dataVersion = 1
aclVersion = 0
ephemeralOwner = 0x0
dataLength = 4
numChildren = 1
```

2. Java API 操作

（1）环境准备。

① 在 IDEA 中创建一个 Maven 工程 zookeeper，并且在 pom 文件中添加以下内容。

```xml
<dependencies>
    <dependency>
        <groupId>junit</groupId>
        <artifactId>junit</artifactId>
        <version>RELEASE</version>
    </dependency>

    <dependency>
        <groupId>org.apache.logging.log4j</groupId>
        <artifactId>log4j-core</artifactId>
        <version>2.8.2</version>
    </dependency>

    <dependency>
        <groupId>org.apache.zookeeper</groupId>
        <artifactId>zookeeper</artifactId>
        <version>3.5.7</version>
    </dependency>
</dependencies>
```

② 在 zookeeper 工程的 src/main/resources 目录下新建 log4j.properties 文件，在该文件中添加以下内容。

```
log4j.rootLogger=INFO, stdout
log4j.appender.stdout=org.apache.log4j.ConsoleAppender
log4j.appender.stdout.layout=org.apache.log4j.PatternLayout
log4j.appender.stdout.layout.ConversionPattern=%d %p [%c] - %m%n
log4j.appender.logfile=org.apache.log4j.FileAppender
log4j.appender.logfile.File=target/spring.log
log4j.appender.logfile.layout=org.apache.log4j.PatternLayout
log4j.appender.logfile.layout.ConversionPattern=%d %p [%c] - %m%n
```

③ 创建 com.atguigu.zk 包，在该包中创建 zkClient 类。

（2）编写初始化方法 init()，在该方法中创建 ZooKeeper 客户端，增加 @Before 注解，在测试其他方法时会首先调用该方法，代码如下：

```java
// 注意：逗号前后不能有空格
private static String connectString =
 "hadoop102:2181,hadoop103:2181,hadoop104:2181";

    private static int sessionTimeout = 2000;
    private ZooKeeper zkClient = null;

    @Before
    public void init() throws Exception {

    zkClient = new ZooKeeper(connectString, sessionTimeout, new Watcher() {

            @Override
            public void process(WatchedEvent watchedEvent) {

                // 收到事件通知后的回调方法（用户的业务逻辑）
```

```
            System.out.println(watchedEvent.getType() + "--" + watchedEvent.getPath());

            // 再次启动监听功能
            try {
                List<String> children = zkClient.getChildren("/", true);
                for (String child : children) {
                    System.out.println(child);
                }
            } catch (Exception e) {
                e.printStackTrace();
            }
        }
    });
}
```

（3）编写创建子节点的方法，代码如下：
```
// 创建子节点
@Test
public void create() throws Exception {

    // 参数1：要创建的节点的路径； 参数2：节点数据； 参数3：节点权限； 参数4：节点的类型
    String nodeCreated = zkClient.create("/atguigu", "shuaige".getBytes(), Ids.OPEN_ACL_UNSAFE, CreateMode.PERSISTENT);
}
```

运行该方法，在 hadoop102 节点服务器的 ZooKeeper 客户端查看创建的节点情况，具体如下：
```
[zk: localhost:2181(CONNECTED) 16] get -s /atguigu
shuaige
```

（4）编写获取子节点并监听节点变化的方法，代码如下：
```
// 获取子节点
@Test
public void getChildren() throws Exception {

    List<String> children = zkClient.getChildren("/", true);

    for (String child : children) {
        System.out.println(child);
    }

    // 延时阻塞
    Thread.sleep(Long.MAX_VALUE);
}
```

运行该方法，在 IDEA 控制台中输出以下节点。
```
zookeeper
sanguo
atguigu
```

在 hadoop102 节点服务器的 ZooKeeper 客户端的根目录下创建一个节点 atguigu1，代码如下：
```
[zk: localhost:2181(CONNECTED) 3] create /atguigu1 "atguigu1"
```

在 IDEA 控制台中输出以下节点。
```
zookeeper
sanguo
```

```
atguigu
atguigu1
```
在 hadoop102 节点服务器的 ZooKeeper 客户端删除节点 atguigu1，代码如下：
```
[zk: localhost:2181(CONNECTED) 4] delete /atguigu1
```
在 IDEA 控制台中输出以下节点。
```
zookeeper
sanguo
atguigu
```
（5）编写判断 ZNode 是否存在的方法，代码如下：
```java
// 判断 ZNode 是否存在
@Test
public void exist() throws Exception {

    Stat stat = zkClient.exists("/atguigu", false);

    System.out.println(stat == null ? "not exist" : "exist");
}
```
运行该方法，输出结果如下：
```
exist
```

7.2 HA 概述

7.2.1 什么是 HA

对一个集群来说，HA（High Availability，高可用）是指可以 7×24 小时不间断地提供服务。实现 HA 的关键在于消除单点故障。由于 Hadoop 的两大重要组件 HDFS 和 YARN 都采用 Master-Slave 架构，因此 Hadoop 的 HA 也应该分为两种，分别为 HDFS HA 和 YARN HA。

HDFS 的 NameNode 主要在以下两个方面影响 HDFS 集群。

- 如果 NameNode 发生意外，如宕机，那么集群会无法使用，直到管理员重启。
- 如果 NameNode 需要升级，包括软件、硬件升级，那么集群也会无法使用。

HDFS HA 功能通过配置多个 NameNode（状态是 Active 或 Standby）解决上述问题。如果出现故障，如节点崩溃、节点需要升级维护，那么可以通过此种方式将 NameNode 很快地切换为另一台节点服务器。

7.2.2 HDFS HA 的工作机制

在配置多个 NameNode 实现 HA 前，HDFS 通过 SecondaryNameNode 机制保障 NameNode 的正常运行及宕机重启。在配置了多个 NameNode 后，我们需要回答以下几个问题。

问题一：如何保证多个 NameNode 的数据一致性？

其中一个 NameNode 负责生成快照文件 FsImage，其他 NameNode 拉取同步。通过引进新的集群角色日志节点 JournalNode，保证多个 NameNode 中的编辑日志文件 EditLog 的数据一致性。当 FsImage 文件与 EditLog 文件都能保持一致时，NameNode 可以提供相同的元数据管理服务。

在备用 NameNode 接管工作后，会加载所有现有 FsImage 文件和 EditLog 文件，实现状态同步。

问题二：如何使多个 NameNode 中的一个处于 Active 状态，其他处于 Standby 状态？

同时有两个 NameNode 处于 Active 状态对 Hadoop 集群来说是致命的，这种现象称为脑裂，应该尽量避免。Hadoop 提供了一种称为故障转移控制器（Failover Controller）的监控进程，它可以在每一个 NameNode 上启动，时刻监控 NameNode 的状态。ZKFC 是一种常用的故障转移控制器，是基于 ZooKeeper 实现的。

当 ZKFC 运行在处于 Active 状态的 NameNode 上时，会在发现其状态不正常时向 ZooKeeper 中写入数据。当 ZKFC 运行在处于 Standby 状态的 NameNode 上时，会从 ZooKeeper 中读取数据，从而感知处于 Active 状态的 NameNode 是否在正常工作，以便顺利完成故障转移工作。备用 NameNode 在感知到处于 Active 状态的 NameNode 出现异常后，并不会立即切换状态，它会先通过 SSH 远程杀死处于 Active 状态的 NameNode 进程。但是这样并不能保证执行成功，进一步的规避机制包括撤销处于 Active 状态的 NameNode 访问共享存储目录的权限、通过远程管理命令屏蔽相应的网络接口，以及通过一个特定的供电单元对相应主机进行断点操作。以上规避行为可以确保在自动故障转移过程中，只有一个 NameNode 处于 Active 状态。

当然，用户也可以在进行日常维护时，手动切换 NameNode 的状态。

问题三：HA 架构中不包含 SecondaryNameNode，那么定期合并 FsImage 文件和 EditLog 文件的工作由谁负责？

备用 NameNode 中包含 SecondaryNameNode 的角色，处于 Standby 状态的 NameNode 会定时为处于 Active 状态的 NameNode 合并 FsImage 文件和 EditLog 文件。

7.3 Hadoop HA 集群的搭建

在了解了 HA 的工作机制后，下面详细讲解如何搭建一个 Hadoop HA 集群。

7.3.1 HDFS HA 手动故障转移

前面已经提到过，在 Hadoop HA 集群配置完成后，如果处于 Active 状态的 NameNode 宕机，那么有两种方式可以完成故障转移工作，将备用 NameNode 的状态切换为 Active，分别是手动故障转移和自动故障转移，本节主要介绍手动故障转移的集群配置。

非高可用 HDFS 集群规划如表 7-2 所示。

表 7-2 非高可用 HDFS 集群规划

hadoop102	hadoop103	hadoop104
NameNode	—	SecondaryNameNode
DataNode	DataNode	DataNode

为了构建 Hadoop HA 集群，需要重新进行集群规划。高可用 HDFS 集群规划如表 7-3 所示，每台节点服务器上都运行了一个 NameNode 进程和一个 JournalNode 进程。

表 7-3 高可用 HDFS 集群规划

hadoop102	hadoop103	hadoop104
NameNode	NameNode	NameNode
JournalNode	JournalNode	JournalNode
DataNode	DataNode	DataNode

在高可用 HDFS 集群规划完成后，执行以下操作。

(1) 在 opt 目录下创建一个 ha 文件夹，用于安装高可用的 Hadoop HA 集群。

```
[atguigu@hadoop102 ~]$ cd /opt
[atguigu@hadoop102 opt]$ sudo mkdir ha
[atguigu@hadoop102 opt]$ sudo chown atguigu:atguigu /opt/ha
```

(2) 将/opt/module/目录下的 hadoop-3.1.3 复制到/opt/ha 目录下，并且删除 data 和 log 目录。

```
[atguigu@hadoop102 opt]$ cp -r /opt/module/hadoop-3.1.3 /opt/ha/
[atguigu@hadoop102 opt]$ rm -r /opt/ha/hadoop-3.1.3/data/
[atguigu@hadoop102 opt]$ rm -r /opt/ha/hadoop-3.1.3/log/
```

(3) 修改配置文件 core-site.xml,代码如下:

```xml
<configuration>
  <!-- 将多个NameNode的地址组装成一个集群mycluster -->
  <property>
    <name>fs.defaultFS</name>
    <value>hdfs://mycluster</value>
  </property>
  <!-- 指定Hadoop在运行时产生文件的存储目录 -->
  <property>
    <name>hadoop.tmp.dir</name>
    <value>/opt/ha/hadoop-3.1.3/data</value>
  </property>
</configuration>
```

(4) 修改配置文件 hdfs-site.xml,将 hadoop102、hadoop103 和 hadoop104 节点服务器中的 NameNode 分别命名为 nn1、nn2 和 nn3,代码如下:

```xml
<configuration>
  <!-- NameNode 数据存储目录 -->
  <property>
    <name>dfs.namenode.name.dir</name>
    <value>file://${hadoop.tmp.dir}/name</value>
  </property>
  <!-- DataNode 数据存储目录 -->
  <property>
    <name>dfs.datanode.data.dir</name>
    <value>file://${hadoop.tmp.dir}/data</value>
  </property>
  <!-- JournalNode 数据存储目录 -->
  <property>
    <name>dfs.journalnode.edits.dir</name>
    <value>${hadoop.tmp.dir}/jn</value>
  </property>
  <!-- 完全分布式集群名称 -->
  <property>
    <name>dfs.nameservices</name>
    <value>mycluster</value>
  </property>
  <!-- 集群中的NameNode都有哪些 -->
  <property>
    <name>dfs.ha.namenodes.mycluster</name>
    <value>nn1,nn2,nn3</value>
  </property>
  <!-- NameNode 的 RPC 通信地址 -->
  <property>
    <name>dfs.namenode.rpc-address.mycluster.nn1</name>
    <value>hadoop102:8020</value>
  </property>
  <property>
    <name>dfs.namenode.rpc-address.mycluster.nn2</name>
    <value>hadoop103:8020</value>
  </property>
  <property>
```

```xml
    <name>dfs.namenode.rpc-address.mycluster.nn3</name>
    <value>hadoop104:8020</value>
</property>
<!-- NameNode 的 HTTP 通信地址 -->
<property>
    <name>dfs.namenode.http-address.mycluster.nn1</name>
    <value>hadoop102:9870</value>
</property>
<property>
    <name>dfs.namenode.http-address.mycluster.nn2</name>
    <value>hadoop103:9870</value>
</property>
<property>
    <name>dfs.namenode.http-address.mycluster.nn3</name>
    <value>hadoop104:9870</value>
</property>
<!-- 指定 NameNode 元数据在 JournalNode 中的存储位置 -->
<property>
    <name>dfs.namenode.shared.edits.dir</name>
    <value>qjournal://hadoop102:8485;hadoop103:8485;hadoop104:8485/mycluster</value>
</property>
<!-- 访问代理类：Client 主要用于确定哪个 NameNode 处于 Active 状态 -->
<property>
    <name>dfs.client.failover.proxy.provider.mycluster</name>
    <value>org.apache.hadoop.hdfs.server.namenode.ha.ConfiguredFailoverProxyProvider</value>
</property>
<!-- 配置隔离机制，即同一时刻只能有一台节点服务器对外响应 -->
<property>
    <name>dfs.ha.fencing.methods</name>
    <value>sshfence</value>
</property>
<!-- 在使用隔离机制时，需要使用 SSH 免密登录-->
<property>
    <name>dfs.ha.fencing.ssh.private-key-files</name>
    <value>/home/atguigu/.ssh/id_rsa</value>
</property>
</configuration>
```

（5）在修改完配置文件后，将配置好的 Hadoop 安装包 hadoop-3.1.3 分发到其他节点中，代码如下：

```
[atguigu@hadoop102 ~]$ xsync /opt/ha/hadoop-3.1.3
```

（6）修改环境变量，将 HADOOP_HOME 环境变量放到 HA 目录下，代码如下：

```
[atguigu@hadoop102 ~]$ sudo vim /etc/profile.d/my_env.sh
```

对 HADOOP_HOME 环境变量的部分修改如下：

```
#HADOOP_HOME
export HADOOP_HOME=/opt/ha/hadoop-3.1.3
export PATH=$PATH:$HADOOP_HOME/bin
export PATH=$PATH:$HADOOP_HOME/sbin
```

使修改后的环境变量生效，代码如下：

```
[atguigu@hadoop102 ~]$source /etc/profile
```

在 hadoop103 和 hadoop104 节点服务器上分别执行修改环境变量和使环境变量生效的操作。

（7）在各个 JournalNode 上执行以下命令，启动 journalnode 服务。

```
[atguigu@hadoop102 ~]$ hdfs --daemon start journalnode
```

```
[atguigu@hadoop103 ~]$ hdfs --daemon start journalnode
[atguigu@hadoop104 ~]$ hdfs --daemon start journalnode
```

（8）在 hadoop102 节点服务器上对 NameNode 进行格式化并启动，代码如下：

```
[atguigu@hadoop102 ~]$ hdfs namenode -format
[atguigu@hadoop102 ~]$ hdfs --daemon start namenode
```

（9）在 hadoop103 和 hadoop104 节点服务器中同步 hadoop102 节点服务器中 NameNode 的元数据信息，代码如下：

```
[atguigu@hadoop103 ~]$ hdfs namenode -bootstrapStandby
[atguigu@hadoop104 ~]$ hdfs namenode -bootstrapStandby
```

（10）分别在 hadoop103 和 hadoop104 节点服务器上启动 NameNode，代码如下：

```
[atguigu@hadoop103 ~]$ hdfs --daemon start namenode
[atguigu@hadoop104 ~]$ hdfs --daemon start namenode
```

（11）分别登录 3 台节点服务器中 NameNode 的 Web 端页面，查看 NameNode 的状态，分别如图 7-10、图 7-11 和图 7-12 所示。可以看到，3 台节点服务器中的 NameNode 都处于 Standby 状态。

图 7-10　hadoop102 节点服务器中的 NameNode 处于 Standby 状态

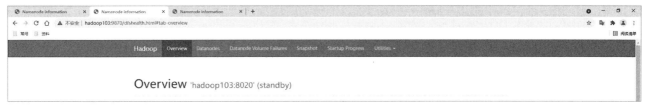

图 7-11　hadoop103 节点服务器中的 NameNode 处于 Standby 状态

图 7-12　hadoop104 节点服务器中的 NameNode 处于 Standby 状态

（12）在所有节点服务器上启动 DataNode，代码如下：

```
[atguigu@hadoop102 ~]$ hdfs --daemon start datanode
[atguigu@hadoop103 ~]$ hdfs --daemon start datanode
[atguigu@hadoop104 ~]$ hdfs --daemon start datanode
```

（13）将 nn1 的状态切换为 Active，代码如下：

```
[atguigu@hadoop102 ~]$ hdfs haadmin -transitionToActive nn1
```

（14）执行以下命令，查看 nn1 是否处于 Active 状态。

```
[atguigu@hadoop102 ~]$ hdfs haadmin -getServiceState nn1
```

7.3.2　HDFS HA 自动故障转移

HDFS HA 自动故障转移过程如图 7-13 所示，该过程为 HDFS 部署了两个新组件，分别为 ZooKeeper 和 ZKFailoverController（ZKFC）进程。其中，ZooKeeper 会维护 NameNode 的状态数据，并且通知 ZKFC

这些数据的变化情况。

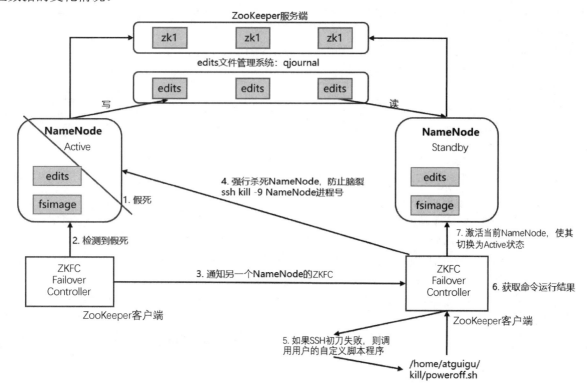

图 7-13　HDFS HA 自动故障转移过程

根据 HDFS HA 自动故障转移过程进行集群规划，如表 7-4 所示。

表 7-4　HDFS HA 自动故障转移集群规划

hadoop102	hadoop103	hadoop104
NameNode	NameNode	NameNode
JournalNode	JournalNode	JournalNode
DataNode	DataNode	DataNode
ZooKeeper	ZooKeeper	ZooKeeper
ZKFC	ZKFC	ZKFC

具体配置步骤如下。

（1）修改配置文件。

在 hdfs-site.xml 文件中添加以下内容，启用 HDFS HA 自动故障转移功能。

```
<!-- 启用 HDFS HA 自动故障转移功能 -->
<property>
    <name>dfs.ha.automatic-failover.enabled</name>
    <value>true</value>
</property>
```

在 core-site.xml 文件中添加以下内容，指定 ZooKeeper 服务器的地址和端口号。

```
<!-- 指定 ZKFC 要连接的 zkServer 地址 -->
<property>
    <name>ha.zookeeper.quorum</name>
    <value>hadoop102:2181,hadoop103:2181,hadoop104:2181</value>
</property>
```

分发修改后的配置文件，代码如下：

```
[atguigu@hadoop102 etc]$ pwd
```

```
/opt/ha/hadoop-3.1.3/etc
[atguigu@hadoop102 etc]$ xsync hadoop/
```

（2）启动。

关闭所有的 HDFS 服务，代码如下：

```
[atguigu@hadoop102 ~]$ stop-dfs.sh
```

启动 ZooKeeper 集群，代码如下：

```
[atguigu@hadoop102 ~]$ zkServer.sh start
[atguigu@hadoop103 ~]$ zkServer.sh start
[atguigu@hadoop104 ~]$ zkServer.sh start
```

在启动 ZooKeeper 集群后，初始化 HA 在 ZooKeeper 集群中的状态，代码如下：

```
[atguigu@hadoop102 ~]$ hdfs zkfc -formatZK
```

再次启动 HDFS 服务，代码如下：

```
[atguigu@hadoop102 ~]$ start-dfs.sh
```

分别登录 3 台节点服务器中 NameNode 的 Web 端页面，查看 NameNode 的状态，分别如图 7-14、图 7-15 和图 7-16 所示。可以看到，hadoop102 节点服务器中的 NameNode 处于 Active 状态，hadoop103 和 hadoop104 节点服务器中的 NameNode 处于 Standby 状态。

图 7-14　hadoop102 节点服务器中的 NameNode 处于 Active 状态

图 7-15　hadoop103 节点服务器中的 NameNode 处于 Standby 状态

图 7-16　hadoop104 节点服务器中的 NameNode 处于 Standby 状态

（3）验证。

将处于 Active 状态的 NameNode 进程关闭，代码如下：

```
[atguigu@hadoop102 ~]$ jps
2529 JournalNode
4027 DFSZKFailoverController
4107 Jps
2061 NameNode
2238 DataNode
3086 QuorumPeerMain
[atguigu@hadoop102 ~]$ kill -9 2061
```

查看 3 台节点服务器中 NameNode 的 Web 端页面，观察 NameNode 的状态变化，如图 7-17 所示。可以看到，hadoop103 节点服务器中的 NameNode 状态切换成功，由 Standby 转换为了 Active。

图 7-17　hadoop103 节点服务器中的 NameNode 状态切换为 Active

7.3.3　YARN HA

YARN 的整体架构与 HDFS 的整体架构类似，都采用 Master-Slaver 架构，所以配置 YARN HA 架构的思路与配置 HDFS HA 架构的思路类似。配置多个 ResourceManager，其中一个 ResourceManager 处于 Active 状态，其他 ResourceManager 处于 Standby 状态，如图 7-18 所示，处于 Active 状态的 ResourceManager 将状态写入 ZooKeeper，当其他备用 ResourceManager 切换状态时，可以直接从 ZooKeeper 中读取，从而继续进行任务和资源调度。

图 7-18　YARN HA 的架构

根据 YARN HA 的工作机制进行集群规划，如表 7-5 所示。

表 7-5　YARN HA 的集群规划

hadoop102	hadoop103	hadoop104
ResourceManager	ResourceManager	ResourceManager
NodeManager	NodeManager	NodeManager
ZooKeeper	ZooKeeper	ZooKeeper

具体配置步骤如下。

（1）修改配置文件 yarn-site.xml，将 hadoop102、hadoop103 和 hadoop104 节点服务器中的 ResourceManager 分别命名为 rm1、rm2 和 rm3。

```
<configuration>

    <property>
        <name>yarn.nodemanager.aux-services</name>
        <value>mapreduce_shuffle</value>
    </property>

    <!-- 启用 resourcemanager ha -->
    <property>
        <name>yarn.resourcemanager.ha.enabled</name>
        <value>true</value>
    </property>

    <!-- 声明 3 个 ResourceManager 的地址 -->
```

```xml
<property>
    <name>yarn.resourcemanager.cluster-id</name>
    <value>cluster-yarn1</value>
</property>
<!--指定 ResourceManager 的逻辑列表-->
<property>
    <name>yarn.resourcemanager.ha.rm-ids</name>
    <value>rm1,rm2,rm3</value>
</property>
<!-- ========== rm1 的配置 ========== -->
<!-- 指定 rm1 的主机名 -->
<property>
    <name>yarn.resourcemanager.hostname.rm1</name>
    <value>hadoop102</value>
</property>
<!-- 指定 rm1 的 Web 端地址 -->
<property>
    <name>yarn.resourcemanager.webapp.address.rm1</name>
    <value>hadoop102:8088</value>
</property>
<!-- 指定 rm1 的内部通信地址 -->
<property>
    <name>yarn.resourcemanager.address.rm1</name>
    <value>hadoop102:8032</value>
</property>
<!-- 指定 ApplicationMaster 向 rm1 申请资源的地址 -->
<property>
    <name>yarn.resourcemanager.scheduler.address.rm1</name>
    <value>hadoop102:8030</value>
</property>
<!-- 指定供 NodeManager 连接的地址 -->
<property>
    <name>yarn.resourcemanager.resource-tracker.address.rm1</name>
    <value>hadoop102:8031</value>
</property>
<!-- ========== rm2 的配置 ========== -->
<!-- 指定 rm2 的主机名 -->
<property>
    <name>yarn.resourcemanager.hostname.rm2</name>
    <value>hadoop103</value>
</property>
<property>
    <name>yarn.resourcemanager.webapp.address.rm2</name>
    <value>hadoop103:8088</value>
</property>
<property>
    <name>yarn.resourcemanager.address.rm2</name>
    <value>hadoop103:8032</value>
</property>
<property>
    <name>yarn.resourcemanager.scheduler.address.rm2</name>
```

```xml
        <value>hadoop103:8030</value>
    </property>
    <property>
        <name>yarn.resourcemanager.resource-tracker.address.rm2</name>
        <value>hadoop103:8031</value>
    </property>
<!-- ========== rm3 的配置 ========== -->
<!-- 指定 rm3 的主机名 -->
    <property>
        <name>yarn.resourcemanager.hostname.rm3</name>
        <value>hadoop104</value>
    </property>
<!-- 指定 rm3 的 Web 端地址 -->
    <property>
        <name>yarn.resourcemanager.webapp.address.rm3</name>
        <value>hadoop104:8088</value>
    </property>
<!-- 指定 rm3 的内部通信地址 -->
    <property>
        <name>yarn.resourcemanager.address.rm3</name>
        <value>hadoop104:8032</value>
    </property>
<!-- 指定 ApplicationMaster 向 rm3 申请资源的地址 -->
    <property>
        <name>yarn.resourcemanager.scheduler.address.rm3</name>
        <value>hadoop104:8030</value>
    </property>
<!-- 指定供 NodeManager 连接的地址 -->
    <property>
        <name>yarn.resourcemanager.resource-tracker.address.rm3</name>
        <value>hadoop104:8031</value>
    </property>
<!-- 指定 ZooKeeper 集群的地址 -->
    <property>
        <name>yarn.resourcemanager.zk-address</name>
        <value>hadoop102:2181,hadoop103:2181,hadoop104:2181</value>
    </property>

<!-- 启用自动恢复功能 -->
    <property>
        <name>yarn.resourcemanager.recovery.enabled</name>
        <value>true</value>
    </property>

<!-- 指定 ResourceManager 的状态信息存储于 ZooKeeper 集群中 -->
    <property>
        <name>yarn.resourcemanager.store.class</name>
        <value>org.apache.hadoop.yarn.server.resourcemanager.recovery.ZKRMStateStore</value>
    </property>
<!-- 环境变量的继承 -->
    <property>
```

```xml
        <name>yarn.nodemanager.env-whitelist</name>
        <value>JAVA_HOME,HADOOP_COMMON_HOME,HADOOP_HDFS_HOME,HADOOP_CONF_DIR,CLASSPATH_PREPEND_DISTCACHE,HADOOP_YARN_HOME,HADOOP_MAPRED_HOME</value>
    </property>

</configuration>
```

（2）将配置文件分发至其他节点服务器中，代码如下：
```
[atguigu@hadoop102 etc]$ xsync hadoop/
```
（3）在 hadoop102 节点服务器中启动 YARN，代码如下：
```
[atguigu@hadoop102 ~]$ start-yarn.sh
```
分别在 hadoop103 和 hadoop104 节点服务器中执行以下命令，启动 ResourceManager。
```
[atguigu@hadoop103 ~]$ start-yarn.sh
[atguigu@hadoop104 ~]$ start-yarn.sh
```
（4）查看 rm1 的服务状态，是 Active 状态。
```
[atguigu@hadoop102 ~]$ yarn rmadmin -getServiceState rm1
active
```
（5）可以在 zkCli.sh 客户端查看 ResourceManager 的状态，即以下加粗内容，可以看到 rm1 处于 Active 状态。
```
[atguigu@hadoop102 ~]$ zkCli.sh
localhost:2181(CONNECTED) 0] get -s /yarn-leader-election/cluster-yarn1/ActiveStandbyElectorLock

cluster-yarn1                                    rm1
cZxid = 0x7300000015
ctime = Sat Nov 13 18:43:49 CST 2021
mZxid = 0x7300000015
mtime = Sat Nov 13 18:43:49 CST 2021
pZxid = 0x7300000015
cversion = 0
dataVersion = 0
aclVersion = 0
ephemeralOwner = 0x2000006cb0a0000
dataLength = 20
numChildren = 0
```
（6）在 YARN 的 Web 端页面查看 hadoop102:8088、hadoop103:8088、hadoop104:8088 的 YARN 的状态，如图 7-19 所示。需要注意的是，无论用户访问哪台节点服务器的 8088 端口，都会自动跳转到处于 Active 状态的节点服务器。

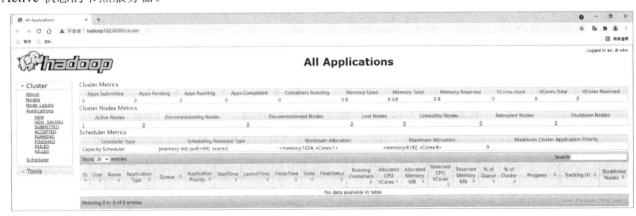

图 7-19　YARN 的 Web 端页面

7.3.4 Hadoop HA 集群规划

在 Hadoop HA 集群搭建完成后，对整个集群进行规划，如表 7-6 所示。

表 7-6 Hadoop HA 集群的完整规划

hadoop102	hadoop103	hadoop104
NameNode	NameNode	NameNode
JournalNode	JournalNode	JournalNode
DataNode	DataNode	DataNode
ZooKeeper	ZooKeeper	ZooKeeper
ZKFC	ZKFC	ZKFC
ResourceManager	ResourceManager	ResourceManager
NodeManager	NodeManager	NodeManager

7.4 本章总结

本章主要讲解了如何搭建一个高可用的 Hadoop 集群，这在企业中是十分必要的。在企业的生产环境中，如果 HDFS 集群的 NameNode 发生宕机无法重启，则会造成严重后果，所以大部分企业都会搭建高可用集群，避免在故障转移过程中造成损失。通过对本章内容的学习，读者应该重点理解 Hadoop 对单点故障问题提供的解决方案，并且熟悉高可用 Hadoop 集群的搭建流程，这些在今后的工作中都是非常重要的。

第8章 生产调优手册

经过前面的学习，读者可以学会如何搭建一套高可用的 Hadoop 集群，了解 Hadoop 底层的运行机制，但是如何保障 Hadoop 集群平稳、高效地运行？如何提高资源利用率？这些问题都将在本章得到解答。在学习各种生产调优手段前，掌握扎实的基础知识是必不可少的，只有了解了底层运行机制，调优手段才能有的放矢。本章主要介绍 HDFS 与 MapReduce 的调优手段，对于 YARN 的生产调优手段，读者可以参考第 6 章。

8.1 HDFS 的核心参数

在 HDFS 的使用过程中，有一些核心参数使用并不是特别频繁，但是经过合理的配置，可以优化 HDFS 的运行。

8.1.1 NameNode 的内存生产配置

通过 NameNode 的内存空间可以计算出一个 Hadoop 集群的数据存储上限。假设将 NameNode 部署在一台内存空间为 128GB 的服务器上，一个数据块的元数据信息大概占用 150Byte 内存空间，那么整个 Hadoop 集群大约可以存储 9.2 亿（128×1024×1024×1024÷150）个数据块。数据块大小一般为 128MB。所以集群中不能存储大量的小文件，小文件过多会大幅降低集群的存储性能。

在 Hadoop 2.x 中，NameNode 的内存空间由 hadoop-env.sh 配置文件中的一个参数指定。在配置这个参数时应该考虑，部署 NameNode 的服务器上是否还运行着其他进程，是否需要为其他进程的运行预留内存空间。为了最大化集群的存储性能，一般会单独部署 NameNode。例如，本书搭建的 Hadoop 集群为每台服务器分配了 4GB 的内存空间，并且在 NameNode 上部署了 DataNode 和 NodeManager。在考虑了这些因素后，即可为 NameNode 分配合理的内存空间，具体配置参数如下，参数默认值为 2000MB。

```
HADOOP_NAMENODE_OPTS=-Xmx2000m
```

在 Hadoop 3.x 中，Hadoop 中 NameNode 的内存空间是动态分配的，在 hadoop-env.sh 配置文件中，动态分配 Hadoop 中 NameNode 内存空间的相关参数及其描述如下：

```
# Specify the JVM options to be used when starting the NameNode.
# These options will be appended to the options specified as HADOOP_OPTS
# and therefore may override any similar flags set in HADOOP_OPTS
#
# a) Set JMX options
# export HDFS_NAMENODE_OPTS="-Dcom.sun.management.jmxremote=true -Dcom.sun.management.jmxremote.authenticate=false -Dcom.sun.management.jmxremote.ssl=false -Dcom.sun.management.jmxremote.port=1026"
#
# b) Set garbage collection logs
```

```
# export HDFS_NAMENODE_OPTS="${HADOOP_GC_SETTINGS}  -Xloggc:${HADOOP_LOG_DIR}/gc-rm.log-
$(date +'%Y%m%d%H%M')"
#
# c) ... or set them directly
# export HDFS_NAMENODE_OPTS="-verbose:gc  -XX:+PrintGCDetails  -XX:+PrintGCTimeStamps -
XX:+PrintGCDateStamps -Xloggc:${HADOOP_LOG_DIR}/gc-rm.log-$(date +'%Y%m%d%H%M')"

# this is the default:
# export HDFS_NAMENODE_OPTS="-Dhadoop.security.logger=INFO,RFAS"

# Specify the JVM options to be used when starting the DataNode.
# These options will be appended to the options specified as HADOOP_OPTS
# and therefore may override any similar flags set in HADOOP_OPTS
#
# This is the default:
# export HDFS_DATANODE_OPTS="-Dhadoop.security.logger=ERROR,RFAS"
```

执行 jmap -heap 命令，可以查看对应进程可能占用的最大堆内存。先执行 jps 命令，查看 NameNode 和 DataNode 的进程号，再执行 jmap -heap 命令。

```
[atguigu@hadoop102 ~]$ jps
3088 NodeManager
2611 NameNode
3271 JobHistoryServer
2744 DataNode
3579 Jps
[atguigu@hadoop102 ~]$ jmap -heap 2611
Heap Configuration:
   MaxHeapSize              = 1031798784 (984.0MB)
[atguigu@hadoop102 ~]$ jmap -heap 2744
Heap Configuration:
   MaxHeapSize              = 1031798784 (984.0MB)
```

可以看到 NameNode 和 DataNode 的最大堆内存都可以占用当前堆内存的最大空闲内存空间。这样的分配是不合理的，在 NameNode 和 DataNode 共同部署的服务器上，很可能出现其中一个进程占用全部内存空间的情况。对此，Cloudera 公司提供了一些参考配置，如图 8-1 所示，NameNode 的最大允许内存空间为 1GB，每增加 1 000 000 个数据块，都会增加 1GB；DataNode 的最大允许内存空间为 4GB，可以满足低于 4 000 000 个数据副本的存储要求，每增加 1 000 000 个副本，都会增加 1GB。

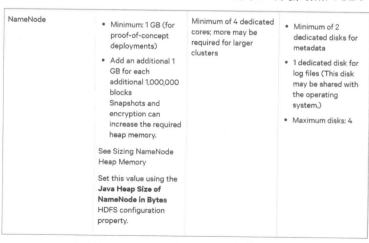

图 8-1　Cloudera 公司提供的参考配置

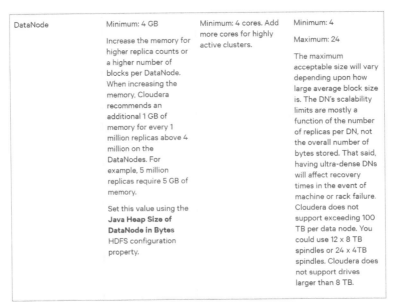

图 8-1 Cloudera 公司提供的参考配置（续）

hadoop-env.sh 文件中的具体配置参数如下，用户可以根据自己集群的具体情况进行配置。

```
export HDFS_NAMENODE_OPTS="-Dhadoop.security.logger=INFO,RFAS -Xmx1024m"

export HDFS_DATANODE_OPTS="-Dhadoop.security.logger=ERROR,RFAS -Xmx1024m"
```

8.1.2 NameNode 心跳并发配置

Hadoop 集群中的 DataNode 会保持与 NameNode 的心跳联系，如图 8-2 所示，还需要处理来自客户端的数据读/写请求，那么 NameNode 准备多少线程合适呢？NameNode 有一个工作线程池，主要用于处理不同 DataNode 的并发心跳及客户端并发的元数据操作。线程池的默认线程数配置是 10，对于大集群或有大量客户端的集群，通常需要增大该参数。

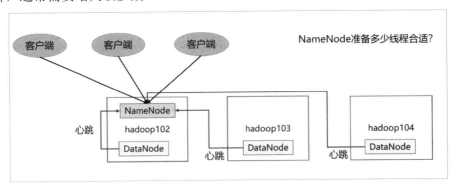

图 8-2 DataNode 与 NameNode 之间的心跳联系

在配置文件 hdfs-site.xml 中进行参数配置，具体参数配置如下：

```
<property>
    <name>dfs.namenode.handler.count</name>
    <value>10</value>
</property>
```

对于该参数，企业实际生产的经验如下：

$$\text{dfs.namenode.handler.count} = 20 \times \log_e^{\text{ClusterSize}}$$

例如，当集群规模（Cluster Size，即 DataNode 数量）为 3 个 DataNode 时，建议将该参数值设置为 21。可以通过简单的 Python 代码计算该值，代码如下：

```
[atguigu@hadoop102 ~]$ sudo yum install -y python
[atguigu@hadoop102 ~]$ python
Python 2.7.5 (default, Apr 11 2018, 07:36:10)
[GCC 4.8.5 20150623 (Red Hat 4.8.5-28)] on linux2
Type "help", "copyright", "credits" or "license" for more information.
>>> import math
>>> print int(20*math.log(3))
21
>>> quit()
```

8.1.3 启用回收站功能

启用 Hadoop 的回收站功能，删除的文件在不超时的情况下可以恢复，起到防止数据误删除的作用。回收站的工作机制如图 8-3 所示。被删除的文件会被放入回收站，可以设置一个文件存活时间，也就是在文件被放入回收站后多久才会被真正删除，进而释放数据块。还可以设置一个检查回收站的间隔时间，每隔一段时间对回收站进行检查，检查回收站中的数据是否到了被删除的时间。

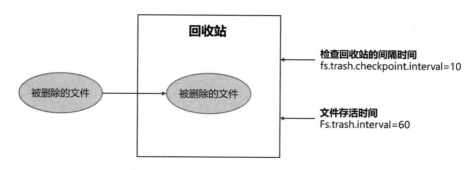

图 8-3　回收站的工作机制

启用回收站功能的参数如下。

- fs.trash.interval：默认值为 0，表示禁用回收站功能，其他值表示文件的存活时间。
- fs.trash.checkpoint.interval：默认值为 0，表示该值和 fs.trash.interval 参数的值相等，其他值表示检查回收站的间隔时间。

要求 fs.trash.checkpoint.interval≤fs.trash.interval。

回收站使用测试如下。

（1）修改配置文件 core-site.xml，启用回收站功能，设置垃圾回收时间为 1 分钟。

```
<property>
    <name>fs.trash.interval</name>
    <value>1</value>
</property>
```

在修改配置文件后，分发配置文件并重启集群，使配置文件生效。

（2）在 HDFS 的 Web 端页面中删除的文件和通过程序删除的文件都不会直接进入回收站，需要调用 moveToTrash()方法才能进入回收站。

```
Trash trash = New Trash(conf);
trash.moveToTrash(path);
```

（3）只有在命令行使用 hadoop fs -rm 命令删除的文件，才能直接进入回收站。

```
[atguigu@hadoop102 hadoop-3.1.3]$ hadoop fs -rm -r /input
2021-07-14 16:13:42,643 INFO fs.TrashPolicyDefault: Moved: 'hdfs://hadoop102:9820/input'
to trash at: hdfs://hadoop102:9820/user/atguigu/.Trash/Current/input
```

在 HDFS 的 Web 端页面中查看进入回收站的文件，如图 8-4 所示。

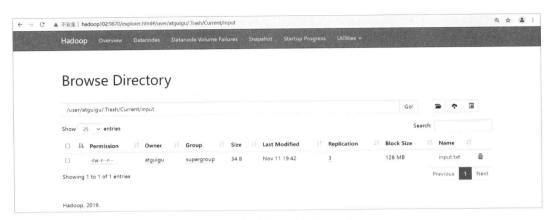

图 8-4 查看进入回收站的文件

（4）执行以下命令，可以恢复回收站数据。
```
[atguigu@hadoop102 hadoop-3.1.3]$ hadoop fs -mv
/user/atguigu/.Trash/Current/input    /input
```

8.2 HDFS 集群压测

在企业的实际生产过程中，企业会非常关心每天从 Java 后台拉取过来的数据需要多久才能上传到集群中，而数据的消费者会关心多久能获取 HDFS 中存储的数据，如图 8-5 所示。为了搞清楚这两个问题，需要对集群进行压力测试，用于了解 HDFS 集群真正的读/写性能。

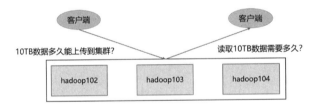

图 8-5 HDFS 的读/写性能

HDFS 的读/写性能主要受网络带宽和磁盘因素的影响。为了方便测试，可以将 hadoop102、hadoop103、hadoop104 节点服务器的网络带宽都设置为 100Mbps，配置方式如图 8-6 所示。

图 8-6 节点服务器的网络带宽配置方式

在图 8-6 中，带宽的配置单位是 Mbps（Million bits per second，每秒传输百万位数据），这里的 bit（比特，1 比特等于 1 位）是表示数字信号数据的最小单位，我们需要将其换算成更直观的 MB/s。对于 MB/s，其中的 MB 表示 Million Bytes（百万字节），Byte（字节）是计算机计量存储容量的一种计量单位。在计算机中，每 8 位为 1 字节，也就是 1Byte＝8bit，当表示网速时，1MB/s=8Mbps，因此可以将 100Mbps 换算成 12.5MB/s。

在带宽配置完毕后，可以运行以下程序，对网速进行测试。

（1）执行以下 Python 命令，开启一个文件下载服务器，默认端口号是 8000。

```
[atguigu@hadoop102 software]$ python -m SimpleHTTPServer
Serving HTTP on 0.0.0.0 port 8000 ...
```

（2）登录带宽测试页面，如图 8-7 所示。

图 8-7　带宽测试页面

（3）单击其中一个文件进行下载，如图 8-8 所示，可以看到下载速度，大约为 10.5MB/s，与配置的带宽接近。

图 8-8　文件下载测试

8.2.1　测试 HDFS 的写性能

HDFS 写性能测试的底层原理如图 8-9 所示，将多个测试文件写入 HDFS 集群，记录每个 MapTask 的数据写入时间和平均速度。

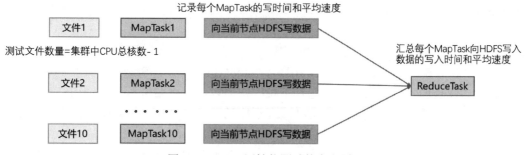

图 8-9　HDFS 写性能测试的底层原理

(1) 运行以下测试程序，测试内容为向 HDFS 集群中写入 10 个 128MB 的文件。

```
[atguigu@hadoop102 mapreduce]$ hadoop jar /opt/module/hadoop-3.1.3/share/hadoop/mapreduce/hadoop-mapreduce-client-jobclient-3.1.3-tests.jar TestDFSIO -write -nrFiles 10 -fileSize 128MB

2021-02-09 10:43:16,853 INFO fs.TestDFSIO: ----- TestDFSIO ----- : write
2021-02-09 10:43:16,854 INFO fs.TestDFSIO:             Date & time: Tue Feb 09 10:43:16 CST 2021
2021-02-09 10:43:16,854 INFO fs.TestDFSIO:         Number of files: 10
2021-02-09 10:43:16,854 INFO fs.TestDFSIO:  Total MBytes processed: 1280
2021-02-09 10:43:16,854 INFO fs.TestDFSIO:       Throughput mb/sec: 1.61
2021-02-09 10:43:16,854 INFO fs.TestDFSIO:  Average IO rate mb/sec: 1.9
2021-02-09 10:43:16,854 INFO fs.TestDFSIO:   IO rate std deviation: 0.76
2021-02-09 10:43:16,854 INFO fs.TestDFSIO:      Test exec time sec: 133.05
2021-02-09 10:43:16,854 INFO fs.TestDFSIO:
```

在上述测试命令中，nrFiles 参数表示尝试上传文件的数量，一般将其设置为可用 CPU 核数-1，可用 CPU 核数可以通过 YARN 的 Web 端页面查看。

在上述测试结果中，参数说明如下。

- Number of files：上传文件的数量，一般是集群中可用 CPU 核数-1。
- Total MBytes processed：任务处理的总文件大小（单位为 MB）。
- Throughput mb/sec：单个文件处理速度，计算方式是任务处理的总文件大小/每个 MapTask 写数据的累加时间。
- Average IO rate mb/sec：每个 MapTask 的平均吞吐量，计算方式是先计算每个 MapTask 处理的文件大小/每个 MapTask 写数据的时间，再计算所有 MapTask 吞吐量的和/MapTask 数量。
- IO rate std deviation：MapTask 平均吞吐量的方差，反映了各个 MapTask 平均吞吐量处理速度的差值，该值越小越均衡。

（2）测试结果分析。

文件副本存储情况如图 8-10 所示。由于文件在上传后生成的 3 个副本中的副本 1 位于本地 DataNode 中，因此该副本不计入测试结果。

图 8-10 文件副本存储情况

计算过程如下。

实际参与测试的文件总数：10 个文件 × 2 个副本 = 20 个。

压测后的单个文件处理速度：1.61 MB/s。

实际处理速度：1.61 MB/s × 20 个文件 ≈ 32MB/s。

3 台节点服务器的带宽：12.5 MB/s + 12.5 MB/s + 12.5MB/s = 37.5 MB/s。

实际处理速度与 3 台节点服务器的总带宽几乎相等，所以所有网络资源都已经用满。

如果实测速度远远小于网络带宽，并且实测速度不能满足工作需求，则可以考虑采用固态硬盘或增加磁盘数量。

（3）如果客户端不在集群节点上，上传 3 个文件副本都需要占用网络带宽资源，那么 3 个文件副本都参与计算，如图 8-11 所示。

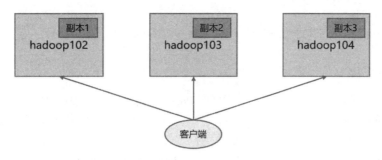

图 8-11　非本地上传文件副本的存储情况

8.2.2　测试 HDFS 的读性能

（1）运行以下测试程序，测试内容为从 HDFS 集群中读取 10 个 128MB 的文件，测试结果如下：

```
[atguigu@hadoop102 mapreduce]$ hadoop jar /opt/module/hadoop-3.1.3/share/hadoop/mapreduce/hadoop-mapreduce-client-jobclient-3.1.3-tests.jar TestDFSIO -read -nrFiles 10 -fileSize 128MB

2021-02-09 11:34:15,847 INFO fs.TestDFSIO: ----- TestDFSIO ----- : read
2021-02-09 11:34:15,847 INFO fs.TestDFSIO:            Date & time: Tue Feb 09 11:34:15 CST 2021
2021-02-09 11:34:15,847 INFO fs.TestDFSIO:        Number of files: 10
2021-02-09 11:34:15,847 INFO fs.TestDFSIO: Total MBytes processed: 1280
2021-02-09 11:34:15,848 INFO fs.TestDFSIO:      Throughput mb/sec: 200.28
2021-02-09 11:34:15,848 INFO fs.TestDFSIO: Average IO rate mb/sec: 266.74
2021-02-09 11:34:15,848 INFO fs.TestDFSIO:  IO rate std deviation: 143.12
2021-02-09 11:34:15,848 INFO fs.TestDFSIO:     Test exec time sec: 20.83
```

上述测试结果中的参数说明与 HDFS 写性能测试结果中的参数说明是相同的。

（2）执行以下命令，将读取的数据清除。

```
[atguigu@hadoop102 mapreduce]$ hadoop jar /opt/module/hadoop-3.1.3/share/hadoop/mapreduce/hadoop-mapreduce-client-jobclient-3.1.3-tests.jar TestDFSIO -clean
```

（3）测试结果分析。

在 Throughput 的测试结果中，读取文件的平均速度为 200.28MB/s，而在 8.2.1 节中，写入文件的平均速度是 1.61MB/s，可以发现，读取文件的速度（数据读取速度）远高于写入文件的速度（数据写入速度）。目前只有 3 台节点服务器，并且有 3 个文件副本，数据读取操作会采用就近原则，相当于读取本地磁盘中的数据，如图 8-12 所示，因此不会占用网络带宽资源。但是这个测试在节点数量较多的实际生产集群环境中是比较实用的，可以了解集群的数据读取速度。

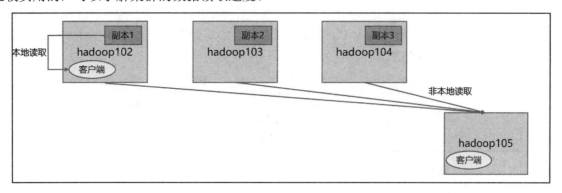

图 8-12　HDFS 的读性能测试

8.3 HDFS 的多目录配置

HDFS 提供了不同集群角色的多目录配置，分别起到了不同的作用。

8.3.1 NameNode 的多目录配置

NameNode 作为 HDFS 的主节点，负责整个集群的元数据管理工作，还要负责响应客户端的数据读/写请求，所以其数据存储的可靠性至关重要。在 HDFS 中，NameNode 的数据存储目录可以配置成多个，并且每个目录中存储的数据是相同的。我们可以选择来自不同磁盘的多个目录，当 NameNode 中一个磁盘因为故障不能使用时，可以使用其他目录中的数据，从而提高数据的可靠性。

具体操作步骤如下。

（1）在配置文件 hdfs-site.xml 中添加以下内容，其中，value 部分的多个目录使用逗号分隔，在实际生产环境中，不同的目录可以来自不同的磁盘。

```xml
<property>
    <name>dfs.namenode.name.dir</name>
    <value>file://${hadoop.tmp.dir}/dfs/name1,file://${hadoop.tmp.dir}/dfs/name2</value>
</property>
```

需要注意的是，此项配置是针对当前 NameNode 生效的，所以不用分发至其他节点服务器中。

（2）在修改 NameNode 的配置后，停止集群，删除 3 台节点服务器中 data 和 logs 目录下的所有数据。

```
[atguigu@hadoop102 hadoop-3.1.3]$ rm -rf data/ logs/
[atguigu@hadoop103 hadoop-3.1.3]$ rm -rf data/ logs/
[atguigu@hadoop104 hadoop-3.1.3]$ rm -rf data/ logs/
```

（3）格式化集群并重新启动，代码如下：

```
[atguigu@hadoop102 hadoop-3.1.3]$ bin/hdfs namenode -format
[atguigu@hadoop102 hadoop-3.1.3]$ sbin/start-dfs.sh
```

（4）查看数据存储路径${hadoop.tmp.dir}/dfs，其中。${hadoop.tmp.dir}是 Hadoop 的临时目录，该属性在 core-site.xml 文件中配置。

```
[atguigu@hadoop102 dfs]$ ll
总用量 12
drwx------. 3 atguigu atguigu 4096 12月 11 08:03 data
drwxrwxr-x. 3 atguigu atguigu 4096 12月 11 08:03 name1
drwxrwxr-x. 3 atguigu atguigu 4096 12月 11 08:03 name2
```

检查 name1 和 name2 目录下的内容，可以发现数据是一样的。

8.3.2 DataNode 的多目录配置

DataNode 可以配置多个数据目录，但是目的与 NameNode 多目录配置的目的不同。DataNode 配置多个数据目录，每个目录下存储的数据是不一样的。也就是说，当 DataNode 的存储空间不足时，可以为服务器挂载磁盘，然后增加目录，从而达到为 DataNode 扩容的目的。为 hadoop102 和 hadoop103 节点服务器分别增加一个磁盘目录，如图 8-13 所示。

每个磁盘存储的数据不一样

图 8-13 增加磁盘目录

具体操作步骤如下。

（1）在配置文件 hdfs-site.xml 中添加以下内容，其中，value 部分的多个目录之间使用逗号分隔。

```
<property>
    <name>dfs.datanode.data.dir</name>
    <value>file://${hadoop.tmp.dir}/dfs/data1,file://${hadoop.tmp.dir}/dfs/data2</value>
</property>
```

需要注意的是，此项配置是针对当前 DataNode 生效的，所以不用分发至其他节点服务器中。

（2）查看${hadoop.tmp.dir}/dfs 路径，可以看到增加了一个数据存储目录。

```
[atguigu@hadoop102 dfs]$ ll
总用量 12
drwx------. 3 atguigu atguigu 4096 4月   4 14:22 data1
drwx------. 3 atguigu atguigu 4096 4月   4 14:22 data2
drwxrwxr-x. 3 atguigu atguigu 4096 12月 11 08:03 name1
drwxrwxr-x. 3 atguigu atguigu 4096 12月 11 08:03 name2
```

（3）向集群中上传一个文件，再次观察 data1 和 data2 目录，可以发现，其中一个目录下有数据，另一个目录下没有数据。因此，DataNode 的多目录之间存储的数据是不同的。

```
[atguigu@hadoop102 hadoop-3.1.3]$ hadoop fs -put wcinput/word.txt /
```

8.3.3　集群数据均衡之磁盘之间的数据均衡

在集群的实际运行过程中，在发现磁盘存储空间不足后，通常需要增加磁盘。在增加磁盘后，我们会在 HDFS 的多目录配置中增加新磁盘的存储路径。但是在增加配置后，新磁盘中是没有数据的，数据负载不均衡，如图 8-14 所示，在这种情况下，可以执行磁盘均衡命令。

图 8-14　数据负载不均衡

（1）执行以下命令，生成磁盘均衡任务。

```
[atguigu@hadoop103 hadoop-3.1.3]$ hdfs diskbalancer -plan hadoop103
```

（2）执行磁盘均衡任务，代码如下：

```
[atguigu@hadoop103 hadoop-3.1.3]$ hdfs diskbalancer -execute hadoop103.plan.json
```

（3）查看磁盘均衡任务的执行情况，代码如下：

```
[atguigu@hadoop103 hadoop-3.1.3]$ hdfs diskbalancer -query hadoop103
```

（4）取消磁盘均衡任务，代码如下：

```
[atguigu@hadoop103 hadoop-3.1.3]$ hdfs diskbalancer -cancel hadoop103.plan.json
```

8.4　HDFS 集群的扩容及缩容

Hadoop 的集群是可灵活扩展的，这种灵活性体现在，当集群产能不足时，可以增加节点服务器；在集群产能过剩时，可以退役节点服务器，即集群的扩容与缩容。

8.4.1　添加白名单

在 Hadoop 集群的安装过程中，可以配置一个白名单，使用白名单中的主机存储数据。在企业中配置白名单，可以尽量防止黑客的恶意访问攻击。如图 8-15 所示，如果将 hadoop102 和 hadoop103 节点服务器放入白名单，那么集群不会再向 hadoop104 节点服务器中存储数据。对于没有被添加到白名单中的节点服务器，虽然不可以存储数据，但是依然可以执行文件上传操作。

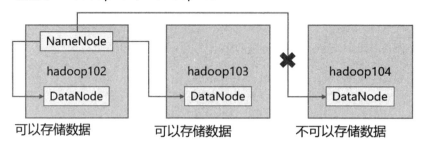

图 8-15　白名单配置示意图

配置白名单的步骤如下。

（1）在 NameNode 的/opt/module/hadoop-3.1.3/etc/hadoop 目录下，分别创建白名单文件 whitelist 和黑名单文件 blacklist。

创建白名单文件 whitelist。

```
[atguigu@hadoop102 hadoop]$ vim whitelist
```

在 whitelist 文件中添加以下主机名，将 hadoop102 和 hadoop103 节点服务器加入白名单。

```
hadoop102
hadoop103
```

创建黑名单文件 blacklist，该文件中暂时不填写任何内容。

```
[atguigu@hadoop102 hadoop]$ touch blacklist
```

（2）在配置文件 hdfs-site.xml 中添加配置参数 dfs.hosts 和 dfs.hosts.exclude。

```xml
<!-- 白名单 -->
<property>
    <name>dfs.hosts</name>
    <value>/opt/module/hadoop-3.1.3/etc/hadoop/whitelist</value>
</property>

<!-- 黑名单 -->
<property>
    <name>dfs.hosts.exclude</name>
    <value>/opt/module/hadoop-3.1.3/etc/hadoop/blacklist</value>
</property>
```

（3）分发配置文件 whitelist 和 hdfs-site.xml。

```
[atguigu@hadoop104 hadoop]$ xsync hdfs-site.xml whitelist
```

（4）由于是第一次添加白名单，因此必须重启集群。如果不是第一次添加白名单，那么只需刷新 NameNode。

```
[atguigu@hadoop102 hadoop-3.1.3]$ myhadoop.sh stop
[atguigu@hadoop102 hadoop-3.1.3]$ myhadoop.sh start
```

（5）查看 HDFS 的 Web 端页面，查看可用的 DataNode 列表，如图 8-16 所示，发现只有 2 个 DataNode 可用。

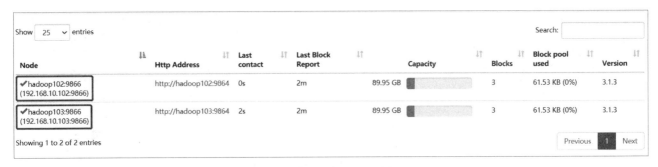

图 8-16　可用的 DataNode 列表（一）

（6）在 hadoop104 节点服务器中上传数据。

```
[atguigu@hadoop104 hadoop-3.1.3]$ hadoop fs -put NOTICE.txt /
2021-11-12 10:38:51,161 INFO sasl.SaslDataTransferClient: SASL encryption trust check:
localHostTrusted = false, remoteHostTrusted = false
```

（7）在数据上传完成后，查看 HDFS 中文件副本的存储情况，文件副本只存在于 hadoop102 和 hadoop103 节点服务器中，如图 8-17 所示。

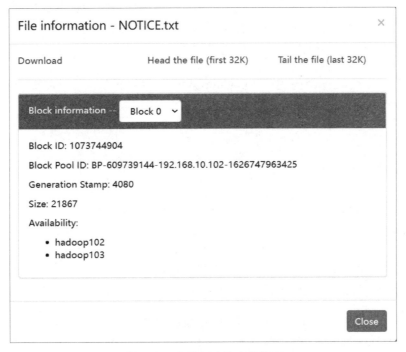

图 8-17　文件副本的存储情况

（8）修改白名单文件 whitelist，在白名单列表中添加 hadoop104 节点服务器。

```
[atguigu@hadoop102 hadoop]$ vim whitelist

hadoop102
hadoop103
hadoop104
```

（9）执行以下命令，刷新 NameNode。

```
[atguigu@hadoop102 hadoop-3.1.3]$ hdfs dfsadmin -refreshNodes
Refresh nodes successful
```

（10）再次查看 HDFS 的 Web 端页面，查看可用的 DataNode 列表，如图 8-18 所示，发现目前有 3 个可用 DataNode。

图 8-18 可用的 DataNode 列表（二）

8.4.2 服役新服务器

随着公司业务的增长，数据量越来越大，原有的数据节点的容量已经不能满足存储数据的需求，需要在原有集群的基础上动态添加新的数据节点，这种操作称为服役新服务器。我们可以在本地搭建的测试节点服务器上做简单操作，如果读者的个人计算机不能满足条件，则可以略过本节操作。

具体操作步骤如下。

（1）首先需要再克隆一台节点服务器，参考 2.4.6 节中的操作步骤，将主机名更改为 hadoop105。

（2）参考 2.4.3 节提供的主机名与 IP 地址映射文件，修改主机名与 IP 地址。

```
[root@hadoop105 ~]# vim /etc/sysconfig/network-scripts/ifcfg-ens33
[root@hadoop105 ~]# vim /etc/hostname
```

（3）将 hadoop102 节点服务器中的/opt/module 目录和环境变量文件/etc/profile.d/my_env.sh 复制到 hadoop105 节点服务器中，并且在 hadoop105 节点服务器中使环境变量生效。

```
[atguigu@hadoop102 opt]$ scp -r module/* atguigu@hadoop105:/opt/module/

[atguigu@hadoop102 opt]$ sudo scp /etc/profile.d/my_env.sh root@hadoop105:/etc/profile.d/my_env.sh

[atguigu@hadoop105 hadoop-3.1.3]$ source /etc/profile
```

（4）删除 hadoop105 节点服务器中 Hadoop 的历史数据 data 和 logs 目录。

```
[atguigu@hadoop105 hadoop-3.1.3]$ rm -rf data/ logs/
```

（5）配置 hadoop102 和 hadoop103 节点服务器到 hadoop105 节点服务器的 SSH 免密登录。

```
[atguigu@hadoop102 .ssh]$ ssh-copy-id hadoop105

[atguigu@hadoop103 .ssh]$ ssh-copy-id hadoop105
```

（6）在 hadoop105 节点服务器上执行以下命令，直接启动 DataNode 和 NodeManager。

```
[atguigu@hadoop105 hadoop-3.1.3]$ hdfs --daemon start datanode
[atguigu@hadoop105 hadoop-3.1.3]$ yarn --daemon start nodemanager
```

在 DataNode 和 NodeManager 启动成功后，查看 HDFS 的 Web 端页面，如图 8-19 所示，可以看到，可用的 DataNode 列表中没有 hadoop105 节点服务器。

图 8-19 可用的 DataNode 列表（一）

（7）在白名单文件 whitelist 中添加 hadoop105 节点服务器。

```
[atguigu@hadoop102 hadoop]$ vim whitelist
修改为以下内容
hadoop102
hadoop103
hadoop104
hadoop105
```

（8）修改分发脚本 xsync，添加 hadoop105 节点服务器，代码如下：

```
[atguigu@hadoop102 bin]$ vim xsync
#!/bin/bash
#获取输入参数的数量，如果没有输入参数，则直接退出
pcount=$#
if((pcount==0)); then
echo no args;
exit;
fi

#获取文件名称
p1=$1
fname=`basename $p1`
echo fname=$fname

#获取上级目录的绝对路径
pdir=`cd -P $(dirname $p1); pwd`
echo pdir=$pdir

#获取当前用户名
user=`whoami`

#循环
for((host=103; host<106; host++)); do
    echo --------------------- hadoop$host ----------------
    rsync -rvl $pdir/$fname $user@hadoop$host:$pdir
done
```

在修改完成后，分发白名单文件 whitelist。

```
[atguigu@hadoop102 hadoop]$ xsync whitelist
```

（9）刷新 NameNode。

```
[atguigu@hadoop102 hadoop-3.1.3]$ hdfs dfsadmin -refreshNodes
Refresh nodes successful
```

（10）重新查看 HDFS 的 Web 端页面，查看可用的 DataNode 列表，发现此时有 4 个可用的 DataNode，如图 8-20 所示。

图 8-20 可用的 DataNode 列表（二）

（11）在 hadoop105 节点服务器上执行文件上传命令。
```
[atguigu@hadoop105 hadoop-3.1.3]$ hadoop fs -put /opt/module/hadoop-3.1.3/LICENSE.txt /
```
在文件上传成功后，查看 HDFS 的 Web 端页面，查看上传文件的副本情况，如图 8-21 所示，其中包含 hadoop105 节点服务器，说明 hadoop105 节点服务器已经正常服役了。

图 8-21　上传文件的副本情况

8.4.3　服务器之间的数据均衡

在企业开发中，Hadoop 集群经常会出现节点服务器之间数据不均衡的情况。如果经常在 hadoop102 和 hadoop104 节点服务器中提交任务，并且将副本数量设置为 2 个，那么遵循数据本地性原则，会导致 hadoop102 和 hadoop104 节点服务器中的数据过多，hadoop103 节点服务器中的数据过少。此外，如果对集群进行扩容，那么新服役服务器中的数据存储量远远小于老服务器中的数据存储量。基于以上情况，我们经常需要执行数据均衡命令。

执行以下命令，启用数据均衡功能。
```
[atguigu@hadoop105 hadoop-3.1.3]$ sbin/start-balancer.sh -threshold 10
```
参数-threshold 10 表示集群中各个节点的磁盘空间利用率相差不超过 10%，可以根据实际情况进行调整。

执行以下命令，即可关闭数据均衡功能。
```
[atguigu@hadoop105 hadoop-3.1.3]$ sbin/stop-balancer.sh
```
注意：由于 HDFS 需要启动单独的 Rebalance Server 执行数据均衡操作，因此尽量不要在 NameNode 上执行以上命令，改为找一台比较空闲的节点服务器。

8.4.4　黑名单退役服务器

与白名单机制相对应，Hadoop 还提供了黑名单机制。如图 8-22 所示，被添加到黑名单中的主机不能存储数据。所以在企业中，通常使用黑名单退役节点服务器。黑名单机制与白名单机制不同，它是一种比较"严格"的机制，被写入黑名单的主机，需要将主机中原先存储的数据转移至正常服役的主机中，所以在实际生产环境中，不要尝试黑名单操作。

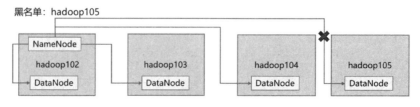

图 8-22　黑名单配置示意图

配置黑名单的步骤如下。

（1）编辑/opt/module/hadoop-3.1.3/etc/hadoop 目录下的 blacklist 文件。

```
[atguigu@hadoop102 hadoop] vim blacklist
```

将即将退役的主机名写入文件。

```
hadoop105
```

（2）修改配置文件 hdfs-site.xml 中的黑名单参数。需要注意的是，配置 dfs.hosts.exclude 参数的前提是已经配置了 dfs.hosts 参数。

```
<!-- 黑名单 -->
<property>
    <name>dfs.hosts.exclude</name>
    <value>/opt/module/hadoop-3.1.3/etc/hadoop/blacklist</value>
</property>
```

（3）分发配置文件 blacklist 和 hdfs-site.xml。

```
[atguigu@hadoop104 hadoop]$ xsync hdfs-site.xml blacklist
```

（4）第一次添加黑名单必须重启集群。由于在讲解白名单时我们已经添加过了，因此只需刷新 NameNode。

```
[atguigu@hadoop102 hadoop-3.1.3]$ hdfs dfsadmin -refreshNodes
Refresh nodes successful
```

（5）节点服务器的状态图标含义如图 8-23 所示。

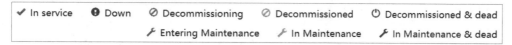

图 8-23　节点服务器的状态图标含义

查看 HDFS 的 Web 端页面，如图 8-24 所示，退役节点服务器的状态为 Decommissioning（退役中），说明 hadoop105 节点服务器正在复制数据块到其他节点服务器中。

图 8-24　hadoop105 节点服务器退役中

（6）等待退役节点服务器的状态为 Decommissioned（已退役），如图 8-25 所示。

图 8-25　hadoop105 节点服务器已退役

停止该节点服务器的 DataNode 及 NodeManager 进程。

```
[atguigu@hadoop105 hadoop-3.1.3]$ hdfs --daemon stop datanode
stopping datanode
[atguigu@hadoop105 hadoop-3.1.3]$ yarn --daemon stop nodemanager
stopping nodemanager
```

在使用黑名单退役节点服务器时，如果退役后的节点服务器数量少于集群配置的副本数量，那么是不能退役成功的。如果集群中有 3 台节点服务器，并且副本数量配置为 3 个，那么如果要退役其中一台节点服务器，则需要事先将集群中的副本数量修改为 2 个。

（7）在节点服务器退役后，如果数据不均衡，则可以执行数据均衡命令。

```
[atguigu@hadoop102 hadoop-3.1.3]$ sbin/start-balancer.sh -threshold 10
```

8.5 HDFS 的存储优化策略

HDFS 作为一个分布式存储框架，在实际生产过程中，如何提升存储空间的利用效率、节省存储空间、保证数据的可靠性非常重要，本节为读者讲解 HDFS 的存储优化策略。

8.5.1 纠删码

在默认情况下，HDFS 会为一个文件存储 3 个副本，这样的机制提高了数据的可靠性，但也带来了 2 倍的数据冗余。Hadoop 3.x 引入了纠删码（Erasure Coding，EC），采用计算的方式，可以节省大约 50%的存储空间。

纠删码是一种数据冗余保护技术，以更少数据冗余的方式提供近似 3 个副本的数据可靠性。如图 8-26 所示，在传统的三副本机制中，一个 300MB 的数据会复制得到 3 个副本，总共占用 900MB 的存储空间。在引入纠删码后，一个 300MB 的文件可以被拆分成 3 个数据单元和 2 个校验单元，这 5 个单元会分别存储在 5 个数据节点中，总共占用 500MB 的存储空间，数据冗余度降低了将近 50%。可以容忍其中 2 个单元的丢失，任意 2 个单元的损坏都可以通过计算的方式还原数据。

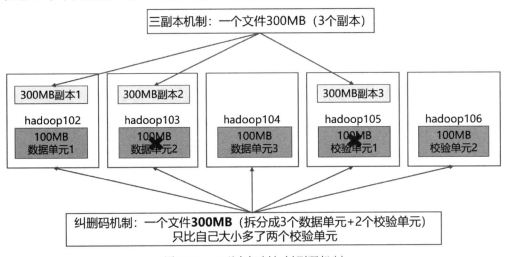

图 8-26　三副本机制与纠删码机制

执行以下命令，可以查看当前 HDFS 支持的纠删码策略。

```
[atguigu@hadoop102 hadoop-3.1.3] hdfs ec -listPolicies

Erasure Coding Policies:
ErasureCodingPolicy=[Name=RS-10-4-1024k, Schema=[ECSchema=[Codec=rs, numDataUnits=10, numParityUnits=4]], CellSize=1048576, Id=5], State=DISABLED

ErasureCodingPolicy=[Name=RS-3-2-1024k, Schema=[ECSchema=[Codec=rs, numDataUnits=3, numParityUnits=2]], CellSize=1048576, Id=2], State=DISABLED

ErasureCodingPolicy=[Name=RS-6-3-1024k, Schema=[ECSchema=[Codec=rs, numDataUnits=6, numParityUnits=3]], CellSize=1048576, Id=1], State=ENABLED

ErasureCodingPolicy=[Name=RS-LEGACY-6-3-1024k, Schema=[ECSchema=[Codec=rs-legacy, numDataUnits=6, numParityUnits=3]], CellSize=1048576, Id=3], State=DISABLED

ErasureCodingPolicy=[Name=XOR-2-1-1024k, Schema=[ECSchema=[Codec=xor, numDataUnits=2, numParityUnits=1]], CellSize=1048576, Id=4], State=DISABLED
```

对以上纠删码策略的说明如下。

- RS-3-2-1024k：使用 RS 编码，每 3 个数据单元生成 2 个校验单元，共 5 个单元。也就是说，在这 5 个单元中，只要有任意 3 个单元存在（不管是数据单元，还是校验单元），就可以得到原始数据。每个单元的大小是 1 024KB（1 048 576B）。
- RS-10-4-1024k：使用 RS 编码，每 10 个数据单元生成 4 个校验单元，共 14 个单元。也就是说，在这 14 个单元中，只要有任意 10 个单元存在（不管是数据单元，还是校验单元），就可以得到原始数据。每个单元的大小是 1 024KB（1 048 576B）。
- RS-6-3-1024k：使用 RS 编码，每 6 个数据单元生成 3 个校验单元，共 9 个单元。也就是说，在这 9 个单元中，只要有任意 6 个单元存在（不管是数据单元，还是校验单元），就可以得到原始数据。每个单元的大小是 1 024KB（1 048 576B）。
- RS-LEGACY-6-3-1024k：和 RS-6-3-1024k 策略类似，只是使用的编码算法是 rs-legacy。
- XOR-2-1-1024k：使用 XOR 编码（速度比 RS 编码快），每 2 个数据单元生成 1 个校验单元，共 3 个单元。也就是说，在这 3 个单元中，只要有任意 2 个单元存在（不管是数据单元，还是校验单元），就可以得到原始数据。每个单元的大小是 1 024KB（1 048 576B）。

根据以上纠删码策略可知，数据单元的数量受集群规模的影响，生成的数据单元越多，需要的集群节点越多。校验单元越多，故障容忍度越高。

接下来进行纠删码案例实操演示。我们选择使用 RS-3-2-1024k 策略，因此需要一个包括 5 台节点服务器的 Hadoop 集群，在 8.4.2 节中包括 4 台节点服务器的集群基础上，我们再克隆一台虚拟机，将其命名为 hadoop106，并且对其进行必要配置，将 hadoop106 节点服务器写入白名单，使其正式服役。

在新增节点服务器后，参考 8.4.2 节中的操作步骤，重新配置 hadoop102～hadoop106 节点服务器的 SSH 免密登录，修改分发脚本 xsync。

修改 workers 配置文件，代码如下：

```
[atguigu@hadoop102 hadoop]$ vim workers
hadoop102
hadoop103
hadoop104
hadoop105
hadoop106
```

个人计算机性能不允许这样操作的读者可以只阅读，或者观看本书附赠的课程视频。

使用 HDFS 的纠删码策略可以对文件系统的具体路径进行配置，在配置完成后，所有存储在此路径下的文件都会执行纠删码策略。默认使用 RS-6-3-1024k 策略，如果要使用其他纠删码策略，则需要自行启用。

（1）执行以下命令，查看纠删码的相关命令。

```
[atguigu@hadoop102 hadoop-3.1.3]$ hdfs ec
Usage: bin/hdfs ec [COMMAND]
       [-listPolicies]
       [-addPolicies -policyFile <file>]
       [-getPolicy -path <path>]
       [-removePolicy -policy <policy>]
       [-setPolicy -path <path> [-policy <policy>] [-replicate]]
       [-unsetPolicy -path <path>]
       [-listCodecs]
       [-enablePolicy -policy <policy>]
       [-disablePolicy -policy <policy>]
       [-help <command-name>].
```

（2）开启对 RS-3-2-1024k 策略的支持。
```
[atguigu@hadoop102 hadoop-3.1.3]$ hdfs ec -enablePolicy -policy RS-3-2-1024k
Erasure coding policy RS-3-2-1024k is enabled
```
（3）在 HDFS 中创建目录/input，并且设置此目录使用 RS-3-2-1024k 策略。
```
[atguigu@hadoop102 hadoop-3.1.3]$ hdfs dfs -mkdir /input

[atguigu@hadoop202 hadoop-3.1.3]$ hdfs ec -setPolicy -path /input -policy RS-3-2-1024k
```
（4）上传文件，并且查看文件编码后的存储情况。
```
[atguigu@hadoop102 hadoop-3.1.3]$ hdfs dfs -put web.log /input
```
注意：上传的文件需要大于 2MB 才能看出效果。如果文件小于 2MB，则只产生 1 个数据单元和 2 个校验单元。

（5）查看存储路径下的数据单元和校验单元，并且对其进行破坏测试。

8.5.2 异构存储

存储于 HDFS 中的数据的使用频率通常差别很大，有的数据可能每天都需要使用，有的数据可能只使用一次；存储介质也有差别，包括内存镜像、固态硬盘、机械硬盘和破旧硬盘，如图 8-27 所示。异构存储（冷热数据分离）可以将不同使用频率的数据存储于适合的存储介质中，从而达到最佳存储性能。

图 8-27　不同类型的存储介质

在 HDFS 中，将存储介质分为 4 种，这 4 种存储介绍按照数据读取速度由高到低的顺序分别如下。
- RAM_DISK：内存镜像。
- SSD：固态硬盘。
- DISK：机械硬盘。
- ARCHIVE：归档存储，读/写速度比机械硬盘慢，但是数据不易丢失。

存储目录属于哪种磁盘，由用户在配置文件中指定。
根据存储文件的磁盘类型，将存储策略分为以下几种。
- lazy_persist：1 个副本存储于内存镜像中，其他副本存储于机械硬盘中。
- All_SSD：所有副本都存储于固态硬盘中。
- One_SSD：1 个副本存储于固态硬盘中，其他副本存储于机械硬盘中。
- HOT：默认配置，所有副本都存储于机械硬盘中。
- WARM：1 个副本存储于机械硬盘中，其他副本归档存储。
- COLD：所有副本都归档存储。

使用异构存储的具体步骤如下。
（1）查看当前有哪些可用的存储策略。
```
[atguigu@hadoop102 hadoop-3.1.3]$ hdfs storagepolicies -listPolicies
Block Storage Policies:
```

```
    BlockStoragePolicy{COLD:2,        storageTypes=[ARCHIVE],        creationFallbacks=[],
replicationFallbacks=[]}
    BlockStoragePolicy{WARM:5, storageTypes=[DISK, ARCHIVE], creationFallbacks=[DISK, ARCHIVE],
replicationFallbacks=[DISK, ARCHIVE]}
    BlockStoragePolicy{HOT:7, storageTypes=[DISK], creationFallbacks=[], replicationFallbacks=
[ARCHIVE]}
    BlockStoragePolicy{ONE_SSD:10, storageTypes=[SSD, DISK], creationFallbacks=[SSD, DISK],
replicationFallbacks=[SSD, DISK]}
    BlockStoragePolicy{ALL_SSD:12,        storageTypes=[SSD],        creationFallbacks=[DISK],
replicationFallbacks=[DISK]}
    BlockStoragePolicy{LAZY_PERSIST:15, storageTypes=[RAM_DISK, DISK], creationFallbacks=
[DISK], replicationFallbacks=[DISK]}
```

（2）为指定路径（数据存储目录）设置存储策略。

```
hdfs storagepolicies -setStoragePolicy -path xxx -policy xxx
```

（3）获取指定路径（数据存储目录或文件）的存储策略。

```
hdfs storagepolicies -getStoragePolicy -path xxx
```

（4）取消存储策略的命令如下。在执行该命令后，该目录或文件的存储策略以其上级目录的存储策略为准，如果是根目录，那么使用 HOT 策略。

```
hdfs storagepolicies -unsetStoragePolicy -path xxx
```

（5）查看数据块副本的分布情况，代码如下：

```
bin/hdfs fsck xxx -files -blocks -locations
```

（6）查看集群节点，代码如下：

```
hadoop dfsadmin -report
```

接下来使用 8.5.1 节中配置的包括 5 台节点服务器的集群，针对 6 种存储策略进行测试。修改集群配置，将副本数量修改为 2 个，以便观察结果。提前在各个节点中创建带有存储类型的目录，集群的总体规划如表 8-1 所示。

表 8-1　集群的总体规划

节　　点	存储类型分配
hadoop102	RAM_DISK、SSD
hadoop103	SSD、DISK
hadoop104	DISK、RAM_DISK
hadoop105	ARCHIVE
hadoop106	ARCHIVE

针对以上集群规划，对配置文件进行修改，并且准备测试数据。

（1）在 hadoop102 节点服务器的 hdfs-site.xml 文件中添加以下信息。

```
<property>
    <name>dfs.replication</name>
    <value>2</value>
</property>
<property>
    <name>dfs.storage.policy.enabled</name>
    <value>true</value>
</property>
<property>
    <name>dfs.datanode.data.dir</name>
    <value>[SSD]file:///opt/module/hadoop-3.1.3/hdfsdata/ssd,[RAM_DISK]file:///opt/module/hadoop-3.1.3/hdfsdata/ram_disk</value>
</property>
```

（2）在 hadoop103 节点服务器的 hdfs-site.xml 文件中添加以下信息。
```xml
<property>
    <name>dfs.replication</name>
    <value>2</value>
</property>
<property>
    <name>dfs.storage.policy.enabled</name>
    <value>true</value>
</property>
<property>
    <name>dfs.datanode.data.dir</name>
    <value>[SSD]file:///opt/module/hadoop-3.1.3/hdfsdata/ssd,[DISK]file:///opt/module/hadoop-3.1.3/hdfsdata/disk</value>
</property>
```

（3）在 hadoop104 节点服务器的 hdfs-site.xml 文件中添加以下信息。
```xml
<property>
    <name>dfs.replication</name>
    <value>2</value>
</property>
<property>
    <name>dfs.storage.policy.enabled</name>
    <value>true</value>
</property>
<property>
    <name>dfs.datanode.data.dir</name>
    <value>[RAM_DISK]file:///opt/module/hdfsdata/ram_disk,[DISK]file:///opt/module/hadoop-3.1.3/hdfsdata/disk</value>
</property>
```

（4）在 hadoop105 节点服务器的 hdfs-site.xml 文件中添加以下信息。
```xml
<property>
    <name>dfs.replication</name>
    <value>2</value>
</property>
<property>
    <name>dfs.storage.policy.enabled</name>
    <value>true</value>
</property>
<property>
    <name>dfs.datanode.data.dir</name>
    <value>[ARCHIVE]file:///opt/module/hadoop-3.1.3/hdfsdata/archive</value>
</property>
```

（5）在 hadoop106 节点服务器的 hdfs-site.xml 文件中添加以下信息。
```xml
<property>
    <name>dfs.replication</name>
    <value>2</value>
</property>
<property>
    <name>dfs.storage.policy.enabled</name>
    <value>true</value>
</property>
<property>
```

```
        <name>dfs.datanode.data.dir</name>
        <value>[ARCHIVE]file:///opt/module/hadoop-3.1.3/hdfsdata/archive</value>
</property>
```
（6）停止集群，删除 5 台节点服务器中 data 和 logs 目录下的所有数据。
```
[atguigu@hadoop102 hadoop-3.1.3]$ rm -rf data/ logs/
[atguigu@hadoop103 hadoop-3.1.3]$ rm -rf data/ logs/
[atguigu@hadoop104 hadoop-3.1.3]$ rm -rf data/ logs/
[atguigu@hadoop105 hadoop-3.1.3]$ rm -rf data/ logs/
[atguigu@hadoop106 hadoop-3.1.3]$ rm -rf data/ logs/
```
格式化 NameNode 并重启集群。
```
[atguigu@hadoop102 hadoop-3.1.3]$ hdfs namenode -format
[atguigu@hadoop102 hadoop-3.1.3]$ myhadoop.sh start
```
（7）在 HDFS 中创建文件目录。
```
[atguigu@hadoop102 hadoop-3.1.3]$ hadoop fs -mkdir /hdfsdata
```
（8）上传测试文件。
```
[atguigu@hadoop102 hadoop-3.1.3]$ hadoop fs -put /opt/module/hadoop-3.1.3/NOTICE.txt /hdfsdata
```

1. HOT 存储策略

（1）在未设置存储策略的情况下，获取该目录的存储策略，发现使用的是 HOT 策略。
```
[atguigu@hadoop102 hadoop-3.1.3]$ hdfs storagepolicies -getStoragePolicy -path /hdfsdata
```
（2）查看上传文件中数据块副本的分布情况。
```
[atguigu@hadoop102 hadoop-3.1.3]$ hdfs fsck /hdfsdata -files -blocks -locations

[DatanodeInfoWithStorage[192.168.10.104:9866,DS-0b133854-7f9e-48df-939b-5ca6482c5afb,DISK],
DatanodeInfoWithStorage[192.168.10.103:9866,DS-ca1bd3b9-d9a5-4101-9f92-3da5f1baa28b,DISK]]
```
在未设置存储策略的情况下，所有数据块副本都存储于 DISK 中，默认的存储策略为 HOT。

2. WARM 存储策略

（1）将存储策略修改为 WARM。
```
[atguigu@hadoop102 hadoop-3.1.3]$ hdfs storagepolicies -setStoragePolicy -path /hdfsdata -policy WARM
```
（2）再次查看数据块副本的分布情况，可以发现，数据块副本依然存储于原处。
```
[atguigu@hadoop102 hadoop-3.1.3]$ hdfs fsck /hdfsdata -files -blocks -locations
```
（3）执行以下命令，使 HDFS 按照存储策略移动数据块。
```
[atguigu@hadoop102 hadoop-3.1.3]$ hdfs mover /hdfsdata
```
（4）再次查看数据块副本的分布情况，可以发现，一个数据块副本存储于 DISK 中，一个数据块副本存储于 ARCHIVE 中，符合 WARM 策略。
```
[atguigu@hadoop102 hadoop-3.1.3]$ hdfs fsck /hdfsdata -files -blocks -locations

[DatanodeInfoWithStorage[192.168.10.105:9866,DS-d46d08e1-80c6-4fca-b0a2-
4a3dd7ec7459,ARCHIVE], DatanodeInfoWithStorage[192.168.10.103:9866,DS-ca1bd3b9-d9a5-4101-
9f92-3da5f1baa28b,DISK]]
```

3. COLD 存储策略

（1）将存储策略修改为 COLD。
```
[atguigu@hadoop102 hadoop-3.1.3]$ hdfs storagepolicies -setStoragePolicy -path /hdfsdata -policy COLD
```

注意：在将目录的存储策略设置为 COLD 且未配置 ARCHIVE 存储目录的情况下，不可以直接向该目录上传文件，会报出异常。

（2）执行以下命令，使 HDFS 按照存储策略移动数据块。
```
[atguigu@hadoop102 hadoop-3.1.3]$ hdfs mover /hdfsdata
```
（3）检查数据块副本的分布情况，可以发现，所有数据块副本都存储于 ARCHIVE 中，符合 COLD 存储策略。
```
[atguigu@hadoop102 hadoop-3.1.3]$ bin/hdfs fsck /hdfsdata -files -blocks -locations

[DatanodeInfoWithStorage[192.168.10.105:9866,DS-d46d08e1-80c6-4fca-b0a2-4a3dd7ec7459,ARCHIVE],
DatanodeInfoWithStorage[192.168.10.106:9866,DS-827b3f8b-84d7-47c6-8a14-0166096f919d,ARCHIVE]]
```

4. One_SSD 存储策略

（1）将存储策略修改为 One_SSD。
```
[atguigu@hadoop102 hadoop-3.1.3]$ hdfs storagepolicies -setStoragePolicy -path /hdfsdata
-policy One_SSD
```
（2）执行以下命令，使 HDFS 按照存储策略移动数据块。
```
[atguigu@hadoop102 hadoop-3.1.3]$ hdfs mover /hdfsdata
```
（3）检查数据块副本的分布情况，可以发现，一个数据块副本存储于 SSD 中，一个数据块副本存储于 DISK 中，符合 One_SSD 存储策略。
```
[atguigu@hadoop102 hadoop-3.1.3]$ bin/hdfs fsck /hdfsdata -files -blocks -locations

[DatanodeInfoWithStorage[192.168.10.104:9866,DS-0b133854-7f9e-48df-939b-
5ca6482c5afb,DISK],   DatanodeInfoWithStorage[192.168.10.103:9866,DS-2481a204-59dd-46c0-
9f87-ec4647ad429a,SSD]]
```

5. All_SSD 存储策略

（1）将存储策略修改为 All_SSD。
```
[atguigu@hadoop102 hadoop-3.1.3]$ hdfs storagepolicies -setStoragePolicy -path /hdfsdata
-policy All_SSD
```
（2）执行以下命令，使 HDFS 按照存储策略移动数据块。
```
[atguigu@hadoop102 hadoop-3.1.3]$ hdfs mover /hdfsdata
```
（3）查看数据块副本的分布情况，可以发现，所有数据块副本都存储于 SSD 中，符合 All_SSD 存储策略。
```
[atguigu@hadoop102 hadoop-3.1.3]$ bin/hdfs fsck /hdfsdata -files -blocks -locations

[DatanodeInfoWithStorage[192.168.10.102:9866,DS-c997cfb4-16dc-4e69-a0c4-9411a1b0c1eb,SSD],
DatanodeInfoWithStorage[192.168.10.103:9866,DS-2481a204-59dd-46c0-9f87-ec4647ad429a,SSD]]
```

6. lazy_persist 存储策略

（1）将存储策略修改为 lazy_persist。
```
[atguigu@hadoop102 hadoop-3.1.3]$ hdfs storagepolicies -setStoragePolicy -path /hdfsdata
-policy lazy_persist
```
（2）执行以下命令，使 HDFS 按照存储策略移动数据块。
```
[atguigu@hadoop102 hadoop-3.1.3]$ hdfs mover /hdfsdata
```
（3）查看数据块副本的分布情况，可以发现，所有数据块副本都存储于 DISK 中，理论上，应该是一个数据块副本存储于 RAM_DISK 中，其他数据块副本存储于 DISK 中，这是因为，我们还需要配置 dfs.datanode.max.locked.memory 和 dfs.block.size 参数。
```
[atguigu@hadoop102 hadoop-3.1.3]$ hdfs fsck /hdfsdata -files -blocks -locations
```

```
[DatanodeInfoWithStorage[192.168.10.104:9866,DS-0b133854-7f9e-48df-939b-
5ca6482c5afb,DISK],    DatanodeInfoWithStorage[192.168.10.103:9866,DS-ca1bd3b9-d9a5-4101-
9f92-3da5f1baa28b,DISK]]
```

在使用 lazy_persist 存储策略时，数据块副本都存储于 DISK 中的原因如下。
- 当客户端所在的 DataNode 中没有 RAM_DISK 时，会将一个数据块副本写入客户端所在的 DataNode 的 DISK，将其他数据块副本写入其他 DataNode 的 DISK。
- 当客户端所在的 DataNode 中有 RAM_DISK，但未设置 dfs.datanode.max.locked.memory 参数的值或该值过小（小于 dfs.block.size 参数的值）时，会将一个数据块副本写入客户端所在的 DataNode 的 DISK，将其他数据块副本写入其他 DataNode 的 DISK。

由于虚拟机的 max locked memory 参数值为 64KB，因此如果设置的 dfs.datanode.max.locked.memory 参数值过大，则会报出以下错误。

```
ERROR org.apache.hadoop.hdfs.server.datanode.DataNode: Exception in secureMain
java.lang.RuntimeException: Cannot start datanode because the configured max locked memory
size (dfs.datanode.max.locked.memory) of 209715200 bytes is more than the datanode's
available RLIMIT_MEMLOCK ulimit of 65536 bytes.
```

可以通过以下命令查询该虚拟机的 max locked memory 参数值，如果该值较小，则无法对 HDFS 的 dfs.datanode.max.locked.memory 参数值做过大幅度的修改。

```
[atguigu@hadoop102 hadoop-3.1.3]$ ulimit -a
max locked memory       (kbytes, -l) 64
```

8.6 HDFS 的故障排除

本节主要讲解如何排除 HDFS 在运行过程中遇到的几种常见故障。为了节省个人计算机资源，读者可以将集群通过快照恢复至 3 台节点服务器的情况。

8.6.1 NameNode 故障处理

在集群正常运行的过程中，NameNode 突然宕机，并且存储的数据丢失了，如图 8-28 所示，在这种情况下，应该如何处理？

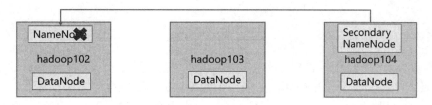

图 8-28 NameNode 故障

1. 故障模拟

执行 kill -9 命令，直接杀死 NameNode 进程，模拟 NameNode 宕机情况。

```
[atguigu@hadoop102 current]$ kill -9 19886
```

删除 NameNode 中存储的数据/opt/module/hadoop-3.1.3/data/dfs/name，模拟数据丢失情况。

```
[atguigu@hadoop102 hadoop-3.1.3]$ rm -rf /opt/module/hadoop-3.1.3/data/dfs/name/*
```

2. 问题解决

将 SecondaryNameNode 中的数据复制到原 NameNode 存储数据的目录下。

```
[atguigu@hadoop102 dfs]$ scp -r atguigu@hadoop104:/opt/module/hadoop-3.1.3/data/dfs/namesecondary/* ./name/
```

重新启动 NameNode。
```
[atguigu@hadoop102 hadoop-3.1.3]$ hdfs --daemon start namenode
```
查看集群运行情况，并且向集群上传一个文件，发现集群可以正常运行了。
```
[atguigu@hadoop102 hadoop-3.1.3]$ hadoop fs -put README.txt
```
该问题解决方案是在没有配置 HDFS 高可用的情况下采用的，在企业的实际生产过程中，通常都会配置高可用，用于避免 NameNode 宕机问题。

8.6.2 集群安全模式&磁盘数据损坏

在集群的安全模式下，文件系统只接受读数据请求，不接受删除、修改等变更请求。集群会在以下两种情况下进入安全模式。
- NameNode 加载镜像文件和编辑日志文件期间。
- 当 NameNode 接收 DataNode 注册时。

集群是否退出安全模式，受以下几个参数影响。
- dfs.namenode.safemode.min.datanodes：最少可用 DataNode 数量，默认值为 0。
- dfs.namenode.safemode.threshold-pct：副本数量达到最小要求的数据块数量占系统总数据块数量的百分比，默认值为 0.999f，也就是说，如果有 1000 个数据块，那么只允许丢失一个数据块。
- dfs.namenode.safemode.extension：稳定时间，默认为 30 000 毫秒，即 30 秒。

当集群处于安全模式时，不能执行数据的写操作。在集群启动完成后，自动退出安全模式。用户可以使用以下命令确认安全模式状态。

```
hdfs dfsadmin -safemode get       功能描述：查看安全模式状态
hdfs dfsadmin -safemode enter     功能描述：进入安全模式状态
hdfs dfsadmin -safemode leave     功能描述：退出安全模式状态
hdfs dfsadmin -safemode wait      功能描述：等待安全模式状态
```

如果集群规模比较大，那么在集群启动之初，DataNode 向 NameNode 上报块信息的过程会比较长，在这种情况下，可以执行退出安全模式命令。
```
hdfs dfsadmin -safemode leave
```
有时，用户期望在执行某条命令前，NameNode 能暂时退出安全模式，特别是在脚本中，可以在脚本中执行等待安全模式命令，如将脚本编辑如下：
```
[atguigu@hadoop102 hadoop-3.1.3]$ vim safemode.sh

#!/bin/bash
hdfs dfsadmin -safemode wait
hdfs dfs -put /opt/module/hadoop-3.1.3/README.txt /
```
有时，存储于磁盘中的数据会发生损坏，如果一个数据块的所有数据副本都损坏了，不能恢复，那么在集群启动，DataNode 上报数据块信息时，永远不能达到退出安全模式的条件，这种问题如何解决呢？

1．故障模拟

分别进入 hadoop102、hadoop103、hadoop104 节点服务器中的 /opt/module/hadoop-3.1.3/data/dfs/data/current/BP-1015489500-192.168.10.102-1611909480872/current/finalized/subdir0/subdir0 目录（DataNode 数据存储目录），统一删除某两个数据块信息。

```
[atguigu@hadoop102 subdir0]$ pwd
/opt/module/hadoop-3.1.3/data/dfs/data/current/BP-1015489500-192.168.10.102-
1611909480872/current/finalized/subdir0/subdir0

[atguigu@hadoop102 subdir0]$ rm -rf blk_1073741847 blk_1073741847_1023.meta
[atguigu@hadoop102 subdir0]$ rm -rf blk_1073741865 blk_1073741865_1042.meta
```

在 hadoop103 和 hadoop104 节点服务器上重复执行以上命令。

重新启动集群。

```
[atguigu@hadoop102 subdir0]$ myhadoop.sh stop
[atguigu@hadoop102 subdir0]$ myhadoop.sh start
```

查看 HDFS 的 Web 端页面，查看集群数据块损坏情况，如图 8-29 所示，数据块数量无法达到要求，无法退出安全模式。

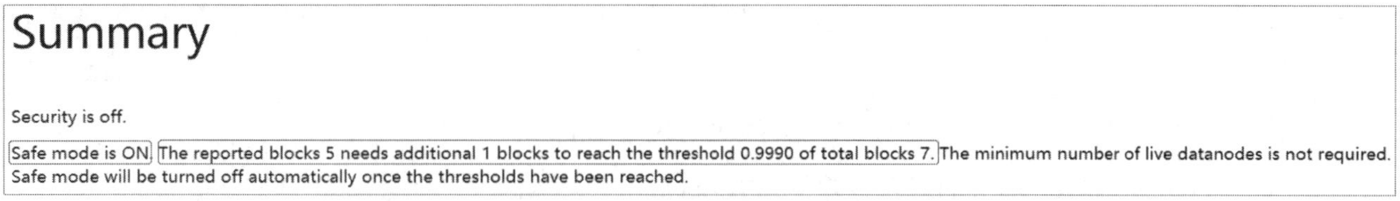

图 8-29　集群数据块损坏情况

2. 问题解决

执行以下命令，强行退出安全模式。

```
[atguigu@hadoop102 subdir0]$ hdfs dfsadmin -safemode get
Safe mode is ON
[atguigu@hadoop102 subdir0]$ hdfs dfsadmin -safemode leave
Safe mode is OFF
```

再次查看 HDFS 的 Web 端页面，如图 8-30 所示，提示两个数据块丢失。在这种情况下，即使暂时退出了安全模式，在下次启动集群时依然会进入安全模式，无法退出。

图 8-30　两个数据块丢失

在 NameNode 的 Web 端页面中将这两个数据块的元数据删除，如图 8-31 和图 8-32 所示。需要注意的是，此项操作无法撤销。

图 8-31　删除数据块的元数据（一）

图 8-32　删除数据块的元数据（二）

再次查看 HDFS 的 Web 端页面，可以看到，集群已经恢复正常。

只有在数据块彻底丢失的情况下，才会使用这种针对数据块损失的解决方案。

8.6.3 慢磁盘监控

慢磁盘是指写入数据非常慢的一类磁盘。其实慢磁盘并不少见，当服务器运行时间长了，上面跑的任务多了，磁盘的读/写性能自然会退化，严重的会出现写入数据延时的问题。

如何发现慢磁盘？在 HDFS 中创建一个目录，通常只需要不到 1s 的时间。如果创建目录超过 1 分钟，并且这种现象并不是每次都有，只是偶尔慢了一下，就很有可能存在慢磁盘。可以采用以下方法找出哪块磁盘慢。

1. 通过心跳未联系时间

出现慢磁盘现象，通常会影响 DataNode 与 NameNode 之间的心跳。在正常情况下，心跳时间间隔是 3s，如果超过 3s，则说明有异常。如图 8-33 所示，通过 NameNode 的 Web 端页面查看所有 DataNode 的上次心跳间隔，如果时间过长，则说明出现了异常。

图 8-33　DataNode 的上次心跳间隔情况

2. 测试磁盘的读/写性能

使用 fio 命令进行顺序读测试。

```
[atguigu@hadoop102 ~]# sudo yum install -y fio
[atguigu@hadoop102 ~]# sudo fio -filename=/home/atguigu/test.log -direct=1 -iodepth 1 -thread -rw=read -ioengine=psync -bs=16k -size=2G -numjobs=10 -runtime=60 -group_reporting -name=test_r

Run status group 0 (all jobs):
   READ: bw=360MiB/s (378MB/s), 360MiB/s-360MiB/s (378MB/s-378MB/s), io=20.0GiB (21.5GB), run=56885-56885msec
```

上述结果显示，磁盘的总体顺序读速度为 360MiB/s。

使用 fio 命令进行顺序写测试。

```
[atguigu@hadoop102 ~]# sudo fio -filename=/home/atguigu/test.log -direct=1 -iodepth 1 -
thread -rw=write -ioengine=psync -bs=16k -size=2G -numjobs=10 -runtime=60 -group_reporting 
-name=test_w
```

```
Run status group 0 (all jobs):
  WRITE: bw=341MiB/s (357MB/s), 341MiB/s-341MiB/s (357MB/s-357MB/s), io=19.0GiB (21.4GB), 
run=60001-60001msec
```

上述结果显示，磁盘的总体顺序写速度为341MiB/s。

使用fio命令进行随机写测试。

```
[atguigu@hadoop102 ~]# sudo fio -filename=/home/atguigu/test.log -direct=1 -iodepth 1 -
thread -rw=randwrite -ioengine=psync -bs=16k -size=2G -numjobs=10 -runtime=60 -
group_reporting -name=test_randw
```

```
Run status group 0 (all jobs):
  WRITE: bw=309MiB/s (324MB/s), 309MiB/s-309MiB/s (324MB/s-324MB/s), io=18.1GiB (19.4GB), 
run=60001-60001msec
```

上述结果显示，磁盘的总体随机写速度为309MiB/s。

使用fio命令进行混合随机读/写测试。

```
[atguigu@hadoop102 ~]# sudo fio -filename=/home/atguigu/test.log -direct=1 -iodepth 1 -
thread -rw=randrw -rwmixread=70 -ioengine=psync -bs=16k -size=2G -numjobs=10 -runtime=60 
-group_reporting -name=test_r_w -ioscheduler=noop
```

```
Run status group 0 (all jobs):
  READ: bw=220MiB/s (231MB/s), 220MiB/s-220MiB/s (231MB/s-231MB/s), io=12.9GiB (13.9GB), 
run=60001-60001msec
  WRITE: bw=94.6MiB/s (99.2MB/s), 94.6MiB/s-94.6MiB/s (99.2MB/s-99.2MB/s), io=5674MiB 
(5950MB), run=60001-60001msec
```

上述测试结果显示，磁盘的总体混合随机读速度为220MiB/s、总体混合随机写速度94.6MiB/s。

8.6.4 小文件存档

在HDFS中，每个文件都按块存储，每个数据块的元数据都存储于NameNode的内存空间中。如图8-34所示，100个1KB的数据块和100个128MB的数据块，产生的元数据大小是相同的。所以大量的小文件会耗尽NameNode中的大部分内存空间，HDFS存储小文件会非常低效。

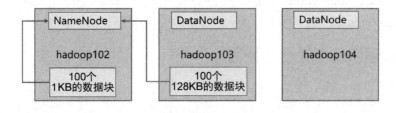

图8-34 小文件元数据存储情况

HDFS存档文件或HAR文件是一个更高效的文件存档工具，它将小文件存入HDFS块，不但可以减少NameNode使用的内存空间，而且允许对文件进行透明访问。具体来说，HDFS存档文件对内是一个个的独立文件，对NameNode而言却是一个整体，减少了NameNode占用的内存空间，如图8-35所示。

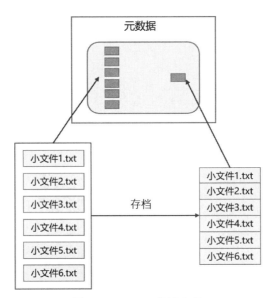

图 8-35　HDFS 存档文件

HDFS 的文件存档操作如下。

（1）保证 HDFS 与 YARN 已经启动。

```
[atguigu@hadoop102 hadoop-3.1.3]$ start-dfs.sh
[atguigu@hadoop103 hadoop-3.1.3]$ start-yarn.sh
```

（2）将/input 目录下的所有文件存档成一个名称为 input.har 的存档文件，并且将存档后的文件存储于/output 目录下。

```
[atguigu@hadoop102 hadoop-3.1.3]$ hadoop archive -archiveName input.har -p /input /output
```

（3）查看存档后的文件。

```
[atguigu@hadoop102 hadoop-3.1.3]$ hadoop fs -ls /output/input.har
[atguigu@hadoop102 hadoop-3.1.3]$ hadoop fs -ls har:///output/input.har
```

（4）解压缩存档文件。

```
[atguigu@hadoop102 hadoop-3.1.3]$ hadoop fs -cp har:///output/input.har/*  /
```

8.7　MapReduce 的生产经验

在 MapReduce 程序运行起来后，我们都希望程序可以运行得快一点。在深入了解了 MapReduce 的底层运行机制后，用户可以基于 MapReduce 的底层运行机制，通过调整相应的参数，使作业运行效率达到最高。本节将简单介绍 MapReduce 在生产环境中的一些基本调优手段。

8.7.1　MapReduce 程序运行较慢的原因

一个 MapReduce 程序运行较慢，主要从以下两方面考虑。

- 计算机性能。集群的计算机性能是有限的，在数据量过于庞大时，计算机的 CPU、内存、磁盘、网络 I/O 都有可能成为程序效率瓶颈，可以通过切分数据集、提升集群性能等措施解决此类问题。
- 用户角度。MapReduce 程序编写得不合理，Hadoop 参数配置不够优化，也有可能导致 MapReduce 程序运行较慢。例如，当某个 Task 运行时间过长时，会造成数据倾斜；当 MapTask 运行时间过长，会导致 ReduceTask 等待时间过久；当小文件过多时，会导致启动 Task 的数量过多，浪费集群资源。在上述情况下，用户应该首要考虑通过调整参数或优化程序提升程序的运行效率。

8.7.2 MapReduce 的常用调优参数

1. MapReduce 的运行全流程

前面讲解过 MapReduce 的运行全流程，如图 8-36 所示，所有的程序调优参数都应该基于 MapReduce 的底层运行机制进行调整。

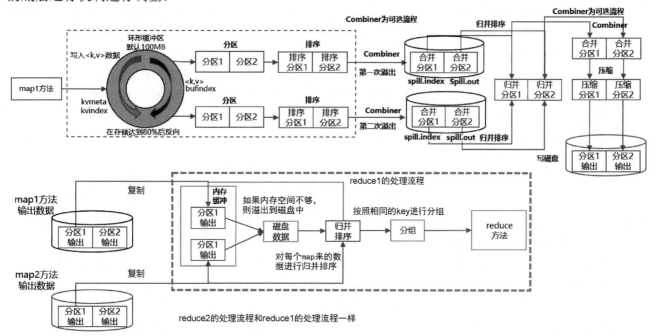

图 8-36　MapReduce 的运行全流程

2. 常用的 MapReduce 参数调优方法

常用的 MapReduce 参数调优方法如下。

1）使用自定义分区器，减少数据倾斜。

自定义分区器，使其继承 Partitioner 接口，重写 getPartition()方法。如果数据集中某个键对应的数据过多，那么使用自定义分区器可以使数据分布更加均匀。

2）减少溢写的次数。

数据在经过 map()方法后，会进入环形缓冲区，环形缓冲区的大小由参数指定。在环形缓冲区内的数据存储量达到的一定阈值（由参数指定）后，会将数据溢写至磁盘中。通过合理地调整参数，可以减少数据溢写次数，减少中间文件数目，提高任务执行效率。

用户可以根据自己作业的特点调整以下参数。

- mapreduce.task.io.sort.mb：环形缓冲区大小，默认值为 100MB，可以增大到 200MB。
- mapreduce.map.sort.spill.percent：环形缓冲区溢出的阈值，默认值为 80%，可以增大到 90%。

3）增加每次 Merge 合并的文件数量。

mapreduce.task.io.sort.factor：在 shuffle 阶段，一次 Merge 合并时的文件数量。默认值为 10，可以适当增大到 20 或更大。

4）提前采用 Combiner。

在不影响业务结果的前提条件下，提前采用 Combiner，可以减少 MapTask 的中间输出结果，降低 ReduceTask 的远程复制数据量，节省 I/O，提高效率。

5）采用 Snappy 或 LZO 压缩格式。

在 MapReduce 程序的适当位置启用压缩功能，可以实现降低 MapTask 的中间输出结果或降低 ReduceTask 的远程复制数据量的效果。但是压缩功能的启用应该慎重，过多使用压缩功能可能会影响计算性能。以 Snappy 为例，可以通过以下代码选择压缩格式。

```
conf.setBoolean("mapreduce.map.output.compress", true);
conf.setClass("mapreduce.map.output.compress.codec",
SnappyCodec.class,CompressionCodec.class);
```

6）mapreduce.map.memory.mb。

mapreduce.map.memory.mb 参数主要用于指定 MapTask 内存空间上限，默认值为 1024MB。可以根据每 128MB 数据对应 1GB 内存空间的原则增大该参数的值。

7）mapreduce.map.java.opts。

mapreduce.map.java.opts 参数主要用于指定 MapTask 的堆内存大小。如果程序在运行过程中报出错误 java.lang.OutOfMemoryError，那么在排查程序没有编写失误的前提下，应该适当增大该参数的值。

8）mapreduce.map.cpu.vcores。

mapreduce.map.cpu.vcores 参数主要用于指定 MapTask 的 CPU 核数，默认为 1 个。在计算密集型任务中，可以适当增加 CPU 核数。

9）异常重试。

mapreduce.map.maxattempts 参数主要用于指定每个 MapTask 的最大失败重试次数，一旦重试次数超过该参数的值，就认为 MapTask 运行失败。默认值为 4，根据机器性能，可以适当增大该参数的值。

10）mapreduce.reduce.shuffle.parallelcopies。

mapreduce.reduce.shuffle.parallelcopies 参数主要用于指定每个 ReduceTask 在 MapTask 结果数据中拉取数据时的并行数，默认值为 5，可以将该值增大到 10。

11）mapreduce.reduce.shuffle.input.buffer.percent。

mapreduce.reduce.shuffle.input.buffer.percent 参数主要用于指定 reduce 阶段 Buffer 大小占 ReduceTask 可用内存空间的比例，默认值为 0.7，可以将该值增大到 0.8，用于减少溢写文件数量。

12）mapreduce.reduce.shuffle.merge.percent。

mapreduce.reduce.shuffle.merge.percent 参数主要用于指定 reduce 阶段 Buffer 中的数据占 ReduceTask 可用内存空间的比例达到多少开始写入磁盘，默认值为 0.66，可以将该值增大到 0.75，用于减少溢写文件数量。

13）mapreduce.reduce.memory.mb。

mapreduce.reduce.memory.mb 参数主要用于指定默认的 ReduceTask 内存空间上限，默认为 1024MB，可以根据每 128MB 数据对应 1GB 内存空间的原则，适当将内存空间扩展到 4~6GB。

14）mapreduce.reduce.java.opts。

mapreduce.reduce.java.opts 参数主要用于控制 ReduceTask 堆内存的大小。如果程序在运行过程中报出错误 java.lang.OutOfMemoryError，那么在排查程序没有编写失误的前提下，应该适当增大该参数的值。

15）mapreduce.reduce.cpu.vcores。

mapreduce.reduce.cpu.vcores 参数主要用于指定 ReduceTask 的 CPU 核数，默认为 1 个，在集群资源允许的前提下，可以适当将其增加到 2~4 个。

16）mapreduce.reduce.maxattempts。

mapreduce.reduce.maxattempts 参数主要用于指定每个 ReduceTask 的最大重试次数，一旦重试次数超过该参数的值，就认为 ReduceTask 运行失败。默认值为 4，可以适当增大该参数的值。

17）mapreduce.job.reduce.slowstart.completedmaps。

在 MapTask 完成的比例达到 mapreduce.job.reduce.slowstart.completedmaps 参数的值后，才会为 ReduceTask 申请资源，默认值为 0.05，可以适当减小该参数的值，以便更早开启 ReduceTask。

18）mapreduce.task.timeout。

如果一个 Task 在一定的时间内没有任何进入，即不会读取新的数据，也没有输出数据，则认为该 Task

处于 Block 状态，可能是卡住了。为了防止因为用户程序永远卡住不退出，通过 mapreduce.task.timeout 参数强制设置了一个超时时间，单位为毫秒，默认值为 600 000（10 分钟）。如果你的程序对每条输入数据的处理时间过长，那么建议将该参数的值调大，从而避免正常运行的程序被杀死。

19）尽量在 MapTask 阶段完成所有数据计算工作，避免使用 ReduceTask。

8.7.3 MapReduce 的数据倾斜

在 MapReduce 程序的运行过程中，经常会出现数据倾斜现象。如图 8-37 所示，MapReduce 任务已经完成了 99%，但是其中两个 ReduceTask 的运行时间过长，拖慢了整体进程，这种现象称为数据倾斜。

```
[2K----------------------------------------------------------------
[2K[31;1mVERTICES: 23/25  [==========================>>-] 99%  0ELAPSED TIME: 5217.72 S
[22;0m[2K---------------------------------------------------------------
[31A[2K---------------------------------------------------------------
[2K[36;1m        VERTICES         MODE        STATUS    TOTAL  COMPLETED  RUNNING  PENDING  FAILED  KILLED
[22;0m[2K---------------------------------------------------------------
[2KMap 6 ..........  container    SUCCEEDED       1          1        0        0       0       0
Map 7 ..........  container    SUCCEEDED      31         31        0        0       0       0
Map 8 ..........  container    SUCCEEDED      31         31        0        0       0       0
Map 9 ..........  container    SUCCEEDED      31         31        0        0       0       0
Map 10 .........  container    SUCCEEDED      31         31        0        0       0       0
Map 11 .........  container    SUCCEEDED      31         31        0        0       0       0
Map 12 .........  container    SUCCEEDED      31         31        0        0       0       0
Map 13 .........  container    SUCCEEDED      31         31        0        0       0       0
Map 14 .........  container    SUCCEEDED      31         31        0        0       0       0
Map 15 .........  container    SUCCEEDED      31         31        0        0       0       0
Map 16 .........  container    SUCCEEDED      31         31        0        0       0       0
Map 17 .........  container    SUCCEEDED      31         31        0        0       0       0
Map 18 .........  container    SUCCEEDED      31         31        0        0       0       0
Map 1 ..........  container    SUCCEEDED      31         31        0        0       0       0
Reducer 5 ......  container    SUCCEEDED      31         31        0        0       0       0
Map 19 .........  container    SUCCEEDED      31         31        0        0       0       0
Map 20 .........  container    SUCCEEDED      31         31        0        0       0       0
Map 21 .........  container    SUCCEEDED      31         31        0        0       0       0
Map 22 .........  container    SUCCEEDED      31         31        0        0       0       0
Map 23 .........  container    SUCCEEDED      31         31        0        0       0       0
Map 24 .........  container    SUCCEEDED      31         31        0        0       0       0
Reducer 2 ......  container    SUCCEEDED      31         31        0        0       0       0
Map 25 .........  container    SUCCEEDED      31         31        0        0       0       0
Reducer 3 ......  container    SUCCEEDED      31         31        0        0       0       0
Reducer 4 ......  container    SUCCEEDED      31         31        0        0       0       0
[2K---------------------------------------------------------------
[2K[31;1mVERTICES: 23/25  [==========================>>-] 99%  0ELAPSED TIME: 5222.73 S
[22;0m[2K---------------------------------------------------------------
[31A[2K---------------------------------------------------------------
```

图 8-37　MapReduce 程序的数据倾斜现象

数据倾斜分为以下两类。
- 数据频率倾斜：某个区域内的数据量远远大于其他区域。例如，在进行词频统计时，以 a~f 开头的单词远远多于其他单词。
- 数据大小倾斜：部分记录的大小远远大于平均值。例如，在进行词频统计时，以 s 开头的单词远远多于其他单词。

针对以上数据倾斜问题，解决方法有以下几种。
- 检查原始数据集，是否由于空值数据过多造成数据倾斜。如果空值无用，则直接过滤掉空值数据。如果要保留空值，则考虑将空值连接随机数，并且配合自定义分区器，对计算结果进行二次聚合处理。
- 如果某个区域内的数据远多于其他区域内的数据，则考虑自定义分区器，对数据进行合理分配。
- 考虑增加 ReduceTask 的数量。
- 重新设计 MapReduce 程序，尽量在 MapTask 中完成数据计算工作。
- 增加 Combiner 组件，将数据在 map 输出阶段进行轻度聚合，避免大量数据流入同一个 ReduceTask。

8.8 Hadoop 的综合调优

为了更好地在 Hadoop 集群上运行程序，应该综合考虑调优策略，本节主要讲解一些综合调优案例。

8.8.1 Hadoop 的小文件优化方法

小文件存储的弊端我们已经讲解过，HDFS 中的每个文件都要在 NameNode 上创建对应的元数据，这样，当小文件比较多时，会产生很多元数据文件，一方面会占用 NameNode 的大量内存空间，另一方面会使寻址索引速度变慢。如果小文件过多，那么在进行 MapReduce 计算时，会生成过多数据切片，需要启动过多的 MapTask。MapTask 处理的数据量小，会导致 MapTask 的处理时间比启动时间还短，白白消耗资源。

针对小文件的问题，我们从以下 4 个方向考虑解决方案。

- 数据源头。在采集数据时，先将小文件或小批数据合成大文件，再将其上传至 HDFS 中。
- 存储方向。将小文件存档放入 HDFS 数据块，能够将多个小文件打包成一个 HAR 文件，从而减少占用的 NameNode 内存空间。
- 使用 CombineTextInputFormat。CombineTextInputFormat 主要用于将多个小文件在切片过程中生成一个单独的数据切片或少量的数据切片。
- 开启 uber 模式，实现 JVM 重用。在默认情况下，每个 Task 都需要启动一个 JVM，用于运行该 Task，如果 Task 计算的数据量很小，则可以让同一个 Job 的多个 Task 运行在一个 JVM 上，不必为每个 Task 都开启一个 JVM。

在以上解决方案中，前 3 种都已经详细讲解过了，下面详细讲解第 4 种解决方案。

（1）在未开启 uber 模式的情况下，在/input 目录下上传多个小文件，并且运行 WordCount 程序。

```
[atguigu@hadoop102 hadoop-3.1.3]$ hadoop jar share/hadoop/mapreduce/hadoop-mapreduce-examples-3.1.3.jar wordcount /input /output2
```

（2）观察控制台中打印的日志。

```
2021-02-14 16:13:50,607 INFO mapreduce.Job: Job job_1613281510851_0002 running in uber mode : false
```

（3）观察 YARN 的 Web 端页面，单击打开任务运行页面，如图 8-38 所示，可以发现，一共使用 5 个 Container。

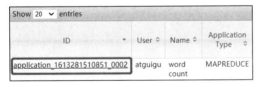 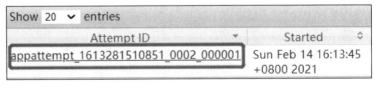

图 8-38　任务运行页面（一）

（4）在 mapred-site.xml 文件中添加以下配置，开启 uber 模式。

```xml
<!-- 开启 uber 模式，默认关闭 -->
<property>
    <name>mapreduce.job.ubertask.enable</name>
    <value>true</value>
</property>
```

```xml
<!-- uber 模式中最大的 MapTask 并行度，只可以将其修改为较小值 -->
<property>
    <name>mapreduce.job.ubertask.maxmaps</name>
    <value>9</value>
</property>
<!-- uber 模式中最大的 ReduceTask 并行度，只可以将其修改为较小值 -->
<property>
    <name>mapreduce.job.ubertask.maxreduces</name>
    <value>1</value>
</property>
<!-- uber 模式中最大的输入数据量，默认使用 dfs.blocksize 参数的值指定，只可以将其修改为较小值 -->
<property>
    <name>mapreduce.job.ubertask.maxbytes</name>
    <value></value>
</property>
```

（5）分发配置文件。

```
[atguigu@hadoop102 hadoop]$ xsync mapred-site.xml
```

（6）再次运行 WordCount 程序。

```
[atguigu@hadoop102 hadoop-3.1.3]$ hadoop jar share/hadoop/mapreduce/hadoop-mapreduce-examples-3.1.3.jar wordcount /input /output2
```

（7）观察控制台中打印的日志，显示已经开启 uber 模式。

```
2021-02-14 16:28:36,198 INFO mapreduce.Job: Job job_1613281510851_0003 running in uber mode : true
```

（8）观察 YARN 的 Web 端页面，单击打开任务运行页面，如图 8-39 所示，可以发现，一共使用 1 个 Container，表示 uber 模式开启成功。

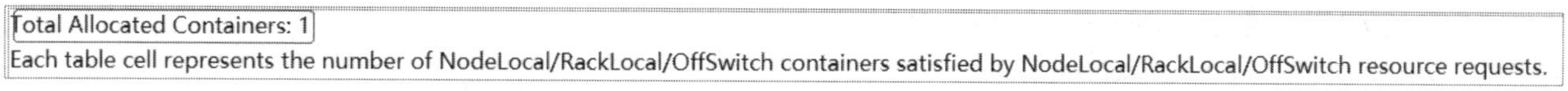

```
Total Allocated Containers: 1
Each table cell represents the number of NodeLocal/RackLocal/OffSwitch containers satisfied by NodeLocal/RackLocal/OffSwitch resource requests.
```

图 8-39 任务运行页面（二）

8.8.2 测试 MapReduce 的计算性能

Hadoop 提供了一个 Sort 程序，用于测试集群的计算性能。如果读者的个人计算机可以分配给虚拟机的磁盘空间不能达到 150GB，则不要运行 Sort 程序。

（1）使用 RandomWriter 程序产生随机数，每个节点都运行 10 个 MapTask，每个 MapTask 都产生大约 1GB 的二进制随机数。

```
[atguigu@hadoop102 mapreduce]$ hadoop jar /opt/module/hadoop-3.1.3/share/hadoop/mapreduce/hadoop-mapreduce-examples-3.1.3.jar randomwriter random-data
```

（2）针对上述程序的数据输出结果运行 Sort 程序。

```
[atguigu@hadoop102 mapreduce]$ hadoop jar /opt/module/hadoop-3.1.3/share/hadoop/mapreduce/hadoop-mapreduce-examples-3.1.3.jar sort random-data sorted-data
```

（3）验证数据排序是否成功。

```
[atguigu@hadoop102 mapreduce]$
hadoop jar /opt/module/hadoop-3.1.3/share/hadoop/mapreduce/hadoop-mapreduce-client-jobclient-3.1.3-tests.jar testmapredsort -sortInput random-data -sortOutput sorted-data
```

用户可以根据程序运行时间了解当前集群的计算性能，在一般情况下，排序程序会在 1 分钟内运行完毕。

8.8.3 企业开发场景案例

1. 需求分析

1) 需求。

使用词频统计程序,统计 1GB 数据中每个单词出现的次数。

2) 服务器现状。

现有服务器 3 台,每台都有 4GB 内存空间、4 核 CPU、4 线程。

3) 具体需求分析。

假设 1GB 的数据不存在小文件问题,数据块大小为 128MB,那么将这 1GB 数据分为 8 个数据块进行存储。在 MapReduce 程序中会产生 8 个 MapTask,需要运行 1 个 ReduceTask,还需要运行 MRAppMaster 进程,即 ApplicationMaster 角色。共计需要运行 10 个进程,为尽量平衡各节点负载,按照(4,3,3)的形式分配进程。

2. HDFS 参数调优

(1) 修改配置文件 hadoop-env.sh,调整 NameNode 和 DataNode 的内存空间配置,代码如下:

```
export HDFS_NAMENODE_OPTS="-Dhadoop.security.logger=INFO,RFAS -Xmx1024m"

export HDFS_DATANODE_OPTS="-Dhadoop.security.logger=ERROR,RFAS -Xmx1024m"
```

(2) 修改配置文件 hdfs-site.xml,增加 NameNode 线程池数量,代码如下:

```xml
<!-- NameNode 处理客户端请求的线程池的线程数量,默认为 10 个 -->
<property>
    <name>dfs.namenode.handler.count</name>
    <value>21</value>
</property>
```

(3) 修改配置文件 core-site.xml,代码如下:

```xml
<!-- 设置垃圾回收时间为 60 分钟 -->
<property>
    <name>fs.trash.interval</name>
    <value>60</value>
</property>
```

(4) 分发以上配置文件,代码如下:

```
[atguigu@hadoop102 hadoop]$ xsync hadoop-env.sh hdfs-site.xml core-site.xml
```

3. MapReduce 参数调优

(1) 修改配置文件 mapred-site.xml,代码如下:

```xml
<!-- 环形缓冲区大小,默认为 100MB -->
<property>
  <name>mapreduce.task.io.sort.mb</name>
  <value>100</value>
</property>

<!-- 环形缓冲区溢写阈值,默认值为 0.8 -->
<property>
  <name>mapreduce.map.sort.spill.percent</name>
  <value>0.80</value>
</property>

<!-- Merge 合并次数,默认为 10 次 -->
<property>
```

```xml
    <name>mapreduce.task.io.sort.factor</name>
    <value>10</value>
</property>

<!--为 MapTask 申请的内存空间，默认为 1GB；如果未配置该参数，或者该参数的值为负数，则通过 mapreduce.map.java.opts 和 mapreduce.job.heap.memory-mb.ratio 计算得到 -->
<property>
    <name>mapreduce.map.memory.mb</name>
    <value>-1</value>
    <description>The amount of memory to request from the scheduler for each  map task. If this is not specified or is non-positive, it is inferred from mapreduce.map.java.opts and mapreduce.job.heap.memory-mb.ratio. If java-opts are also not specified, we set it to 1024.
    </description>
</property>

<!-- MapTask 的 CPU 核数，默认为 1 个 -->
<property>
    <name>mapreduce.map.cpu.vcores</name>
    <value>1</value>
</property>

<!-- MapTask 失败重试次数，默认为 4 次 -->
<property>
    <name>mapreduce.map.maxattempts</name>
    <value>4</value>
</property>

<!-- ReduceTask 在 MapTask 结果数据中拉取数据时的并行数，默认值为 5 -->
<property>
    <name>mapreduce.reduce.shuffle.parallelcopies</name>
    <value>5</value>
</property>

<!-- Buffer 大小占 ReduceTask 可用内存空间的比例，默认值为 0.7 -->
<property>
    <name>mapreduce.reduce.shuffle.input.buffer.percent</name>
    <value>0.70</value>
</property>

<!-- Buffer 中的数据占 ReduceTask 可用内存空间的比例达到多少开始写入磁盘，默认值为 0.66 -->
<property>
    <name>mapreduce.reduce.shuffle.merge.percent</name>
    <value>0.66</value>
</property>

<!--为 ReduceTask 申请的内存空间，默认为 1GB；如果未配置该参数，或者该参数的值为负数，则通过 mapreduce.map.java.opts 和 mapreduce.job.heap.memory-mb.ratio 计算得到 -->
<property>
    <name>mapreduce.reduce.memory.mb</name>
    <value>-1</value>
```

```xml
    <description>The amount of memory to request from the scheduler for each    reduce task.
If this is not specified or is non-positive, it is inferred
     from mapreduce.reduce.java.opts and mapreduce.job.heap.memory-mb.ratio.
     If java-opts are also not specified, we set it to 1024.
  </description>
</property>

<!-- ReduceTask 的 CPU 核数，默认为 1 个 -->
<property>
  <name>mapreduce.reduce.cpu.vcores</name>
  <value>2</value>
</property>

<!-- ReduceTask 失败重试次数，默认为 4 次 -->
<property>
  <name>mapreduce.reduce.maxattempts</name>
  <value>4</value>
</property>

<!-- 当 MapTask 完成的比例达到该值后才会为 ReduceTask 申请资源，默认值为 0.05 -->
<property>
  <name>mapreduce.job.reduce.slowstart.completedmaps</name>
  <value>0.05</value>
</property>

<!-- 如果程序在规定的默认 10 分钟内没有读取数据，则会强制超时退出 -->
<property>
  <name>mapreduce.task.timeout</name>
  <value>600000</value>
</property>
```

（2）分发配置文件，代码如下：

```
[atguigu@hadoop102 hadoop]$ xsync mapred-site.xml
```

4．YARN 参数调优

（1）修改配置文件 yarn-site.xml，代码如下：

```xml
<!-- 选择调度器，默认容量 -->
<property>
    <description>The class to use as the resource scheduler.</description>
    <name>yarn.resourcemanager.scheduler.class</name>
    <value>org.apache.hadoop.yarn.server.resourcemanager.scheduler.capacity.CapacityScheduler</value>
</property>

<!-- ResourceManager 处理调度器请求的线程数量，默认值为 50；如果提交的任务数大于 50，则可以增加该值，但是不能超过 12（3 台 × 4 线程 = 12 线程），去除其他应用程序，实际不能超过 8 -->
<property>
    <description>Number of threads to handle scheduler interface.</description>
    <name>yarn.resourcemanager.scheduler.client.thread-count</name>
    <value>8</value>
</property>
```

```xml
<!-- 是否让 YARN 自动检测硬件进行配置，默认值为 false，如果该节点上有其他应用程序，则建议手动配置；如果该节点上没有其他应用程序，则可以采用自动配置 -->
<property>
    <description>Enable auto-detection of node capabilities such as
    memory and CPU.
    </description>
    <name>yarn.nodemanager.resource.detect-hardware-capabilities</name>
    <value>false</value>
</property>

<!-- 是否将虚拟核数作为 CPU 核数，默认值为 false，采用物理 CPU 核数 -->
<property>
    <description>Flag to determine if logical processors(such as
    hyperthreads) should be counted as cores. Only applicable on Linux
    when yarn.nodemanager.resource.cpu-vcores is set to -1 and
    yarn.nodemanager.resource.detect-hardware-capabilities is true.
    </description>
    <name>yarn.nodemanager.resource.count-logical-processors-as-cores</name>
    <value>false</value>
</property>

<!-- 虚拟核数和物理核数的乘数，默认值为 1.0 -->
<property>
    <description>Multiplier to determine how to convert phyiscal cores to
    vcores. This value is used if yarn.nodemanager.resource.cpu-vcores
    is set to -1(which implies auto-calculate vcores) and
    yarn.nodemanager.resource.detect-hardware-capabilities is set to true. The number of
vcores will be calculated as number of CPUs * multiplier.
    </description>
    <name>yarn.nodemanager.resource.pcores-vcores-multiplier</name>
    <value>1.0</value>
</property>

<!-- NodeManager 使用的内存空间，默认为 8GB，将其修改为 4GB -->
<property>
    <description>Amount of physical memory, in MB, that can be allocated
    for containers. If set to -1 and
    yarn.nodemanager.resource.detect-hardware-capabilities is true, it is
    automatically calculated(in case of Windows and Linux).
    In other cases, the default is 8192MB.
    </description>
    <name>yarn.nodemanager.resource.memory-mb</name>
    <value>4096</value>
</property>

<!-- NodeManager 的 CPU 核数，当不按照硬件环境自动设置时，默认为 8 个，将其修改为 4 个 -->
<property>
    <description>Number of vcores that can be allocated
    for containers. This is used by the RM scheduler when allocating
    resources for containers. This is not used to limit the number of
    CPUs used by YARN containers. If it is set to -1 and
```

```xml
    yarn.nodemanager.resource.detect-hardware-capabilities is true, it is
    automatically determined from the hardware in case of Windows and Linux.
    In other cases, number of vcores is 8 by default.</description>
    <name>yarn.nodemanager.resource.cpu-vcores</name>
    <value>4</value>
</property>

<!-- 容器最小内存空间，默认为 1GB -->
<property>
    <description>The minimum allocation for every container request at the RM in MBs.
Memory requests lower than this will be set to the value of this property. Additionally,
a node manager that is configured to have less memory than this value will be shut down
by the resource manager.
    </description>
    <name>yarn.scheduler.minimum-allocation-mb</name>
    <value>1024</value>
</property>

<!-- 容器最大内存空间，默认为 8GB，将其修改为 2GB -->
<property>
    <description>The maximum allocation for every container request at the RM in MBs.
Memory requests higher than this will throw an InvalidResourceRequestException.
    </description>
    <name>yarn.scheduler.maximum-allocation-mb</name>
    <value>2048</value>
</property>

<!-- 容器最小 CPU 核数，默认为 1 个 -->
<property>
    <description>The minimum allocation for every container request at the RM in terms of
virtual CPU cores. Requests lower than this will be set to the value of this property.
Additionally, a node manager that is configured to have fewer virtual cores than this
value will be shut down by the resource manager.
    </description>
    <name>yarn.scheduler.minimum-allocation-vcores</name>
    <value>1</value>
</property>

<!-- 容器最大 CPU 核数，默认为 4 个，将其修改为 2 个 -->
<property>
    <description>The maximum allocation for every container request at the RM in terms of
virtual CPU cores. Requests higher than this will throw an
    InvalidResourceRequestException.</description>
    <name>yarn.scheduler.maximum-allocation-vcores</name>
    <value>2</value>
</property>

<!-- 虚拟内存空间检查限制，默认为开启，将其修改为关闭 -->
<property>
    <description>Whether virtual memory limits will be enforced for
    containers.</description>
```

```xml
    <name>yarn.nodemanager.vmem-check-enabled</name>
    <value>false</value>
</property>

<!-- 设置虚拟内存空间与物理内存空间的比值，默认值为 2.1 -->
<property>
    <description>Ratio between virtual memory to physical memory when setting  memory
limits for containers. Container allocations are expressed in terms of physical memory,
and virtual memory usage is allowed to exceed this allocation by this ratio.
    </description>
    <name>yarn.nodemanager.vmem-pmem-ratio</name>
    <value>2.1</value>
</property>
```

（2）分发配置文件，代码如下：

```
[atguigu@hadoop102 hadoop]$ xsync yarn-site.xml
```

5. 执行程序

（1）重启 YARN 集群，代码如下：

```
[atguigu@hadoop102 hadoop-3.1.3]$ sbin/stop-yarn.sh
[atguigu@hadoop103 hadoop-3.1.3]$ sbin/start-yarn.sh
```

（2）执行 WordCount 程序，代码如下：

```
[atguigu@hadoop102 hadoop-3.1.3]$ hadoop jar share/hadoop/mapreduce/hadoop-mapreduce-examples-3.1.3.jar wordcount /input /output
```

（3）查看 YARN 的任务运行页面。

8.9 本章总结

本章从不同层面讲解了在使用 Hadoop 的过程中应该采取哪些措施进行调优，而在实际的生产环境中，调优措施并不局限于本章讲解的内容。在实际应用 Hadoop 时，如果遇到 HDFS 存储故障、MapReduce 运行效率低等问题，那么用户应该结合 Hadoop 的底层运行机制，灵活运用调优措施。本章讲解的内容为 Hadoop 的生产环境调优提供了广泛的解决思路，如果读者可以灵活掌握现有的调优策略，那么在遇到实际问题时，即可快速打开思路，找到最优解决方法。

第9章 源码解析

Hadoop 是大数据中的基础框架，为众多大数据处理工具提供了海量数据存储及资源调度管理服务。但是在使用大规模的 Hadoop 集群时，可能会遇到以下问题。
- 随着数据存储量越来越大，元数据会越来越多，NameNode 占用的内存空间也会越来越大，启动和维护都会变得异常困难。
- 如何保证规模庞大的 HDFS 集群的高可用？
- 在 NameNode 中发生长时间的 GC，导致 NameNode 进程退出，应该如何避免？
- 如何优化 DataNode 的锁粒度，使其更加高效？

以上问题都需要 Hadoop 的使用者阅读源码，甚至修改 Hadoop 源码才能解决。所以，阅读源码虽然辛苦，但也是我们必须要做的一件事。

9.1 RPC 通信原理

RPC（Remote Procedure Call，远程过程调用）协议是一种通过网络从远程计算机程序中请求服务，而不需要了解底层网络技术的协议。也就是说，客户端在不知道调用细节的情况下，调用存储于远程计算机中的某个过程或方法，就像调用本地应用程序中的过程或方法一样。

一个完整的 RPC 架构中包含 4 个核心组件，分别是客户端（Client）、客户端存根（Client Stub）、服务器端（Server）及服务器端存根（Server Stub）。
- 客户端：服务的调用器端。
- 客户端存根：存储服务器端的地址消息，将客户端的请求参数打包成网络消息，并且通过网络远程发送给服务方。
- 服务器端：真正的服务提供者。
- 服务器端存根：接收客户端发送过来的消息，将消息解包，并且调用本地的方法。

RPC 协议的调用过程如图 9-1 所示。

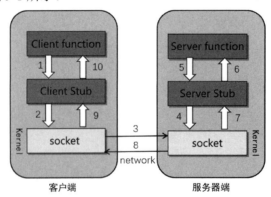

图 9-1 RPC 协议的调用过程

（1）客户端使用本地调用方式（以接口的方式）调用服务。

（2）客户端存根在接收到调用请求后，将方法、参数等组装成能够进行网络传输的消息（将消息对象序列化为二进制数）。

（3）客户端通过 socket 将消息发送给服务器端。

（4）服务器端存根在收到消息后对其进行解码（将消息对象反序列化）。

（5）服务器端存根根据解码结果调用本地的服务。

（6）执行本地服务并将结果返回给服务器端存根。

（7）服务器端存根将返回结果打包成消息（将结果消息对象序列化）。

（8）服务器端通过 socket 将消息发送给客户端。

（9）客户端存根在收到结果消息后对其进行解码（将结果消息对象反序列化）。

（10）客户端得到最终结果。

RPC 协议的目标是将步骤（2）、（3）、（4）、（7）、（8）和（9）封装起来。

后面我们会对 Hadoop 的相关源码进行分析，很多场景都会用到 RPC 协议。HDFS、YARN、MapReduce 的各个组件本质上都是进程，它们之间的通信（如 NameNode 与 DataNode 之间的通信、ResourceManager 与 NodeManager 之间的通信、ResourceManager 与 MRAppMaster 之间的通信）基本都是通过 RPC 协议完成的。

下面通过一段简单的测试代码讲解 RPC 协议的使用方法。测试代码由接口协议、服务器端与客户端组成，如图 9-2 所示。接口协议是指提前约定的一个通信规范，客户端与服务器端之间的通信就是通过接口协议完成的。

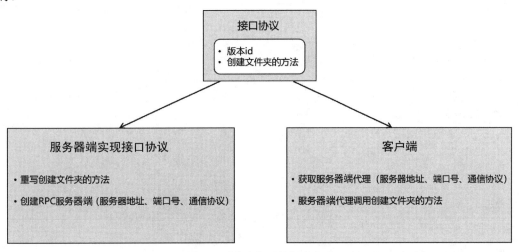

图 9-2　测试代码的组成部分

编写测试代码需要引入 hadoop-client 依赖，在 pom.xml 文件中写入以下代码。

```
<dependencies>
    <dependency>
        <groupId>org.apache.hadoop</groupId>
        <artifactId>hadoop-client</artifactId>
        <version>3.1.3</version>
    </dependency>
</dependencies>
```

首先创建 RPC 接口协议 RPCProtocol，其中必须包含一个与版本有关的属性，并且约定一些通信方法，如创建目录的方法。

```
package com.atguigu.rpc;
```

```java
public interface RPCProtocol {
    // 确定协议版本，自定义版本id
    long versionID = 666;

    // 约定通信方法
    void mkdirs(String path);
}
```

创建 RPC 服务器端，需要实现 RPCProtocol 接口，并且实现约定的通信方法。

```java
package com.atguigu.rpc;

import org.apache.hadoop.conf.Configuration;
import org.apache.hadoop.ipc.RPC;
import org.apache.hadoop.ipc.Server;

import java.io.IOException;

public class RPCServer implements RPCProtocol {
    @Override
    public void mkdirs(String path) {
        System.out.println("服务器端：创建路径" + path);
    }
```

在 main() 方法中，先以 new RPC.Builder() 的形式创建 RPC 服务器端对象，需要传入配置对象、服务器端所在主机的 IP、端口号、RPC 接口协议文件类型、绑定实现类对象等，再通过 RPC 服务器端对象启动 RPC 服务器端。

```java
    public static void main(String[] args) throws IOException {
        // 创建RPC服务器端对象
        Server server = new RPC.Builder(new Configuration())
                .setBindAddress("localhost")
                .setPort(8888)
                .setProtocol(RPCProtocol.class)
                .setInstance(new RPCServer())
                .build();
        //启动RPC服务器端
        server.start();
    }
}
```

运行服务器端代码，等待客户端连接。

创建 RPC 客户端，首先获取 RPC 代理，然后需要提供接口协议类型、接口协议版本 id、用于连接的 InetSocketAddress 对象与配置对象，最后需要调用接口协议中约定的方法进行测试。

```java
package com.atguigu.rpc;

import org.apache.hadoop.conf.Configuration;
import org.apache.hadoop.ipc.RPC;

import java.io.IOException;
import java.net.InetSocketAddress;

public class RPCClient {
    public static void main(String[] args) throws IOException {
        //创建RPC客户端对象，并且连接RPC服务器端
```

```
    RPCProtocol client = RPC.getProxy(
        RPCProtocol.class,
        RPCProtocol.versionID,
        new InetSocketAddress("localhost", 8888),
        new Configuration()
    );

    //客户端调用约定的方法进行测试
    client.mkdirs("/atguigu");
  }
}
```

在运行客户端代码后，到服务器端查看，可以发现 mkdirs() 方法被调用的结果。

9.2 NameNode 启动源码解析

前面讲解过 NameNode 的工作机制，本节将深入解析 NameNode 的启动源码。NameNode 的启动过程主要包括以下几步。

（1）启动 9870 端口服务。
（2）加载镜像文件和编辑日志文件。
（3）初始化 RPC 服务器端。
（4）检查资源。
（5）检查心跳信息并进行超时判断。
（6）退出安全模式。

9.2.1 查看源码的准备工作

为了查看 NameNode 的启动源码，需要在工程中添加 hadoop-client、hadoop-hdfs 与 hadoop-hdfs-client 的相关依赖，版本统一使用 3.1.3。

要启动 NameNode，需要先创建 NameNode 对象，再使用 NameNode 对象调用相关方法，从而进行业务处理。NameNode 的启动流程如图 9-3 所示。

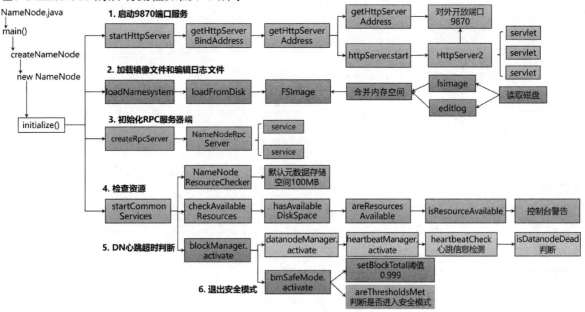

图 9-3　NameNode 的启动流程

根据图 9-3 可知，重要的源码都在 initialize()方法中，为了查看各个流程的源码，首先需要进入 initialize() 方法。进入 initialize()方法的流程如下：先找到 NameNode 类文件，再在其中找到 main()方法，主要的创建与启动流程都是在 main()方法中实现的。main()方法中包含创建 NameNode 对象的相关方法，进入相关方法，找到实例化 NameNode 对象的位置，即可进入 NameNode 类内部，具体流程如下。

（1）按 Ctrl+N 快捷键，全局查找 NameNode 类。

（2）按 Ctrl+F 快捷键，在弹出的搜索框中输入"main"，查找对应的 main()方法，该方法为启动 NameNode 的主程序入口。main()方法中的代码如下：

```java
public static void main(String argv[]) throws Exception {
    if (DFSUtil.parseHelpArgument(argv, NameNode.USAGE, System.out, true)) {
        System.exit(0);
    }

    try {
        StringUtils.startupShutdownMessage(NameNode.class, argv, LOG);
        NameNode namenode = createNameNode(argv, null); //创建 NameNode 对象
        if (namenode != null) {
            namenode.join();
        }
    } catch (Throwable e) {
        LOG.error("Failed to start namenode.", e);
        terminate(1, e);
    }
}
```

（3）按住 Ctrl 键，单击代码"NameNode namenode = createNameNode(argv, null);"中的"createName-Node"，进入该方法内部，关键代码如下。该方法主要用于创建 NameNode 对象。

```java
public static NameNode createNameNode(String argv[], Configuration conf) throws IOException {
    ...
    switch (startOpt) {
    ...
    default:
        DefaultMetricsSystem.initialize("NameNode");
        return new NameNode(conf);
    }
}
```

（4）按住 Ctrl 键，单击代码"return new NameNode(conf);"中的"NameNode"，进入该类内部，代码如下：

```java
public NameNode(Configuration conf) throws IOException {
    this(conf, NamenodeRole.NAMENODE);
}
```

（5）按住 Ctrl 键，单击代码"this(conf, NamenodeRole.NAMENODE);"中的"this"，进入该类内部，关键代码如下：

```java
protected NameNode(Configuration conf, NamenodeRole role) throws IOException {
    ...
    try {
        initializeGenericKeys(conf, nsId, namenodeId);
        initialize(getConf());
        ...
    }
```

```
    ...
}
```

（6）此时可以在 NameNode 类中看到 initialize()方法，NameNode 启动的重要流程源码都在其中。按住 Ctrl 键，单击"initialize"，进入该方法内部，代码如下：

```
protected void initialize(Configuration conf) throws IOException {
    if (conf.get(HADOOP_USER_GROUP_METRICS_PERCENTILES_INTERVALS) == null) {
        String intervals = conf.get(DFS_METRICS_PERCENTILES_INTERVALS_KEY);
        if (intervals != null) {
            conf.set(HADOOP_USER_GROUP_METRICS_PERCENTILES_INTERVALS, intervals);
        }
    }

    UserGroupInformation.setConfiguration(conf);
    loginAsNameNodeUser(conf);

    NameNode.initMetrics(conf, this.getRole());
    StartupProgressMetrics.register(startupProgress);

    pauseMonitor = new JvmPauseMonitor();
    pauseMonitor.init(conf);
    pauseMonitor.start();
    metrics.getJvmMetrics().setPauseMonitor(pauseMonitor);

    if (NamenodeRole.NAMENODE == role) {
        startHttpServer(conf);
    }

    loadNamesystem(conf);
    startAliasMapServerIfNecessary(conf);

    rpcServer = createRpcServer(conf);

    initReconfigurableBackoffKey();

    if (clientNamenodeAddress == null) {
        // This is expected for MiniDFSCluster. Set it now using
        // the RPC server's bind address.
        clientNamenodeAddress = NetUtils.getHostPortString(getNameNodeAddress());
        LOG.info("Clients are to use " + clientNamenodeAddress + " to access" + " this namenode/service.");
    }
    if (NamenodeRole.NAMENODE == role) {
      httpServer.setNameNodeAddress(getNameNodeAddress());
      httpServer.setFSImage(getFSImage());
    }

    startCommonServices(conf);
    startMetricsLogger(conf);
}
```

9.2.2 启动 9870 端口服务

在配置 Hadoop 集群时，在 hdfs-site.xml 配置文件中，设置 dfs.namenode.http-address 属性的值为 hadoop102:9870，该配置项主要用于设置 NameNode 的 Web 端访问地址，使用的端口号为 9870。在启动 NameNode 前，需要先启动 9870 端口所对应的服务。在 initialize()方法内部调用 startHttpServer()方法，用于启动 Web 端；访问 NameNode 的服务。具体流程如图 9-4 所示。

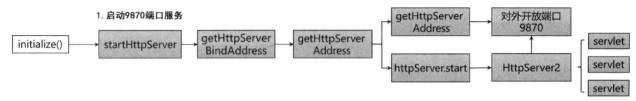

图 9-4　启动 9870 端口服务的具体流程

（1）查看 initialize()方法，关键代码如下：

```
protected void initialize(Configuration conf) throws IOException {
    ...
    if (NamenodeRole.NAMENODE == role) {
        startHttpServer(conf);
    }

    loadNamesystem(conf);
    ...

    rpcServer = createRpcServer(conf);

    ...

    startCommonServices(conf);
    ...
}
```

（2）按住 Ctrl 键，单击"startHttpServer"，进入该方法内部，代码如下：

```
private void startHttpServer(final Configuration conf) throws IOException {
    httpServer = new NameNodeHttpServer(conf, this, getHttpServerBindAddress(conf));
    httpServer.start();
    httpServer.setStartupProgress(startupProgress);
}
```

（3）在 startHttpServer()方法内部，通过代码 "new NameNodeHttpServer(conf, this, getHttpServerBindAddress(conf))" 创建一个 httpServer 对象，即 Web 端访问 NameNode 服务的对象。构造器中的 getHttpServerBindAddress(conf)方法主要用于获取服务绑定的 IP 地址与端口号 Port。按住 Ctrl 键，依次单击 "getHttpServerBindAddress" → "getHttpServerAddress" → "getHttpAddress"，进入 getHttpAddress()方法内部，代码如下：

```
public static InetSocketAddress getHttpAddress(Configuration conf) {
    return NetUtils.createSocketAddr(
        conf.getTrimmed(DFS_NAMENODE_HTTP_ADDRESS_KEY, DFS_NAMENODE_HTTP_ADDRESS_DEFAULT));
}
```

在 getHttpAddress()方法中，关键代码为 "conf.getTrimmed(DFS_NAMENODE_HTTP_ADDRESS_KEY, DFS_NAMENODE_HTTP_ADDRESS_DEFAULT))"。单击第一个参数 DFS_NAMENODE_HTTP_ADDRESS_KEY，查看该参数的相关代码，可以发现，该参数的最终值为 dfs.namenode.http-address，也就是说，默认获

取配置中的 dfs.namenode.http-address 属性值,并且将其作为 NameNode 的 IP 地址与端口号 Port。如果没有配置该属性值,那么使用第二个参数绑定的数据。单击查看第二个参数 DFS_NAMENODE_HTTP_ADDRESS_DEFAULT 的相关代码,可以发现绑定的 IP 地址默认使用 0.0.0.0,端口号为 DFS_NAMENODE_HTTP_PORT_DEFAULT,单击查看端口号,可以发现该值为 9870。

(4)返回 startHttpServer()方法内部,代码如下。按住 Ctrl 键,单击代码"httpServer.start();"中的"start",进入该方法内部。

```
private void startHttpServer(Configuration conf) throws IOException {
    httpServer = new NameNodeHttpServer(conf, this, this.getHttpServerBindAddress(conf));
    httpServer.start();
    httpServer.setStartupProgress(startupProgress);
}
```

(5) start()方法的关键代码如下:

```
void start() throws IOException {
    ...
    HttpServer2.Builder builder = DFSUtil.httpServerTemplateForNNAndJN(conf,
        httpAddr, httpsAddr, "hdfs",
        DFSConfigKeys.DFS_NAMENODE_KERBEROS_INTERNAL_SPNEGO_PRINCIPAL_KEY,
        DFSConfigKeys.DFS_NAMENODE_KEYTAB_FILE_KEY);

    ...

    httpServer = builder.build();

    ...
    setupServlets(httpServer, conf);
    httpServer.start();

    ...
}
```

在上述代码中,首先通过 HttpServer2 构建关联 Web 的服务对象,HttpServer2 是对 httpServer 的再一次封装;然后调用 setupServlets(httpServer, conf)方法,对 httpServer 进行初始化;最后通过 httpServer.start()方法启动 9870 端口服务。

9.2.3 加载镜像文件和编辑日志文件

在讲解 NameNode 的工作机制时曾经介绍过,在启动 NameNode 时,需要将 FsImage 文件加载到内存中,并且将 EditLog 文件中记录的操作在内存中执行一遍,得到最新的 NameNode 元数据。下面看一下源码是如何实现的,具体流程如图 9-5 所示。

图 9-5 加载镜像文件和编辑日志文件的具体流程

(1)查看 initialize()方法,关键代码如下:

```
protected void initialize(Configuration conf) throws IOException {
    ...
    if (NamenodeRole.NAMENODE == role) {
```

```
        startHttpServer(conf);
    }
    
    loadNamesystem(conf);
    ...
    
    rpcServer = createRpcServer(conf);
    
    ...
    
    startCommonServices(conf);
    ...
}
```

（2）在 initialize()方法中，loadNamesystem(conf)方法为加载镜像文件与编辑日志文件的方法。按住 Ctrl 键，单击"loadNamesystem"，进入该方法内部，代码如下：

```
protected void loadNamesystem(Configuration conf) throws IOException {
    this.namesystem = FSNamesystem.loadFromDisk(conf);
}
```

（3）在 loadNamesystem(conf)方法中，loadFromDisk()方法主要用于从磁盘中加载文件。按住 Ctrl 键，单击"loadFromDisk"，进入该方法内部，关键代码如下：

```
static FSNamesystem loadFromDisk(Configuration conf) throws IOException {
    
    ...
    FSImage fsImage = new FSImage(conf,
        FSNamesystem.getNamespaceDirs(conf),
        FSNamesystem.getNamespaceEditsDirs(conf));
    FSNamesystem namesystem = new FSNamesystem(conf, fsImage, false);
    StartupOption startOpt = NameNode.getStartupOption(conf);
    if (startOpt == StartupOption.RECOVER) {
      namesystem.setSafeMode(SafeModeAction.SAFEMODE_ENTER);
    }
    
    ...
    try {
      namesystem.loadFSImage(startOpt);
    }
    ...
    namesystem.getFSDirectory().createReservedStatuses(namesystem.getCTime());
    return namesystem;
}
```

在上述代码中，创建的 FSImage 对象 fsImage 中包含 FsImage 文件与 EditLog 文件，二者会被加载到内存元数据中。

9.2.4 初始化 RPC 服务器端

NameNode 是一个服务器端，需要开启 RPC 服务，允许客户端访问。既然是一个服务器端，就需要在源码中找到 RPC 服务器端的特征，即创建一个 RPC.Builder 对象，提供相关的参数，具体流程如图 9-6 所示。

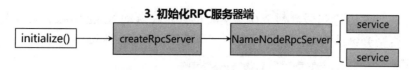

图 9-6　初始化 RPC 服务器端的具体流程

（1）进入 initialize()方法内部，关键代码如下：

```
protected void initialize(Configuration conf) throws IOException {
    ...
    if (NamenodeRole.NAMENODE == role) {
        startHttpServer(conf);
    }

    loadNamesystem(conf);
    ...

    rpcServer = createRpcServer(conf);

    ...

    startCommonServices(conf);
    ...
}
```

（2）其中，createRpcServer()方法为创建 RPC 服务器端的方法。按住 Ctrl 键，单击"createRpcServer"，进入该方法内部，代码如下：

```
protected NameNodeRpcServer createRpcServer(Configuration conf) throws IOException {
    return new NameNodeRpcServer(conf, this);
}
```

（3）按住 Ctrl 键，单击"NameNodeRpcServer"，进入该类内部，这里会返回一个 RPC 服务器端对象。下面深入 NameNodeRpcServer 类，探寻创建 RPC 服务器端的源码，关键代码如下：

```
public NameNodeRpcServer(Configuration conf, NameNode nn) throws IOException {
    ...

clientRpcServer = new RPC.Builder(conf)
    //RPC 接口协议
        .setProtocol(org.apache.hadoop.hdfs.protocolPB.ClientNamenodeProtocolPB.class)
        .setInstance(clientNNPbService)
        .setBindAddress(bindHost)     //RPC 服务器端绑定的 IP
        .setPort(rpcAddr.getPort())  //RPC 服务器端绑定的 Port
        .setNumHandlers(handlerCount)
        .setVerbose(false)
        .setSecretManager(namesystem.getDelegationTokenSecretManager())
        .setAlignmentContext(new GlobalStateIdContext(namesystem))
        .build();

    ...
}
```

9.2.5　检查资源

在启动 NameNode 后，需要保证磁盘存储空间充足，能将后续的操作记录正常写入 EditLog 文件，所以在启动 NameNode 时，需要检测磁盘剩余存储空间是否允许启动 NameNode，如果磁盘剩余存储空间不足，则会抛出异常，否则会继续启动 NameNode。检查资源的具体流程如图 9-7 所示。

图 9-7　检查资源的具体流程

（1）进入 initialize()方法内部，关键代码如下：

```
protected void initialize(Configuration conf) throws IOException {
  ...
  if (NamenodeRole.NAMENODE == role) {
    startHttpServer(conf);
  }

  loadNamesystem(conf);
  ...

  rpcServer = createRpcServer(conf);

  ...

  startCommonServices(conf);
  ...
}
```

（2）按住 Ctrl 键，单击"startCommonServices"，进入该方法内部，关键代码如下：

```
private void startCommonServices(Configuration conf) throws IOException {
  namesystem.startCommonServices(conf, haContext);
  ...
}
```

（3）按住 Ctrl 键，单击代码"namesystem.startCommonServices(conf, haContext);"中的"startCommonServices"，进入该方法内部，资源检查的源码均在该方法中，关键代码如下：

```
void startCommonServices(Configuration conf, HAContext haContext) throws IOException {
  ...

  try {
    nnResourceChecker = new NameNodeResourceChecker(conf);
    checkAvailableResources();
    ...
    // 安全模式
    prog.setTotal(Phase.SAFEMODE,
        STEP_AWAITING_REPORTED_BLOCKS,
        completeBlocksTotal);
    // 启动块服务
    blockManager.activate(conf, completeBlocksTotal);
  }
  ...
}
```

（4）按住 Ctrl 键，单击"NameNodeResourceChecker"，进入该方法内部，关键代码如下。在该方法中会检查是否有足够的磁盘存储元数据，其中，FsImage 文件默认占用 100MB 存储空间，EditLog 文件默认占用 100MB 存储空间。
```
public NameNodeResourceChecker(Configuration conf) throws IOException {
    ...
    duReserved = conf.getLong(DFSConfigKeys.DFS_NAMENODE_DU_RESERVED_KEY,
        DFSConfigKeys.DFS_NAMENODE_DU_RESERVED_DEFAULT);
    ...
}
```

在上述代码中，DFS_NAMENODE_DU_RESERVED_DEFAULT 的值为 1024×1024×100，表示 100MB。

（5）返回 startCommonServices() 方法内部，关键代码如下：

```
void startCommonServices(Configuration conf, HAContext haContext) throws IOException {
    ...
    try {
        nnResourceChecker = new NameNodeResourceChecker(conf);
        checkAvailableResources();
        ...
        // 安全模式
        prog.setTotal(Phase.SAFEMODE,
                STEP_AWAITING_REPORTED_BLOCKS,
                completeBlocksTotal);
        // 启动块服务
        blockManager.activate(conf, completeBlocksTotal);
    }
    ...
}
```

（6）按住 Ctrl 键，单击 "checkAvailableResources"，进入该方法内部，代码如下，在该方法中进行正式的磁盘检测。

```
void checkAvailableResources() {
    long resourceCheckTime = monotonicNow();
    Preconditions.checkState(nnResourceChecker != null,
        "nnResourceChecker not initialized");
    hasResourcesAvailable = nnResourceChecker.hasAvailableDiskSpace();
    resourceCheckTime = monotonicNow() - resourceCheckTime;
    NameNode.getNameNodeMetrics().addResourceCheckTime(resourceCheckTime);
}
```

（7）按住 Ctrl 键，依次单击 "hasAvailableDiskSpace" → "areResourcesAvailable"，进入 areResources-Available() 方法内部，关键代码如下。在该方法中找到 isResourceAvailable() 方法。isResourceAvailable() 方法主要用于判断资源是否充足，它是一个接口中的方法。

```
static boolean areResourcesAvailable(
    ...
    for (CheckableNameNodeResource resource : resources) {
        if (!resource.isRequired()) {
            redundantResourceCount++;
            if (!resource.isResourceAvailable()) {
                disabledRedundantResourceCount++;
            }
        } else {
            requiredResourceCount++;
            if (!resource.isResourceAvailable()) {
                // Short circuit - a required resource is not available.
```

```
        return false;
      }
    }
  }
  ...
}
```

（8）将鼠标指针放到 isResourceAvailable()方法上，按 Ctrl+Alt+B 快捷键，在弹出的快捷菜单中选择 CheckedVolume 命令，可以看到该方法的具体实现，代码如下。可以看到，如果资源不足，则会在控制台中打印相应的信息。

```
@Override
public boolean isResourceAvailable() {
  long availableSpace = df.getAvailable();
  if (LOG.isDebugEnabled()) {
    LOG.debug("Space available on volume '" + volume + "' is "
        + availableSpace);
  }
  if (availableSpace < duReserved) {
    LOG.warn("Space available on volume '" + volume + "' is "
        + availableSpace +
        ", which is below the configured reserved amount " + duReserved);
    return false;
  } else {
    return true;
  }
}
```

9.2.6 检测心跳信息并进行超时判断

NameNode 在启动后会接受 DataNode 的心跳信息，在默认情况下，如果超过 10 分钟 30 秒没有收到 DataNode 的心跳信息，那么 NameNode 会认为 DataNode 下线。下面我们探究一下检测心跳信息并进行超时判断在源码中是如何实现的，具体流程如图 9-8 所示。

图 9-8 检测心跳信息并进行超时判断的具体流程

（1）进入 initialize()方法内部，关键代码如下：

```
protected void initialize(Configuration conf) throws IOException {
  ...
  if (NamenodeRole.NAMENODE == role) {
    startHttpServer(conf);
  }

  loadNamesystem(conf);
  ...

  rpcServer = createRpcServer(conf);
  ...

  startCommonServices(conf);
```

```
   ...
}
```

(2) 按住 Ctrl 键，单击 "startCommonServices"，进入该方法内部，关键代码如下：

```
private void startCommonServices(Configuration conf) throws IOException {
   namesystem.startCommonServices(conf, haContext);
   ...
}
```

(3) 按住 Ctrl 键，单击代码 "namesystem.startCommonServices(conf, haContext)" 中的 "startCommon-Services"，进入该方法内部，心跳超时判断源码在该方法内部，关键代码如下：

```
void startCommonServices(Configuration conf, HAContext haContext) throws IOException {
   ...
   try {
      nnResourceChecker = new NameNodeResourceChecker(conf);
      checkAvailableResources();
      ...
      // 安全模式
      prog.setTotal(Phase.SAFEMODE,
            STEP_AWAITING_REPORTED_BLOCKS,
            completeBlocksTotal);
      // 启动块服务
      blockManager.activate(conf, completeBlocksTotal);
   }
   ...
}
```

(4) 按住 Ctrl 键，单击代码 "blockManager.activate(conf, completeBlocksTotal)" 中的 "activate"，进入该方法内部，代码如下：

```
public void activate(Configuration conf, long blockTotal) {
   pendingReconstruction.start();
   datanodeManager.activate(conf);
   this.redundancyThread.setName("RedundancyMonitor");
   this.redundancyThread.start();
   storageInfoDefragmenterThread.setName("StorageInfoMonitor");
   storageInfoDefragmenterThread.start();
   this.blockReportThread.start();
   mxBeanName = MBeans.register("NameNode", "BlockStats", this);
   bmSafeMode.activate(blockTotal);
}
```

(5) 在 activate() 方法中，按住 Ctrl 键，单击代码 "datanodeManager.activate(conf)" 中的 "activate"，进入该方法内部，该方法是管理 DataNode 与 NameNode 通信的管理者。按住 Ctrl 键，再次单击 "activate"，进入该方法内部，该方法中有一个心跳管理者，代码如下：

```
void activate(final Configuration conf) {
   datanodeAdminManager.activate(conf);
   heartbeatManager.activate();//心跳管理者
}
```

(6) 按住 Ctrl 键，单击代码 "heartbeatManager.activate()" 中的 "activate"，进入该方法内部，代码如下：

```
void activate() {
   heartbeatThread.start();
}
```

（7）此时会启动一个线程，我们需要找到这个线程运行的方法。按 Ctrl+F 快捷键，在弹出的搜索框中输入"run"，查找对应的 run()方法，该方法中的关键代码如下：

```
public void run() {
    while(namesystem.isRunning()) {
        restartHeartbeatStopWatch();
        try {
            final long now = Time.monotonicNow();
            if (lastHeartbeatCheck + heartbeatRecheckInterval < now) {
                heartbeatCheck();
                lastHeartbeatCheck = now;
            }
            ...
        }
        ...
    }
}
```

（8）其中，heartbeatCheck()方法为心跳检测方法。按住 Ctrl 键，单击"heartbeatCheck"，进入该方法内部，关键代码如下：

```
void heartbeatCheck() {
    ...
    while (!allAlive) {
        ...
        synchronized(this) {
            //遍历每个DataNode，判断其是否活着
            for (DatanodeDescriptor d : datanodes) {
                ...
                if (dead == null && dm.isDatanodeDead(d)) {
                    stats.incrExpiredHeartbeats();
                    dead = d;
                }
                ...
            }
            ...
        }
        ...
    }
}
```

（9）上述代码会遍历每个 DataNode，判断其是否活着，其中，isDatanodeDead()方法为判断 DataNode 是否活着的方法。按住 Ctrl 键，单击"isDatanodeDead"，进入该方法内部，代码如下：

```
boolean isDatanodeDead(DatanodeDescriptor node) {
    return (node.getLastUpdateMonotonic() <
        (monotonicNow() - heartbeatExpireInterval));
}
```

判断 DataNode 是否活着的逻辑为，如果上一次获取的时间 node.getLastUpdateMonotonic()小于当前时间 monotonicNow()与 heartbeatExpireInterval 属性值的差，则表示活着，否则表示已经死去。heartbeatExpireInterval 属性值的计算方法如下：

$$heartbeatExpireInterval = 2 \times heartbeatRecheckInterval + 10 \times 1000 \times heartbeatIntervalSeconds$$

其中，heartbeatRecheckInterval 属性的值为配置文件中 dfs.namenode.heartbeat.recheck-interval 配置项的值，如果未配置，则采用默认值 5×60×1000，即 5 分钟；heartbeatIntervalSeconds 属性的值为配置文件中 dfs.heartbeat.interval 配置项的值，如果未配置，则采用默认值 3，即 3 秒。所以在默认情况下，heartbeatExpireInterval 属性表示的时间为 10 分钟 30 秒。

9.2.7 退出安全模式

如果符合标准，则可以退出安全模式，保证集群正常可访问。下面看一下退出安全模式的具体流程，如图 9-9 所示。

图 9-9　退出安全模式的具体流程

（1）查看 initialize() 方法，关键代码如下：

```
protected void initialize(Configuration conf) throws IOException {
    ...
    if (NamenodeRole.NAMENODE == role) {
        startHttpServer(conf);
    }

    loadNamesystem(conf);
    ...

    rpcServer = createRpcServer(conf);

    ...

    startCommonServices(conf);
    ...
}
```

（2）按住 Ctrl 键，单击"startCommonServices"，进入该方法内部，关键代码如下：

```
private void startCommonServices(Configuration conf) throws IOException {
    namesystem.startCommonServices(conf, haContext);
    ...
}
```

（3）按住 Ctrl 键，单击代码"namesystem.startCommonServices(conf, haContext)"中的"startCommonServices"，进入该方法内部，退出安全模式的源码在该方法内部，关键代码如下：

```
void startCommonServices(Configuration conf, HAContext haContext) throws IOException {
    ...
    try {
        nnResourceChecker = new NameNodeResourceChecker(conf);
        checkAvailableResources();
        ...
        // 安全模式
        prog.setTotal(Phase.SAFEMODE,
                STEP_AWAITING_REPORTED_BLOCKS,
```

```
        completeBlocksTotal);
    // 启动块服务
    blockManager.activate(conf, completeBlocksTotal);
  }
  ...
}
```

(4) 按住 Ctrl 键, 单击代码 "blockManager.activate(conf, completeBlocksTotal)" 中的 "activate", 进入该方法内部, 代码如下：

```
public void activate(Configuration conf, long blockTotal) {
    pendingReconstruction.start();
    datanodeManager.activate(conf);
    this.redundancyThread.setName("RedundancyMonitor");
    this.redundancyThread.start();
    storageInfoDefragmenterThread.setName("StorageInfoMonitor");
    storageInfoDefragmenterThread.start();
    this.blockReportThread.start();
    mxBeanName = MBeans.register("NameNode", "BlockStats", this);
    bmSafeMode.activate(blockTotal);
}
```

(5) 按住 Ctrl 键, 单击代码 "bmSafeMode.activate(blockTotal)" 中的 "activate", 进入该方法内部, 查看退出安全模式的源码, 关键代码如下：

```
void activate(long total) {
    ...
    setBlockTotal(total);
    if (areThresholdsMet()) {
        boolean exitResult = leaveSafeMode(false);
        Preconditions.checkState(exitResult, "Failed to leave safe mode.");
    }
    ...
}
```

(6) 其中, leaveSafeMode()方法主要用于退出安全模式, 退出安全模式的条件是 areThresholdsMet()方法返回 true。使用 areThresholdsMet()方法可以验证已注册的数据块是否达到最小副本阈值。最小副本阈值是指文件系统中 99.9%的文件已经达到最小副本数量。最小副本数量可以在配置文件中进行修改, 默认的最小副本数量为 1 个。这个最小副本阈值会通过 setBlockTotal()方法传递给系统, 按住 Ctrl 键, 单击 "setBlockTotal", 进入该方法内部, 代码如下：

```
void setBlockTotal(long total) {
    assert namesystem.hasWriteLock();
    synchronized (this) {
        this.blockTotal = total;
        this.blockThreshold = (long) (total * threshold);
    }
    this.blockReplQueueThreshold = (long) (total * replQueueThreshold);
}
```

在上述代码中, 最小副本阈值通过公式 total × threshold 计算得到, 其中, total 为总量, threshold 为百分比。用户可以通过修改配置文件中的 dfs.namenode.safemode.threshold-pct 配置项, 对百分比进行修改, 如果未配置, 则采用默认值 0.999f。

(7) 回到步骤 (5) 中的代码, 按住 Ctrl 键, 单击 "areThresholdsMet", 进入该方法内部, 代码如下。

使用该方法可以判断 DataNode 和块信息是否达到退出安全模式的标准。

```
private boolean areThresholdsMet() {
    assert namesystem.hasWriteLock();
    int datanodeNum = 0;
    if (datanodeThreshold > 0) {
        datanodeNum = blockManager.getDatanodeManager().getNumLiveDataNodes();
    }
    synchronized (this) {
        //已经正常注册的数据块数 >= 数据块的最小副本阈值 >= 最小可用 DataNode
        return blockSafe >= blockThreshold && datanodeNum >= datanodeThreshold;
    }
}
```

9.3 DataNode 启动源码解析

4.5.4 节讲解过 DataNode 的工作机制，本节将深入解析 DataNode 的启动源码，DataNode 的启动过程主要包括以下几步。

（1）初始化 DataXceiverServer。
（2）初始化 HTTP 服务。
（3）初始化 RPC 服务器端。
（4）DataNode 向 NameNode 注册。
（5）DataNode 向 NameNode 发送心跳信息。

9.3.1 查看源码的准备工作

为了查看 DataNode 的启动源码，需要在工程中添加相关的依赖，具体如下：

```xml
<dependencies>
    <dependency>
        <groupId>org.apache.hadoop</groupId>
        <artifactId>hadoop-client</artifactId>
        <version>3.1.3</version>
    </dependency>
    <dependency>
        <groupId>org.apache.hadoop</groupId>
        <artifactId>hadoop-hdfs</artifactId>
        <version>3.1.3</version>
    </dependency>
    <dependency>
        <groupId>org.apache.hadoop</groupId>
        <artifactId>hadoop-hdfs-client</artifactId>
        <version>3.1.3</version>
        <scope>provided</scope>
    </dependency>
</dependencies>
```

DataNode 的启动流程如图 9-10 所示，可以看到，要启动 DataNode，需要创建 DataNode 对象。在创建 DataNode 对象后需要对该对象进行一系列的初始化操作，即初始化 DataXceiverServer、初始化 HTTP 服务、初始化 RPC 服务器端、DataNode 向 NameNode 注册等操作。

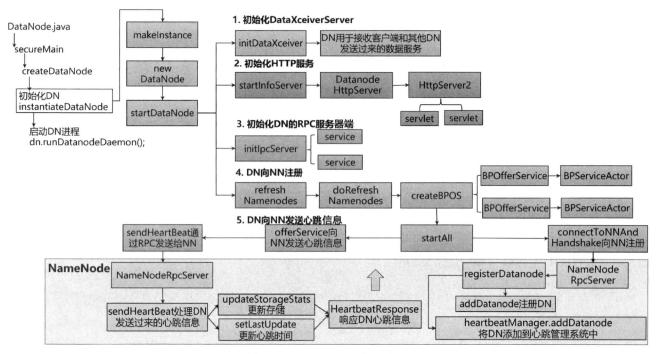

图 9-10 DataNode 的启动流程

DataNode 的主要启动流程都存储于 DataNode 的 startDataNode()方法中，所以我们需要根据特定流程先进入该方法内部，查看 DataNode 的启动源码，具体流程如下。

（1）按 Ctrl+N 快捷键，全局查找 DataNode 类。

（2）按 Ctrl+F 快捷键，在弹出的搜索框中输入"main"，查找对应的 main()方法，该方法中的代码如下：

```
public static void main(String args[]) {
   if (DFSUtil.parseHelpArgument(args, DataNode.USAGE, System.out, true)) {
      System.exit(0);
   }
   secureMain(args, null);
}
```

（3）按住 Ctrl 键，单击"secureMain"，进入该方法内部，关键代码如下：

```
public static void secureMain(String args[], SecureResources resources) {
   int errorCode = 0;
   try {
      StringUtils.startupShutdownMessage(DataNode.class, args, LOG);
      DataNode datanode = createDataNode(args, null, resources);
      ...
   }
   ...
}
```

（4）按住 Ctrl 键，单击"createDataNode"，进入该方法内部，代码如下。createDataNode()方法主要用于创建 DataNode 对象。

```
public static DataNode createDataNode(String args[], Configuration conf, SecureResources
resources) throws IOException {
   DataNode dn = instantiateDataNode(args, conf, resources);
   if (dn != null) {
      dn.runDatanodeDaemon();
   }
```

```
    return dn;
}
```

（5）按住 Ctrl 键，单击"instantiateDataNode"，进入该方法内部，代码如下。instantiateDataNode()方法主要用于初始化一个 DataNode 对象，详细的初始化流程均在该方法内部。在初始化 DataNode 对象后，调用 runDatanodeDaemon()方法运行该 DataNode 对象。

```
public static DataNode instantiateDataNode(String args [], Configuration conf,
SecureResources resources) throws IOException {
    ...
    return makeInstance(dataLocations, conf, resources);
}
```

（6）按住 Ctrl 键，单击"makeInstance"，进入该方法内部，在该方法内部会创建一个 DataNode 对象，关键代码如下：

```
static DataNode makeInstance(Collection<StorageLocation> dataDirs, Configuration conf,
SecureResources resources) throws IOException {
    ...
    return new DataNode(conf, locations, storageLocationChecker, resources);
}
```

（7）按住 Ctrl 键，单击"DataNode"，进入该方法内部，该方法是真正的初始化方法，在该方法内部找到代码"startDataNode(dataDirs, resources)"，按住 Ctrl 键，单击"startDataNode"，进入该方法内部，关键代码如下。所有的初始化操作都在 startDataNode()方法中，本节后续的源码讲解都将围绕该方法展开。

```
void startDataNode(List<StorageLocation> dataDirectories, SecureResources resources)
throws IOException {
    ...
    initDataXceiver();
    startInfoServer();
    ...
    initIpcServer();
    ...
    blockPoolManager.refreshNamenodes(getConf());
    ...
}
```

9.3.2 初始化 DataXceiverServer

DataXceiverServer 服务会启动一个线程，用于接收客户端和其他 DataNode 发送来的数据。初始化 DataXceiverServer 的具体流程如图 9-11 所示。

1. 初始化DataXceiverServer

startDataNode → initDataXceiver → DN用于接收客户端和其他DN发送过来的数据服务

图 9-11 初始化 DataXceiverServer 的具体流程

（1）进入 startDataNode()方法内部，关键代码如下：

```
void startDataNode(List<StorageLocation> dataDirectories, SecureResources resources)
throws IOException {
    ...
    initDataXceiver();
    startInfoServer();
    ...
```

```
    initIpcServer();
    ...
    blockPoolManager.refreshNamenodes(getConf());
    ...
}
```

（2）按住 Ctrl 键，单击"initDataXceiver"，进入该方法内部，关键代码如下。其中，代码"new Daemon(...)"主要用于启动一个线程，从而启动 DataXceiverServer 服务。

```
private void initDataXceiver() throws IOException {
    ...
    if (getConf().getBoolean(
        ...
        if (domainPeerServer != null) {
            this.localDataXceiverServer = new Daemon(threadGroup,
                new DataXceiverServer(domainPeerServer, getConf(), this));
            ...
        }
    }
    ...
}
```

9.3.3 初始化 HTTP 服务

DataNode 也有类似于 NameNode 的 Web 端页面，只不过用得比较少，也需要在启动时开启 HTTP 服务。初始化 HTTP 服务的具体流程如图 9-12 所示。

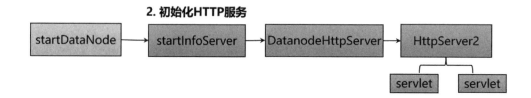

图 9-12 初始化 HTTP 服务的具体流程

（1）进入 startDataNode()方法内部，关键代码如下：

```
void startDataNode(List<StorageLocation> dataDirectories, SecureResources resources)
throws IOException {
    ...
    initDataXceiver();
    startInfoServer();
    ...
    initIpcServer();
    ...
    blockPoolManager.refreshNamenodes(getConf());
    ...
}
```

（2）按住 Ctrl 键，单击"startInfoServer"，进入该方法内部，关键代码如下。其中，代码"new DatanodeHttpServer(...)"主要用于创建一个 HTTP 服务对象。

```
private void startInfoServer() throws IOException {
  ...
  httpServer = new DatanodeHttpServer(getConf(), this, httpServerChannel);
  httpServer.start();
  ...
}
```

9.3.4 初始化 RPC 服务器端

DataNode 是一个服务器端，需要开启 RPC 服务，允许客户端访问。既然是一个服务器端，就需要在源码中找到 RPC 服务器端的特征，即创建一个 RPC.Builder 对象，提供相关的参数，具体流程如图 9-13 所示。

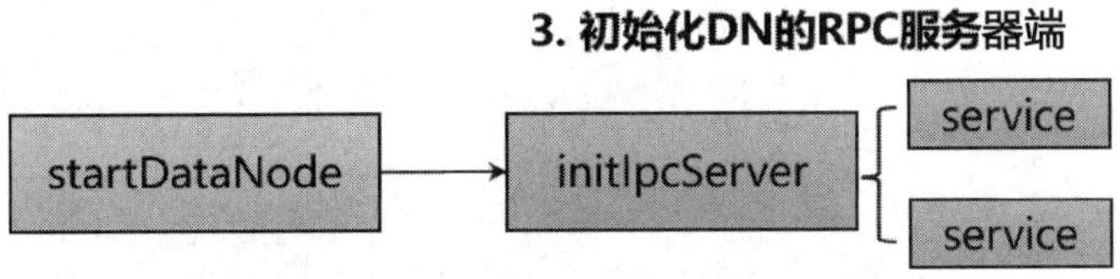

3. 初始化DN的RPC服务器端

startDataNode → initIpcServer ⎱ service
                                ⎰ service

图 9-13　初始化 RPC 服务器端的具体流程

（1）查看 startDataNode()方法，关键代码如下：

```
void startDataNode(List<StorageLocation> dataDirectories, SecureResources resources)
throws IOException {
  ...
  initDataXceiver();
  startInfoServer();

  initIpcServer();

  ...
  blockPoolManager.refreshNamenodes(getConf());

  ...
}
```

（2）按住 Ctrl 键，单击"initIpcServer"，进入该方法内部，在该方法中会创建 RPC 服务器端对象，关键代码如下：

```
private void initIpcServer() throws IOException {
  ...
  ipcServer = new RPC.Builder(getConf())
    .setProtocol(ClientDatanodeProtocolPB.class)
    .setInstance(service)
    .setBindAddress(ipcAddr.getHostName())
    .setPort(ipcAddr.getPort())
    .setNumHandlers(
        getConf().getInt(DFS_DATANODE_HANDLER_COUNT_KEY,
        DFS_DATANODE_HANDLER_COUNT_DEFAULT)).setVerbose(false)
    .setSecretManager(blockPoolTokenSecretManager).build();
```

```
...
}
```

9.3.5　DataNode 向 NameNode 注册

DataNode 在启动时需要向 NameNode 注册信息，让 NameNode 知道自己的存在，并且在双方之间建立通信渠道，具体流程如图 9-14 所示。

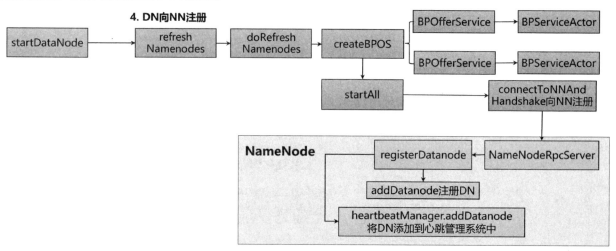

图 9-14　DataNode 向 NameNode 注册的具体流程

（1）查看 startDataNode()方法，关键代码如下：
```
void startDataNode(List<StorageLocation> dataDirectories, SecureResources resources)
throws IOException {
  ...
  initDataXceiver();
  startInfoServer();
  ...
  initIpcServer();

  ...
  blockPoolManager.refreshNamenodes(getConf());

  ...
}
```

（2）按住 Ctrl 键，单击"refreshNamenodes"，进入该方法内部，在该方法中会刷新 NameNode，也就是 NameNode 向 NameNode 注册，关键代码如下：
```
void refreshNamenodes(Configuration conf) throws IOException {
  ...

  synchronized (refreshNamenodesLock) {
    doRefreshNamenodes(newAddressMap, newLifelineAddressMap);
  }
}
```

（3）按住 Ctrl 键，单击"doRefreshNamenodes"，进入该方法内部，进行刷新操作，关键代码如下：
```
private void doRefreshNamenodes(
    Map<String, Map<String, InetSocketAddress>> addrMap,
    Map<String, Map<String, InetSocketAddress>> lifelineAddrMap)
    throws IOException {
```

```
    ...
    synchronized (this) {
        // Step 3. Start new nameservices
        if (!toAdd.isEmpty()) {
            ...
            for (String nsToAdd : toAdd) {
                ...
                BPOfferService bpos = createBPOS(nsToAdd, addrs, lifelineAddrs);
                bpByNameserviceId.put(nsToAdd, bpos);
                offerServices.add(bpos);
            }
        }
        startAll();
    }
    ...
}
```

DataNode 需要和 NameNode 进行通信，在以后的开发过程中，我们会使用高可用的 Hadoop 集群，所以会存在多个 NameNode。DataNode 需要和每个 NameNode 进行通信，所以在代码中使用 for 循环遍历多个 NameNode，通过 createBPOS()方法创建 BPOS 对象，给每个 NameNode 添加一个 BPOS 对象。

（4）按住 Ctrl 键，单击"createBPOS"，进入该方法内部，代码如下：

```
protected BPOfferService createBPOS(
        final String nameserviceId,
        List<InetSocketAddress> nnAddrs,
        List<InetSocketAddress> lifelineNnAddrs) {
    return new BPOfferService(nameserviceId, nnAddrs, lifelineNnAddrs, dn);
}
```

（5）按住 Ctrl 键，单击"BPOfferService"，进入该方法内部，查看如何创建 BPOfferService 对象，关键代码如下。在该方法内部通过 for 循环创建与 NameNode 数量相同的 BPServiceActor 对象。

```
BPOfferService(
        final String nameserviceId,
        List<InetSocketAddress> nnAddrs,
        List<InetSocketAddress> lifelineNnAddrs,
        DataNode dn) {
    ...

    for (int i = 0; i < nnAddrs.size(); ++i) {
      this.bpServices.add(new BPServiceActor(nnAddrs.get(i),
        lifelineNnAddrs.get(i), this));
    }
}
```

（6）回到步骤（3）中的 doRefreshNamenodes()方法内部，在所有的 BPServiceActor 对象创建完毕后，需要将其开启，其中，startAll()方法主要用于开启相关服务。

```
private void doRefreshNamenodes(
        Map<String, Map<String, InetSocketAddress>> addrMap,
        Map<String, Map<String, InetSocketAddress>> lifelineAddrMap)
        throws IOException {
    ...

    synchronized (this) {
```

```
        // Step 3. Start new nameservices
        if (!toAdd.isEmpty()) {
            ...
            for (String nsToAdd : toAdd) {
                ...
                BPOfferService bpos = createBPOS(nsToAdd, addrs, lifelineAddrs);
                bpByNameserviceId.put(nsToAdd, bpos);
                offerServices.add(bpos);
            }
        }
        startAll();
    }
    ...
}
```

（7）按住 Ctrl 键，单击"startAll"，进入该方法内部，关键代码如下。在 for 循环内部启动所有的 BPOS 对象服务。

```
synchronized void startAll() throws IOException {
    try {
        UserGroupInformation.getLoginUser().doAs(new PrivilegedExceptionAction<Object>() {
            @Override
            public Object run() throws Exception {
                for (BPOfferService bpos : offerServices) {
                    bpos.start();
                }
                return null;
            }
        });
    }
    ...
}
```

（8）按住 Ctrl 键，单击"start"，进入 start()方法内部，按照顺序查看 bpos.start()→actor.start()方法，最终看到的代码如下：

```
void start() {
    ...
    bpThread.start();
    ...
}
```

（9）要找到 bpThread.start()方法运行的方法，需要按 Ctrl+F 快捷键，在弹出的搜索框中输入"run"，查找相应的 run()方法，该方法是真正运行的逻辑代码，该方法中的关键代码如下：

```
@Override
public void run() {
    ...
    try {
        while (true) {
            try {
                connectToNNAndHandshake();
                break;
            }
            ...
        }
```

```
            while (shouldRun()) {
                try {
                    offerService();
                }
                ...
            }
            runningState = RunningState.EXITED;
        }
        ...
    }
```

（10）按住 Ctrl 键，单击 "connectToNNAndHandshake"，进入该方法内部，关键代码如下。该方法主要用于与 NameNode 进行通信，bpNamenode 为获取的 RPC 客户端对象。

```
private void connectToNNAndHandshake() throws IOException {
    // 获取 NameNode 的 RPC 客户端对象
    bpNamenode = dn.connectToNN(nnAddr);

    ...
    register(nsInfo);
}
```

（11）按住 Ctrl 键，依次单击 "connectToNN" → "DatanodeProtocolClientSideTranslatorPB" → "createNamenode"，进入 createNamenode()方法内部，查看 RPC 客户端对象与返回的源码，关键代码如下：

```
private static DatanodeProtocolPB createNamenode(
    InetSocketAddress nameNodeAddr, Configuration conf,
    UserGroupInformation ugi) throws IOException {
    return RPC.getProxy(DatanodeProtocolPB.class,
        RPC.getProtocolVersion(DatanodeProtocolPB.class), nameNodeAddr, ugi,
        conf, NetUtils.getSocketFactory(conf, DatanodeProtocolPB.class));
}
```

（12）回到步骤（10）中的 connectToNNAndHandshake()方法内部，其中，register()方法为注册方法。按住 Ctrl 键，依次单击 "register" → "registerDatanode" → "registerDatanode"，发现无法找到 registerDatanode()方法的具体实现。需要注意的是，RPC 客户端主要负责调用 registerDatanode()方法；RPC 服务器端主要负责执行 registerDatanode()方法，也就是在 NameNode 上执行 registerDatanode()方法。

（13）按 Ctrl+N 快捷键，全局查找 NameNodeRpcServer 类。

（14）在 NameNodeRpcServer 类中，按 Ctrl+F 快捷键，查找 registerDatanode()方法，该方法主要用于在 RPC 客户端与 RPC 服务器端之间进行通信，代码如下：

```
public DatanodeRegistration registerDatanode(DatanodeRegistration nodeReg) throws IOException {
    checkNNStartup();
    verifySoftwareVersion(nodeReg);
    namesystem.registerDatanode(nodeReg);
    return nodeReg;
}
```

（15）按住 Ctrl 键，依次单击 "namesystem.registerDatanode" → "blockManager.registerDatanode" → "datanodeManager.registerDatanode"，进入 registerDatanode()方法内部，该方法主要用于将 DataNode 注册到 NameNode 中，并且将其添加到心跳管理系统中，关键代码如下：

```
public void registerDatanode(DatanodeRegistration nodeReg) throws DisallowedDatanodeException,
UnresolvedTopologyException {
    ...
```

```
    try {
        ...
        try {
            ...
            
            // 注册新的 DataNode
            addDatanode(nodeDescr);
            blockManager.getBlockReportLeaseManager().register(nodeDescr);
            
            // 将 DataNode 添加到心跳管理系统中
            heartbeatManager.addDatanode(nodeDescr);
            heartbeatManager.updateDnStat(nodeDescr);
            ...
        }
        ...
    }
    ...
}
```

9.3.6 DataNode 向 NameNode 发送心跳信息

在将 DataNode 注册到 NameNode 中后，二者之间会频繁进行通信，DataNode 会向 NameNode 发送心跳信息，具体流程如图 9-15 所示。

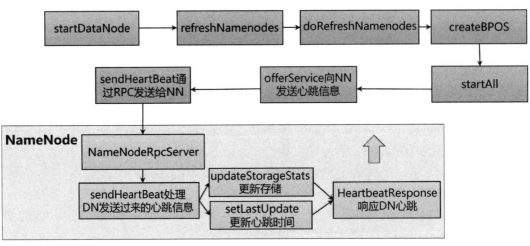

图 9-15　DataNode 向 NameNode 发送心跳信息的具体流程

（1）进入 9.3.5 节中步骤（9）的 bpThread.start()方法运行的 run()方法内部，关键代码如下。在 DataNode 注册成功后，DataNode 需要通过 offerService()方法提供服务。

```
@Override
public void run() {
    ...
    try {
        while (true) {
            try {
                connectToNNAndHandshake();
                break;
            }
```

```
      ...
    }
    while (shouldRun()) {
      try {
        offerService();
      }
      ...
    }
    runningState = RunningState.EXITED;
  }
  ...
}
```

（2）按住 Ctrl 键，依次单击"offerService"→"sendHeartBeat"，进入 sendHeartBeat()方法内部，发现 DataNode 会通过 RPC 客户端给 NameNode 发送信息，关键代码如下：

```
HeartbeatResponse sendHeartBeat(boolean requestBlockReportLease) throws IOException {
  ...
  HeartbeatResponse response = bpNamenode.sendHeartbeat(bpRegistration,
    reports,
    dn.getFSDataset().getCacheCapacity(),
    dn.getFSDataset().getCacheUsed(),
    dn.getXmitsInProgress(),
    dn.getXceiverCount(),
    numFailedVolumes,
    volumeFailureSummary,
    requestBlockReportLease,
    slowPeers,
    slowDisks);
  ...
}
```

其中，bpNamenode 就是 NameNode 对象，也就是说，NameNode 中一定会有一个 sendHeartbeat()方法。

（3）按 Ctrl+N 快捷键，全局查找 NameNodeRpcServer 类。

（4）在 NameNodeRpcServer 类中按 Ctrl+F 快捷键，查找 sendHeartbeat()方法，该方法主要用于在 RPC 客户端与 RPC 服务器端之间发送心跳信息，代码如下：

```
public HeartbeatResponse sendHeartbeat(DatanodeRegistration nodeReg,
    StorageReport[] report, long dnCacheCapacity, long dnCacheUsed,
    int xmitsInProgress, int xceiverCount,
    int failedVolumes, VolumeFailureSummary volumeFailureSummary,
    boolean requestFullBlockReportLease,
    @Nonnull SlowPeerReports slowPeers,
    @Nonnull SlowDiskReports slowDisks) throws IOException {
  checkNNStartup();
  verifyRequest(nodeReg);
  return namesystem.handleHeartbeat(nodeReg, report,
    dnCacheCapacity, dnCacheUsed, xceiverCount, xmitsInProgress,
    failedVolumes, volumeFailureSummary, requestFullBlockReportLease,
    slowPeers, slowDisks);
}
```

（5）按住 Ctrl 键，依次单击"handleHeartbeat"→"handleHeartbeat"→"updateHeartbeat"→"updateHeartbeat"→"updateHeartbeat"→"updateHeartbeatState"，进入 updateHeartbeatState()方法内部，该方法主要用于更新心跳的相关信息，代码如下：

```
void updateHeartbeatState(StorageReport[] reports, long cacheCapacity,
    long cacheUsed, int xceiverCount, int volFailures,
    VolumeFailureSummary volumeFailureSummary) {

//更新存储状态的相关信息
updateStorageStats(reports, cacheCapacity, cacheUsed, xceiverCount,
    volFailures, volumeFailureSummary);

//设置当前时间为上一次更新心跳的时间
setLastUpdate(Time.now());
setLastUpdateMonotonic(Time.monotonicNow());
rollBlocksScheduled(getLastUpdateMonotonic());
}
```

（6）在步骤（4）的 sendHeartbeat()方法内部，handleHeartbeat()方法最后会给 DataNode 返回一个 Heartbeat Response 心跳响应信息。按住 Ctrl 键，单击"handleHeartbeat"，进入该方法内部，关键代码如下：

```
HeartbeatResponse handleHeartbeat(DatanodeRegistration nodeReg,
    StorageReport[] reports, long cacheCapacity, long cacheUsed,
    int xceiverCount, int xmitsInProgress, int failedVolumes,
    VolumeFailureSummary volumeFailureSummary,
    boolean requestFullBlockReportLease,
    @Nonnull SlowPeerReports slowPeers,
    @Nonnull SlowDiskReports slowDisks) throws IOException {
...
try {
    ...
    DatanodeCommand[] cmds = blockManager.getDatanodeManager().handleHeartbeat(
        nodeReg, reports, getBlockPoolId(), cacheCapacity, cacheUsed,
        xceiverCount, maxTransfer, failedVolumes, volumeFailureSummary,
        slowPeers, slowDisks);
    ...
    return new HeartbeatResponse(cmds, haState, rollingUpgradeInfo,
      blockReportLeaseId);
}
...
}
```

9.4 HDFS 写数据流程的源码解析

前面介绍过 HDFS 的写数据流程，可以将 HDFS 的写数据流程分为两个阶段，一个是客户端与 NameNode 交互阶段，另一个是客户端向 DataNode 写数据阶段，前者称为 create 阶段，后者称为 write 阶段。本节将对 HDFS 写数据流程的源码进行解析，如图 9-16 所示。

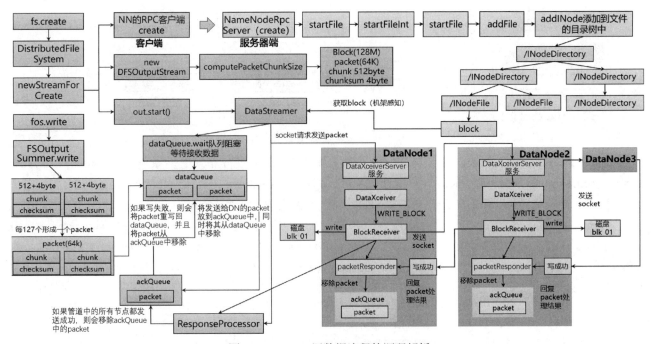

图 9-16 HDFS 写数据流程的源码解析

create 阶段的主要流程如下。

（1）Client 向 NameNode 发起写请求。

（2）NameNode 处理 Client 的写请求。

（3）DataStreamer 启动流程。

write 阶段的主要流程如下。

（1）向 DataStreamer 的队列中写数据。

（2）建立管道之机架感知（块存储位置）。

（3）建立管道之 socket 发送。

（4）建立管道之 socket 接收。

（5）客户端接收 DataNode 的写数据响应。

9.4.1　查看源码的准备工作

继续使用 4.3.2 节中的案例，HdfsClientDemo 工程需要有 hadoop-client、hadoop-hdfs、hadoop-hdfs-client、junit 与 slf4j-log4j12 依赖，并且添加一个上传测试方法，将其作为查看源码的入口，代码如下：

```
@Test
public void testPut2() throws IOException {
// create 阶段
    FSDataOutputStream fos = fs.create(new Path("/input"));

// write 阶段
    fos.write("sunck good".getBytes());
}
```

9.4.2　Client 向 NameNode 发起写请求

Client 向 NameNode 发起写请求的具体流程如图 9-17 所示。

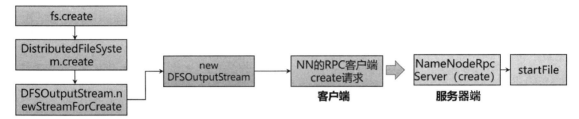

图 9-17　Client 向 NameNode 发起写请求的具体流程

在图 9-17 中，首先使用 fs.create()方法发起写数据请求，然后通过分布式文件系统 DistributedFileSystem 创建一个 StreamForCreate 流，最后客户端通过 RPC 协议与服务器端进行通信，发送 create 请求。

找到客户端向服务器端发送 RPC 请求的位置，从代码"FSDataOutputStream fos = fs.create(new Path("/input"))"处的 create()方法一直追踪深入各种 create()方法，直到遇到 abstract FSDataOutputStream create()方法。按 Ctrl+Alt+B 快捷键，找到 DistributedFileSystem 抽象类的实现类中的 create()实现方法。继续深入该 create()方法，直到找到 DFSOutputStream.newStreamForCreate()方法，进入 newStreamForCreate() 方法内部，查看 DFSOutputStream 对象的创建过程，并且向 RPC 服务器端发送请求，关键代码如下：

```
static DFSOutputStream newStreamForCreate(DFSClient dfsClient, String src,
    FsPermission masked, EnumSet<CreateFlag> flag, boolean createParent,
    short replication, long blockSize, Progressable progress,
    DataChecksum checksum, String[] favoredNodes, String ecPolicyName)
    throws IOException {
    try (TraceScope ignored = dfsClient.newPathTraceScope("newStreamForCreate", src)) {
    ...
        while (shouldRetry) {
            shouldRetry = false;
            try {
                // 创建 DFSOutputStream 对象，并且向 RPC 服务器端发送请求
                stat = dfsClient.namenode.create(src, masked, dfsClient.clientName,
                    new EnumSetWritable<>(flag), createParent, replication,
                    blockSize, SUPPORTED_CRYPTO_VERSIONS, ecPolicyName);
                break;
            }
            ...
        }
    ...
    // 开启线程
    out.start();
    return out;
    }
}
```

9.4.3　NameNode 处理 Client 的写请求

NameNode 的 RPC 服务器端在接收到写请求后，会使用 addINode()方法检查文件上传路径并将其添加到自己的目录树中。在 9.4.2 节代码中的 create()方法处，按 Ctrl+Alt+B 快捷键，定位到 NameNodeRpcServer 实现类中的 create()实现方法。

检测 NameNode 的启动状态，按住 Ctrl 键，依次单击"namesystem.startFile"→"startFileInt"→"FSDirWriteFileOp.startFile"→"addFile"，进入 addFile()方法内部，可以看到 addINode()方法，该方法主要用于将数据写入 INode 的目录树，代码如下：

```
INodesInPath addINode(INodesInPath existing, INode child, FsPermission modes)
```

```
    throws QuotaExceededException, UnresolvedLinkException {
  cacheName(child);
  writeLock();
  try {
    // 将数据写入 INode 的目录树
    return addLastINode(existing, child, modes, true);
  } finally {
    writeUnlock();
  }
}
```

HDFS 与 Linux 一样，也对文件采用树状结构存储方式，INode 是一个抽象类，用于存储文件、目录的公共基本属性。

9.4.4 DataStreamer 启动流程

客户端会创建 DFSOutputStream 数据流，用于计算 packet、chunk 等的相关信息，在一般情况下，数据块大小为 128MB、packet 大小为 64KB、chunk 大小为 512Byte、checksum 大小为 4Byte。通过 out 流对象启动一个 DataStreamer 线程，其中有一个 dataQueue 队列，用于存储读取的本地文件的 packet 数据，当没有数据可以读取时，该队列会阻塞等待。DataStreamer 启动流程如图 9-18 所示。

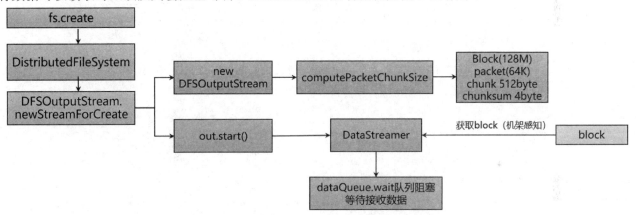

图 9-18　DataStreamer 启动流程

（1）进入 9.4.2 节中的 newStreamForCreate()方法内部，关键代码如下：

```
static DFSOutputStream newStreamForCreate(DFSClient dfsClient, String src,
    FsPermission masked, EnumSet<CreateFlag> flag, boolean createParent,
    short replication, long blockSize, Progressable progress,
    DataChecksum checksum, String[] favoredNodes, String ecPolicyName)
    throws IOException {
  try (TraceScope ignored =
      dfsClient.newPathTraceScope("newStreamForCreate", src)) {
    ...
    while (shouldRetry) {
      shouldRetry = false;
      try {
        // 创建 DFSOutputStream 对象，并且向 RPC 服务器端发送请求
        stat = dfsClient.namenode.create(src, masked, dfsClient.clientName,
            new EnumSetWritable<>(flag), createParent, replication,
            blockSize, SUPPORTED_CRYPTO_VERSIONS, ecPolicyName);
        break;
      }
```

```
        ...
    }
    if(stat.getErasureCodingPolicy() != null) {
        out = new DFSStripedOutputStream(dfsClient, src, stat,
        flag, progress, checksum, favoredNodes);
    } else {
        //创建输出流
        out = new DFSOutputStream(dfsClient, src, stat,
        flag, progress, checksum, favoredNodes, true);
    }
    // 开启线程
    out.start();
    return out;
    }
}
```

（2）按住 Ctrl 键，单击代码 "new DFSOutputStream()" 中的 "DFSOutputStream"，进入该方法内部，代码如下：

```
protected DFSOutputStream(DFSClient dfsClient, String src,
    HdfsFileStatus stat, EnumSet<CreateFlag> flag, Progressable progress,
    DataChecksum checksum, String[] favoredNodes, boolean createStreamer) {
  this(dfsClient, src, flag, progress, stat, checksum);
  this.shouldSyncBlock = flag.contains(CreateFlag.SYNC_BLOCK);

  computePacketChunkSize(dfsClient.getConf().getWritePacketSize(),
    bytesPerChecksum);

  if (createStreamer) {
    streamer = new DataStreamer(stat, null, dfsClient, src, progress,
        checksum, cachingStrategy, byteArrayManager, favoredNodes,
        addBlockFlags);
  }
}
```

按住 Ctrl 键，单击 "computePacketChunkSize"，进入该方法内部，代码如下。可以看到，chunkSize 的值等于 chunk 的值与 checksum 的值的和。

```
protected void computePacketChunkSize(int psize, int csize) {
  final int bodySize = psize - PacketHeader.PKT_MAX_HEADER_LEN;
  final int chunkSize = csize + getChecksumSize();
  chunksPerPacket = Math.max(bodySize/chunkSize, 1);
  packetSize = chunkSize*chunksPerPacket;
  DFSClient.LOG.debug("computePacketChunkSize: src={}, chunkSize={}, "
      + "chunksPerPacket={}, packetSize={}",
    src, chunkSize, chunksPerPacket, packetSize);
}
```

（3）通过 out 流对象启动一个 DataStreamer 线程，找到 DataStreamer 线程中的 run()方法，该方法中的代码如下。其中有一个 dataQueue 队列，用于存储读取的本地文件的 packet 数据，当没有数据可以读取时，该队列会阻塞等待，至此，create 阶段结束。此外，DataStreamer 线程还需要与 NameNode 进行通信，获取可上传数据的 DataNode 信息，建立数据管道，并且启动 ResponseProcessor，用于监听 packet 是否发送成功。当 dataQueue 队列中有数据时，会调用 writeTo()方法，将数据写出到 DataNode 中。

```
@Override
public void run() {
```

```
...
while (!streamerClosed && dfsClient.clientRunning) {
  ...
  synchronized (dataQueue) {
    ...
    try {
      // 如果 dataQueue 队列中没有数据，那么代码会阻塞在这里，等待接收 notify 消息
      dataQueue.wait(timeout);
    }
    ...

    // 队列不为空，从队列中取出 packet
    one = dataQueue.getFirst();
    ...
  }

  ...
  if (stage == BlockConstructionStage.PIPELINE_SETUP_CREATE) {
    LOG.debug("Allocating new block: {}", this);
    // 1 向 NameNode 申请 block 并建立数据管道
    setPipeline(nextBlockOutputStream());
    // 2 启动 ResponseProcessor，用于监听 packet 是否发送成功
    initDataStreaming();
  }
  ...

  synchronized (dataQueue) {

    ...
    // 3 从 dataQueue 队列中将要发送的 packet 移除
    dataQueue.removeFirst();
    // 4 然后在 ackQueue 队列中添加这个 packet
    ackQueue.addLast(one);

  }

  LOG.debug("{} sending {}", this, one);

  // write out data to remote datanode
  try (TraceScope ignored = dfsClient.getTracer().
      newScope("DataStreamer#writeTo", spanId)) {
    // 将数据写出
    one.writeTo(blockStream);
    blockStream.flush();
  } catch (IOException e) {
    errorState.markFirstNodeIfNotMarked();
    throw e;
  }
  ...
}
```

9.4.5 向 DataStreamer 的队列中写数据

在 write 阶段，使用 FSOutputSummer 将本地文件写出，将 chunk+checksum 作为单位写出，每 127 个 chunk+checksum 形成一个 packet，将 packet 写入 dataQueue 队列，如图 9-19 所示。

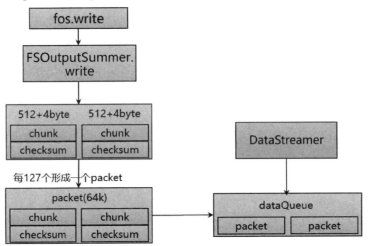

图 9-19 向 dataQueue 队列中写入数据

（1）将鼠标指针放在用户代码"fos.write("sunck good".getBytes())"中的 write()方法处，然后按 Ctrl+Alt+B 快捷键，定位到 FSOutputSummer 实现类中的 write()实现方法，代码如下：

```
public synchronized void write(int b) throws IOException {
    buf[count++] = (byte)b;
    if(count == buf.length) {
        flushBuffer();
    }
}
```

（2）进入 flushBuffer()方法内部，找到 writeChecksumChunks()方法，该方法中的关键代码如下。writeChecksumChunks()方法主要用于写出 chunk+checksum，在该方法中会计算 chunk 的校验和，并且按照 chunk 的大小读取文件并将其写出。

```
private void writeChecksumChunks(byte b[], int off, int len) throws IOException {
    //计算 chunk 的校验和
    sum.calculateChunkedSums(b, off, len, checksum, 0);
    TraceScope scope = createWriteTraceScope();
    try {
        // 按照 chunk 的大小遍历数据
        for (int i = 0; i < len; i += sum.getBytesPerChecksum()) {
            int chunkLen = Math.min(sum.getBytesPerChecksum(), len - i);
            int ckOffset = i / sum.getBytesPerChecksum() * getChecksumSize();
            //逐个 chunk 地将数据写出
            writeChunk(b, off + i, chunkLen, checksum, ckOffset, getChecksumSize());
        }
    }
    ...
}
```

（3）选中步骤（2）中的"writeChunk"，按 Ctrl+Alt+B 快捷键，找到 writeChunk()方法在 DFSOutputStream 实现类中的具体实现过程，代码如下。在该方法中，使用 writeChecksum()方法向 packet 中写入 chunk 的校验和（4Byte），再使用 writeData()方法向 packet 中写入一个 chunk（512Byte），并且通过 incNumChunks()方法

记录写入 packet 的 chunk 数量，累计 127 个 chuck，这个 packet 就满了。此时，使用 enqueueCurrentPacketFull() 方法处理 packet。

```
@Override
protected synchronized void writeChunk(byte[] b, int offset, int len,
        byte[] checksum, int ckoff, int cklen) throws IOException {
    writeChunkPrepare(len, ckoff, cklen);

    // 在 packet 中写入 chunk 的校验和（4Byte）
    currentPacket.writeChecksum(checksum, ckoff, cklen);
    // 向 packet 中写入一个 chunk（512Byte）
    currentPacket.writeData(b, offset, len);
    // 记录写入 packet 的 chunk 数量
    currentPacket.incNumChunks();
    getStreamer().incBytesCurBlock(len);

    // If packet is full, enqueue it for transmission
    if (currentPacket.getNumChunks() == currentPacket.getMaxChunks() ||
            getStreamer().getBytesCurBlock() == blockSize) {
        // packet 满了后的处理方法
        enqueueCurrentPacketFull();
    }
}
```

（4）按住 Ctrl 键，依次单击"enqueueCurrentPacketFull"→"enqueueCurrentPacket"→"waitAndQueuePacket"，进入 waitAndQueuePacket() 方法内部，查看数据写入流程，首先判断 dataQueue 队列是否满了，如果满了，则阻塞等待；如果没有满，则调用 queuePacket() 方法，将 packet 数据写入 dataQueue 队列，关键代码如下：

```
void waitAndQueuePacket(DFSPacket packet) throws IOException {
    synchronized (dataQueue) {
        try {
            try {
                while (!streamerClosed && dataQueue.size() + ackQueue.size() >
dfsClient.getConf().getWriteMaxPackets()) {
                    ...
                    try {
                        // 如果 dataQueue 队列满了，则阻塞等待
                        dataQueue.wait();
                    }
                    ...
                }
            }
            ...
            // 如果 dataQueue 队列没有满，那么将 packet 数据写入 dataQueue 队列
            queuePacket(packet);
        }
        ...
    }
}
```

（5）进入 queuePacket() 方法内部，查看实现过程，其中，dataQueue 队列对象会调用 addLast() 方法，将 packet 数据添加到 dataQueue 队列的末尾，在添加成功后会通知 dataQueue 队列。

```
void queuePacket(DFSPacket packet) {
    synchronized (dataQueue) {
```

```
            if (packet == null) return;
            packet.addTraceParent(Tracer.getCurrentSpanId());
            // 在 dataQueue 队列中添加 packet 数据
            dataQueue.addLast(packet);
            lastQueuedSeqno = packet.getSeqno();
            LOG.debug("Queued {}, {}", packet, this);
            // 通知 dataQueue 队列 packet 数据添加完成
            dataQueue.notifyAll();
        }
}
```

9.4.6 建立管道之机架感知

在启动 DataStreamer 线程时，dataQueue 队列处于无数据阻塞状态，在 dataQueue 队列中有 packet 数据后，会将 packet 数据发送到 DataNode 中。但是在这之前，需要先与 NameNode 通信，获取可上传的 DataNode 信息，如图 9-20 所示。

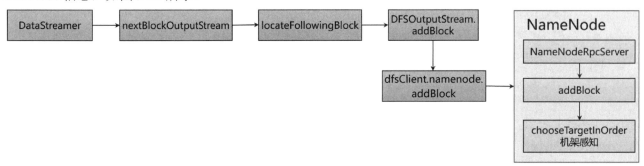

图 9-20 向 NameNode 请求 DataNode 信息

（1）按 Ctrl+N 快捷键，全局查找 DataStreamer 类，在该类中查找 run()方法。当 dataQueue 队列中没有数据时，run()方法会阻塞等待；当 dataQueue 队列中有数据时，run()方法会向 DataNode 上传数据。在 run()方法内部，使用 nextBlockOutputStream()方法向 NameNode 申请 block，并且建立数据管道。进入 nextBlockOutputStream()方法内部，查看具体实现流程，关键代码如下：

```
protected LocatedBlock nextBlockOutputStream() throws IOException {
    ...
    do {
    ...
        // 向 NameNode 获取向哪个 DataNode 写入数据
        lb = locateFollowingBlock(
            excluded.length > 0 ? excluded : null, oldBlock);

        // 建立数据管道
        success = createBlockOutputStream(nodes, nextStorageTypes, nextStorageIDs, 0L, false);

        ...
    } while (!success && --count >= 0);

    ...
    return lb;
}
```

（2）进入 locateFollowingBlock()方法内部，该方法主要用于获取上传的 DataNode 信息，在该方法中可以看到 DFSOutputStream.addBlock()方法，进入 addBlock()方法内部，可以看到 RPC 协议的相关代码。

addBlock()方法中的关键代码如下:
```
static LocatedBlock addBlock(DatanodeInfo[] excludedNodes,
    DFSClient dfsClient, String src, ExtendedBlock prevBlock, long fileId,
        String[] favoredNodes, EnumSet<AddBlockFlag> allocFlags) throws IOException {
    ...
    while (true) {
    try {
        //RPC 客户端向 NameNode 发送写请求
        return     dfsClient.namenode.addBlock(src,     dfsClient.clientName,     prevBlock,
excludedNodes, fileId, favoredNodes, allocFlags);
    }
    ...
    }
}
```

（3）选中步骤（2）中的"addBlock"，按 Ctrl+Alt+B 快捷键，查看 addBlock()方法的具体实现方法，然后单击"NameNodeRpcServer"，进入 NameNodeRpcServer 类内部，在该类中查看 NameNode 是如何处理请求的，代码如下：

```
public LocatedBlock addBlock(String src, String clientName,
    ExtendedBlock previous, DatanodeInfo[] excludedNodes, long fileId,
    String[] favoredNodes, EnumSet<AddBlockFlag> addBlockFlags)
    throws IOException {
  checkNNStartup();
  LocatedBlock locatedBlock = namesystem.getAdditionalBlock(src, fileId,
      clientName, previous, excludedNodes, favoredNodes, addBlockFlags);
  if (locatedBlock != null) {
    metrics.incrAddBlockOps();
  }
  return locatedBlock;
}
```

（4）按住 Ctrl 键，依次单击"getAdditionalBlock"→"chooseTargetForNewBlock"→"chooseTarget4NewBlock"→"chooseTarget"，进入 chooseTarget()方法内部，发现该方法是一个抽象方法。按 Ctrl+Alt+B 快捷键，查看 BlockPlacementPolicyDefault 实现类中的 chooseTarget()实现方法，在该方法中找到 chooseTargetInOrder()方法。chooseTargetInOrder()方法为获取 DataNode 的真正方法，代码如下：

```
protected Node chooseTargetInOrder(int numOfReplicas,
                    Node writer,
                    final Set<Node> excludedNodes,
                    final long blocksize,
                    final int maxNodesPerRack,
                    final List<DatanodeStorageInfo> results,
                    final boolean avoidStaleNodes,
                    final boolean newBlock,
                    EnumMap<StorageType, Integer> storageTypes)
                    throws NotEnoughReplicasException {
  final int numOfResults = results.size();
  if (numOfResults == 0) {
    // 第一个数据块副本存储于当前节点中
    DatanodeStorageInfo storageInfo = chooseLocalStorage(writer,
        excludedNodes, blocksize, maxNodesPerRack, results, avoidStaleNodes,
        storageTypes, true);
```

```
    writer = (storageInfo != null) ? storageInfo.getDatanodeDescriptor()
                                   : null;

  if (--numOfReplicas == 0) {
    return writer;
  }
}
final DatanodeDescriptor dn0 = results.get(0).getDatanodeDescriptor();
// 第二个数据块副本存储于另一个机架中
if (numOfResults <= 1) {
  chooseRemoteRack(1, dn0, excludedNodes, blocksize, maxNodesPerRack,
      results, avoidStaleNodes, storageTypes);
  if (--numOfReplicas == 0) {
    return writer;
  }
}
if (numOfResults <= 2) {
  final DatanodeDescriptor dn1 = results.get(1).getDatanodeDescriptor();
// 如果第一个数据块副本和第二个数据块副本存储于同一个机架中,那么第三个数据块副本存储于其他机架中
  if (clusterMap.isOnSameRack(dn0, dn1)) {
    chooseRemoteRack(1, dn0, excludedNodes, blocksize, maxNodesPerRack,
        results, avoidStaleNodes, storageTypes);
  } else if (newBlock){
// 如果是新数据块,那么将其和第二个数据块副本存储于同一个机架中
    chooseLocalRack(dn1, excludedNodes, blocksize, maxNodesPerRack,
        results, avoidStaleNodes, storageTypes);
  } else {
// 如果不是新数据块,那么将其存储于当前机架中
    chooseLocalRack(writer, excludedNodes, blocksize, maxNodesPerRack,
        results, avoidStaleNodes, storageTypes);
  }
  if (--numOfReplicas == 0) {
    return writer;
  }
}
chooseRandom(numOfReplicas, NodeBase.ROOT, excludedNodes, blocksize,
    maxNodesPerRack, results, avoidStaleNodes, storageTypes);
return writer;
}
```

前面已经介绍过 NameNode 选择可用 DataNode 的原则,如果客户端在 DataNode 上,那么将第一个 DataNode 设置为客户端所在节点的 DataNode,否则随机选取一个 DataNode。将第二个 DataNode 设置为除第一个 NameNode 所在机架外的其他机架中的一个随机 DataNode,将第三个 DataNode 设置为与第二个 DataNode 同机架的 DataNode,最终会通知客户端。

9.4.7　建立管道之 socket 发送

将 dataQueue 队列中的 packet 以 socket 方式发送给 DataNode,并且将该 packet 写入一个 ackQueue 队列,这个 ackQueue 队列主要用于保障将 packet 正确写入 DataNode,整体流程如图 9-21 所示。

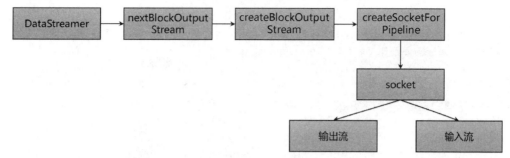

图 9-21　建立管道之 socket 发送的整体流程

（1）按 Ctrl+N 快捷键，全局查找 DataStreamer 类，在该类中查找 run()方法，当 dataQueue 队列中没有数据时，run()方法会阻塞等待；当 dataQueue 队列中有数据时，run()方法会向 DataNode 上传数据。在 run()方法内部，使用 nextBlockOutputStream()方法向 NameNode 申请 block，并且建立数据管道。进入 nextBlockOutputStream()方法内部，查看具体实现流程，关键代码如下：

```
protected LocatedBlock nextBlockOutputStream() throws IOException {
    ...
    do {
        ...
        // 向 NameNode 获取向哪个 DataNode 写入数据
        lb = locateFollowingBlock(
            excluded.length > 0 ? excluded : null, oldBlock);

        // 建立数据管道
        success = createBlockOutputStream(nodes, nextStorageTypes, nextStorageIDs, 0L, false);

        ...
    } while (!success && --count >= 0);

    ...
    return lb;
}
```

（2）进入 createBlockOutputStream()方法内部，查看客户端与 DataNode 建立数据管道的流程，关键代码如下。首先使用 createSocketForPipeline()方法创建 socket，然后创建输出流与输入流，最后调用 writeBlock()方法发送数据。

```
boolean createBlockOutputStream(DatanodeInfo[] nodes,
    StorageType[] nodeStorageTypes, String[] nodeStorageIDs,
    long newGS, boolean recoveryFlag) {
    ...
    // 创建 socket
    s = createSocketForPipeline(nodes[0], nodes.length, dfsClient);

    // 获取输出流，用于将数据写入 DataNode
    OutputStream unbufOut = NetUtils.getOutputStream(s, writeTimeout);
    // 获取输入流，用于读取将数据写入 DataNode 的结果
    InputStream unbufIn = NetUtils.getInputStream(s, readTimeout);

    IOStreamPair saslStreams = dfsClient.saslClient.socketSend(s,
        unbufOut, unbufIn, dfsClient, accessToken, nodes[0]);
    unbufOut = saslStreams.out;
    unbufIn = saslStreams.in;
```

```
  out = new DataOutputStream(new BufferedOutputStream(unbufOut,
      DFSUtilClient.getSmallBufferSize(dfsClient.getConfiguration()))));
  blockReplyStream = new DataInputStream(unbufIn);

  // 发送数据
  new Sender(out).writeBlock(blockCopy, nodeStorageTypes[0], accessToken,
        dfsClient.clientName, nodes, nodeStorageTypes, null, bcs,
        nodes.length, block.getNumBytes(), bytesSent, newGS,
        checksum4WriteBlock, cachingStrategy.get(), isLazyPersistFile,
        (targetPinnings != null && targetPinnings[0]), targetPinnings,
        nodeStorageIDs[0], nodeStorageIDs);
  ...
}
```

9.4.8 建立管道之 socket 接收

DataNode 的 DataXceiver 服务器端在收到客户端的请求后，首先通过 BlockReceiver 服务器端将数据写入磁盘，然后将数据发送给下一个 DataNode 的 DataXceiver 服务器端，最后将数据写入本节点中的一个 ackQueue 队列，依次类推，进行数据传递。

（1）按 Ctrl+N 快捷键，全局查找 DataXceiverServer 类，在该类中查找 run()方法，该方法中的代码如下。使用 run()方法可以接收客户端的 socket 请求，每个 socket 请求都会启动一个 DataXceiver 线程，用于处理 block。

```
public void run() {
  Peer peer = null;
  while (datanode.shouldRun && !datanode.shutdownForUpgrade) {
    try {
    // 接收 socket 请求
      peer = peerServer.accept();

      // Make sure the xceiver count is not exceeded
      int curXceiverCount = datanode.getXceiverCount();
      if (curXceiverCount > maxXceiverCount) {
        throw new IOException("Xceiver count " + curXceiverCount
            + " exceeds the limit of concurrent xcievers: "
            + maxXceiverCount);
      }
    // 客户端每发送一个block，都会启动一个DataXceiver线程，用于处理block
      new Daemon(datanode.threadGroup,
          DataXceiver.create(peer, datanode, this))
          .start();
    } catch (SocketTimeoutException ignored) {
      ...
    }
  }
  ...
}
```

（2）进入 DataXceiver 类内部，在该类中查找 run()方法，该方法中的代码如下。使用 run()方法可以获取当前数据的操作类型，并且根据操作类型处理数据。

```
public void run() {
```

```
int opsProcessed = 0;
Op op = null;

try {
  synchronized(this) {
    xceiver = Thread.currentThread();
  }

  ...

  super.initialize(new DataInputStream(input));

  do {
    updateCurrentThreadName("Waiting for operation #" + (opsProcessed + 1));

    try {
      ...
      // 获取当前数据的操作类型
      op = readOp();
    }
    ...

    opStartTime = monotonicNow();
    // 根据操作类型处理数据
    processOp(op);
    ++opsProcessed;
  } while ((peer != null) &&
    (!peer.isClosed() && dnConf.socketKeepaliveTimeout > 0));
}
...
}
```

（3）首先进入 processOp()方法内部，然后进入一系列 writeBlock()方法，最后按 Ctrl+Alt+B 快捷键，查看 DataXceiver 实现类中的 writeBlock()实现方法。DataXceiver 实现类中的 writeBlock()实现方法可以获取一个 BlockReceiver 对象，用于将数据写入本地磁盘并向下一个 DataNode 发送 socket 请求，从而进行数据传递，关键代码如下：

```
public void writeBlock(...) throws IOException {
  ...
  try {
    final Replica replica;
    if (isDatanode ||
      stage != BlockConstructionStage.PIPELINE_CLOSE_RECOVERY) {

      // 获取一个BlockReceiver对象
      setCurrentBlockReceiver(getBlockReceiver(block, storageType, in,
        peer.getRemoteAddressString(),
        peer.getLocalAddressString(),
        stage, latestGenerationStamp, minBytesRcvd, maxBytesRcvd,
        clientname, srcDataNode, datanode, requestedChecksum,
        cachingStrategy, allowLazyPersist, pinning, storageId));
      replica = blockReceiver.getReplica();
    } else {
```

```java
      replica = datanode.data.recoverClose(
          block, latestGenerationStamp, minBytesRcvd);
    }
    storageUuid = replica.getStorageUuid();
    isOnTransientStorage = replica.isOnTransientStorage();

    //
    // Connect to downstream machine, if appropriate
    // 继续连接下游的节点服务器
    if (targets.length > 0) {
      InetSocketAddress mirrorTarget = null;
      // Connect to backup machine
      mirrorNode = targets[0].getXferAddr(connectToDnViaHostname);
      LOG.debug("Connecting to datanode {}", mirrorNode);
      mirrorTarget = NetUtils.createSocketAddr(mirrorNode);

      // 向下一个 DataNode 发送 socket 请求
      mirrorSock = datanode.newSocket();
      try {

        ...
        if (targetPinnings != null && targetPinnings.length > 0) {
          // 向下游 socket 发送数据
          new Sender(mirrorOut).writeBlock(originalBlock, targetStorageTypes[0],
              blockToken, clientname, targets, targetStorageTypes,
              srcDataNode, stage, pipelineSize, minBytesRcvd, maxBytesRcvd,
              latestGenerationStamp, requestedChecksum, cachingStrategy,
              allowLazyPersist, targetPinnings[0], targetPinnings,
              targetStorageId, targetStorageIds);
        } else {
          new Sender(mirrorOut).writeBlock(originalBlock, targetStorageTypes[0],
              blockToken, clientname, targets, targetStorageTypes,
              srcDataNode, stage, pipelineSize, minBytesRcvd, maxBytesRcvd,
              latestGenerationStamp, requestedChecksum, cachingStrategy,
              allowLazyPersist, false, targetPinnings,
              targetStorageId, targetStorageIds);
        }

        mirrorOut.flush();

        DataNodeFaultInjector.get().writeBlockAfterFlush();

        // read connect ack (only for clients, not for replication req)
        if (isClient) {
          BlockOpResponseProto connectAck =
            BlockOpResponseProto.parseFrom(PBHelperClient.vintPrefixed(mirrorIn));
          mirrorInStatus = connectAck.getStatus();
          firstBadLink = connectAck.getFirstBadLink();
          if (mirrorInStatus != SUCCESS) {
            LOG.debug("Datanode {} got response for connect" +
                "ack  from downstream datanode with firstbadlink as {}",
```

```
            targets.length, firstBadLink);
      }
    }
    ...
    //update metrics
    datanode.getMetrics().addWriteBlockOp(elapsed());
    datanode.getMetrics().incrWritesFromClient(peer.isLocal(), size);
}
```

9.4.9 客户端接收 DataNode 的写数据响应

由最后一个 DataNode 向前报告 packet 是否写入成功，如果写入成功，那么前一个 DataNode 通过 packetResponder 移除本地 ackQueue 队列中的 packet，并且继续向前一个 DataNode 进行响应汇报。最后由第一个 DataNode 通知客户端 packet 的处理结果。客户端请求由 ResponseProcessor 接收并处理，如果写入成功，那么将该 packet 对应于客户端 ackQueue 队列中的 packet 移除；如果写入失败，那么将该 packet 放到 dataQueue 队列中重新发送，具体流程如图 9-22 所示。

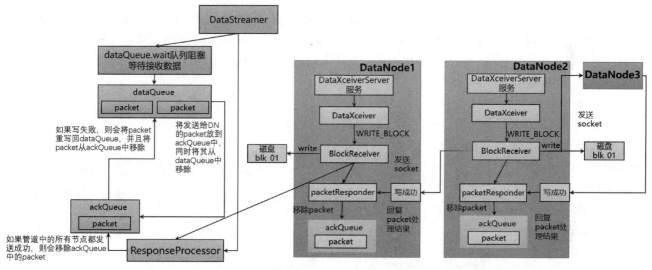

图 9-22　客户端接收 DataNode 写数据响应的具体流程

（1）按 Ctrl+N 快捷键，全局查找 DataStreamer 类，在该类中查找 run()方法，该方法中的关键代码如下。run()方法中的 initDataStreaming()方法可以初始化一个 ResponseProcessor，用于接收 DataNode 的写数据响应。

```
@Override
public void run() {
  ...
  while (!streamerClosed && dfsClient.clientRunning) {
    ...
    if (stage == BlockConstructionStage.PIPELINE_SETUP_CREATE) {
      LOG.debug("Allocating new block: {}", this);
      // 1 向 NameNode 申请 block，并且建立数据管道
      setPipeline(nextBlockOutputStream());
      // 2 初始化 ResponseProcessor，用于接收 DataNode 的写数据响应
```

```
    initDataStreaming();
}
...
}
```

（2）进入 initDataStreaming()方法内部，再进入 ResponseProcessor 类内部，在该类中查找 run()方法，该方法中的关键代码如下。在 run()方法中，可以查看 ResponseProcessor 的运行代码。如果 packet 写入成功，那么将 packet 从 ackQueue 队列中移除；否则将其移入 dataQueue 队列，等待再次发送。

```
public void run() {
    ...
    ackQueue.removeFirst();
    packetSendTime.remove(seqno);
    dataQueue.notifyAll();
    ...
}
```

9.5 YARN 源码解析

5.4.2 节介绍过 Job 提交流程的源码，我们已经了解了，在提交 Job 前，需要获取数据切片信息、配置信息、jar 包信息等，客户端会将这些文件上传到 ResourceManager 指定的位置。6.1.2 节介绍过 YARN 的工作机制，我们已经了解了，在客户端提交 Job 后，会申请执行一个 ApplicationMaster，并且与 ResourceManager 等进行通信，从而进行任务调度，本章将对该过程的源码进行深入解析，该过程的具体步骤如下。

（1）创建 YARN 客户端并提交任务。
（2）启动 MRAppMaster。
（3）调度器任务执行（YarnChild）。

YARN 源码解析的总流程如图 9-23 所示。

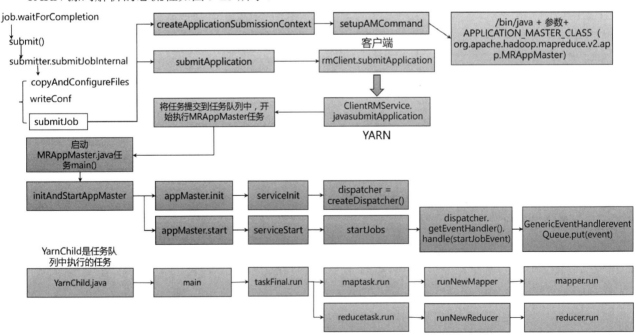

图 9-23　YARN 源码解析的总流程

9.5.1　查看源码的准备工作

查看 5.2.3 节中的 WordCount 程序源码，并且添加一些必要的依赖，具体如下：

```xml
<dependency>
    <groupId>org.apache.hadoop</groupId>
    <artifactId>hadoop-mapreduce-client-app</artifactId>
    <version>3.1.3</version>
</dependency>
```

首先进入 WordCountDriver 类中的 waitForCompletion()方法内部；然后进入 submit()方法内部，进行提交操作；最后进入 submitJobInternal()方法内部，进行内部提交操作。submitJobInternal()方法中的关键代码如下：

```java
JobStatus submitJobInternal(Job job, Cluster cluster) throws ClassNotFoundException,
InterruptedException, IOException {
    //检查输出路径是否存在
    checkSpecs(job);
    
    ...
    
    //获取JobId的值
    JobID jobId = submitClient.getNewJobID();
    job.setJobID(jobId);
    
    try {
        ...
        
        // 复制配置信息
        copyAndConfigureFiles(job, submitJobDir);
        
        // 获取Job切片信息
        int maps = writeSplits(job, submitJobDir);
        ...
        
        // 向资源目录提交配置文件
        writeConf(conf, submitJobFile);
        
        ...
        // 提交Job
        status = submitClient.submitJob(
            jobId, submitJobDir.toString(), job.getCredentials());
        ...
    }
    ...
}
```

submitClient.submitJob()方法位于一个接口文件中，需要找到该方法的具体实现。将鼠标指针放在 submitJob()方法上，按 Ctrl+Alt+B 快捷键，查看 YARNRunner 实现类中的 submit()实现方法，关键代码如下：

```java
@Override
public JobStatus submitJob(JobID jobId, String jobSubmitDir, Credentials ts) throws IOException,
InterruptedException {
    ...
    // 创建提交环境
    ApplicationSubmissionContext appContext =
        createApplicationSubmissionContext(conf, jobSubmitDir, ts);
```

```
try {
    // 向 ResourceManager 提交一个应用程序
    ApplicationId applicationId =
        resMgrDelegate.submitApplication(appContext);

    // 获取提交响应
    ApplicationReport appMaster = resMgrDelegate
        .getApplicationReport(applicationId);

    ...
}
...
}
```

在 submitJob()方法内部，先使用 createApplicationSubmissionContext()方法创建一个提交环境对象 appContext，该对象内部封装了启动 MRAppMaster 和运行 Container 的命令；再使用 resMgrDelegate.submitApplication()方法向 ResourceManager 提交一个应用程序。

9.5.2 创建 YARN 客户端并提交任务

YARN 客户端在向资源目录提交相应的文件后，即可通知 ResourceManager，使其运行 MRAppMaster。但是在这之前，需要提前准备好提交环境，在 submitJob()方法中使用 createApplicationSubmissionContext()方法完成，具体流程如图 9-24 所示。

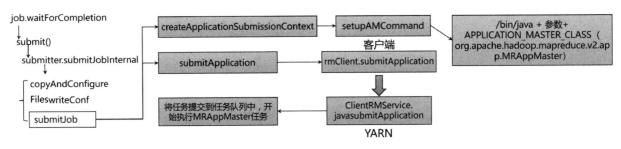

图 9-24　创建 YARN 客户端并提交任务的具体流程

（1）查看 submitJob()方法，关键代码如下：

```
@Override
public JobStatus submitJob(JobID jobId, String jobSubmitDir, Credentials ts) throws
IOException, InterruptedException {
    ...
    // 创建提交环境
    ApplicationSubmissionContext appContext =
        createApplicationSubmissionContext(conf, jobSubmitDir, ts);

    try {
        // 向 ResourceManager 提交一个应用程序
        ApplicationId applicationId =
            resMgrDelegate.submitApplication(appContext);

        // 获取提交响应
        ApplicationReport appMaster = resMgrDelegate
```

```
            .getApplicationReport(applicationId);
    ...
    }
    ...
}
```

（2）按住 Ctrl 键，单击 "createApplicationSubmissionContext"，进入该方法内部，该方法会返回一个上下文相关的环境对象，关键代码如下：

```
public ApplicationSubmissionContext createApplicationSubmissionContext(
    Configuration jobConf, String jobSubmitDir, Credentials ts)
        throws IOException {
ApplicationId applicationId = resMgrDelegate.getApplicationId();

// 封装了本地资源的相关路径
Map<String, LocalResource> localResources =
    setupLocalResources(jobConf, jobSubmitDir);

...

// 封装了启动 MRAppMaster 和运行 Container 的命令
List<String> vargs = setupAMCommand(jobConf);
ContainerLaunchContext amContainer = setupContainerLaunchContextForAM(
    jobConf, localResources, securityTokens, vargs);

...
return appContext;
}
```

（3）按住 Ctrl 键，单击 "setupAMCommand"，进入该方法内部，关键代码如下。该方法中封装了启动 MRAppMaster 的命令，这个命令由 ResourceManager 执行。

```
private List<String> setupAMCommand(Configuration jobConf) {
    List<String> vargs = new ArrayList<>(8);
    // Java 进程启动命令开始
    vargs.add(MRApps.crossPlatformifyMREnv(jobConf, Environment.JAVA_HOME) + "/bin/java");

    ...

    // 用户命令参数
    String mrAppMasterUserOptions = conf.get(MRJobConfig.MR_AM_COMMAND_OPTS,
        MRJobConfig.DEFAULT_MR_AM_COMMAND_OPTS);
    ...

    // 封装了要启动的 MRAppMaster 全类名
    // org.apache.hadoop.mapreduce.v2.app.MRAppMaster
    vargs.add(MRJobConfig.APPLICATION_MASTER_CLASS);
    vargs.add("1>" + ApplicationConstants.LOG_DIR_EXPANSION_VAR +
        Path.SEPARATOR + ApplicationConstants.STDOUT);
    vargs.add("2>" + ApplicationConstants.LOG_DIR_EXPANSION_VAR +
        Path.SEPARATOR + ApplicationConstants.STDERR);
    return vargs;
}
```

在该方法的最后添加了要启动的 MRAppMaster 全类名，即 org.apache.hadoop.mapreduce.v2.app.MRAppMaster。

（4）在提交环境创建完成后，即可向 ResourceManager 提交应用程序。返回步骤（1）中的 submitJob() 方法内部，按住 Ctrl 键，单击"submitApplication"，进入该方法内部，查看如何将应用程序提交到 ResourceManager 中，代码如下：

```
@Override
 public ApplicationId submitApplication(ApplicationSubmissionContext appContext)
        throws YarnException, IOException {
    return client.submitApplication(appContext);
}
```

（5）在 submitApplication() 方法中，按住 Ctrl 键，单击"submitApplication"，发现它是一个抽象方法。按 Ctrl+Alt+B 快捷键，查看 YarnClientImpl 类中的 submitApplication() 实现方法，该方法中的关键代码如下：

```
@Override
public ApplicationId submitApplication(ApplicationSubmissionContext appContext) throws
YarnException, IOException {
    ...
    //客户端提交 Application
    rmClient.submitApplication(request);

    ...

    return applicationId;
}
```

（6）按 Ctrl+Alt+B 快捷键，找到 YarnClientImpl 的实现类 ApplicationClientProtocolPBClientImpl，在该类中找到真正调用的 submitApplication() 方法，该方法中的代码如下：

```
@Override
public SubmitApplicationResponse submitApplication(
    SubmitApplicationRequest request) throws YarnException,
    IOException {
  SubmitApplicationRequestProto requestProto =
    ((SubmitApplicationRequestPBImpl) request).getProto();
  try {
    return new SubmitApplicationResponsePBImpl(proxy.submitApplication(null,
      requestProto));
  } catch (ServiceException e) {
    RPCUtil.unwrapAndThrowException(e);
    return null;
  }
}
```

使用 ResourceManager 代理对象提交一个应用程序，即发送一个 Java 指令，指令格式如下：

```
java-jar jar 包名 org.apache.hadoop.mapreduce.v2.app.MRAppMaster 参数
```

proxy.submitApplication() 方法中的 proxy 是一个 RPC 客户端对象，使用它可以与服务器端进行通信。

```
proxy = RPC.getProxy(ApplicationClientProtocolPB.class, clientVersion, addr, conf);
```

9.5.3 启动 MRAppMaster

MRAppMaster 的启动过程主要包括初始化与启动两部分。

初始化的主要目的是创建一个 dispatcher 调度者，因为 MRAppMaster 是整个 Job 作业中所有任务的管理者，并且可以创建提交路径与 RPC 客户端。

启动的主要目的是将相关任务添加到 YARN 集群的任务资源调度器队列中，以便执行任务。

启动 MRAppMaster 的具体流程如图 9-25 所示。

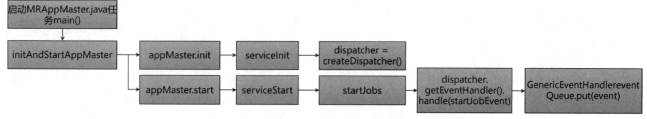

图 9-25 启动 MRAppMaster 的具体流程

（1）按 Ctrl+N 快捷键，全局查找 MRAppMaster。

（2）按 Ctrl+F 快捷键，在弹出的搜索框中输入"main"，查找相应的 main()方法，关键代码如下。在 main()方法中会创建相应的 Container 对象和 MRAppMaster 对象，用于进行初始化与启动操作。

```
public static void main(String[] args) {
    try {
    ...

    //创建 Container 对象
    ContainerId containerId = ContainerId.fromString(containerIdStr);
    ApplicationAttemptId applicationAttemptId =
        containerId.getApplicationAttemptId();
    if (applicationAttemptId != null) {
      CallerContext.setCurrent(new CallerContext.Builder(
         "mr_appmaster_" + applicationAttemptId.toString()).build());
    }
    long appSubmitTime = Long.parseLong(appSubmitTimeStr);

    //创建 MRAppMaster 对象
    MRAppMaster appMaster =
        new MRAppMaster(applicationAttemptId, containerId, nodeHostString,
            Integer.parseInt(nodePortString),
            Integer.parseInt(nodeHttpPortString), appSubmitTime);
    ShutdownHookManager.get().addShutdownHook(
      new MRAppMasterShutdownHook(appMaster), SHUTDOWN_HOOK_PRIORITY);
    JobConf conf = new JobConf(new YarnConfiguration());
    conf.addResource(new Path(MRJobConfig.JOB_CONF_FILE));

    ...
    //初始化与启动 MRAppMaster
    initAndStartAppMaster(appMaster, conf, jobUserName);
    }
    ...
}
```

（3）进入 initAndStartAppMaster()方法内部，查看初始化与启动的流程，关键代码如下。其中，init()方法为初始化方法，start()方法为启动方法。

```
protected static void initAndStartAppMaster(final MRAppMaster appMaster,
    final JobConf conf, String jobUserName) throws IOException,
    InterruptedException {
  ...
```

```
appMasterUgi.doAs(new PrivilegedExceptionAction<Object>() {
    @Override
    public Object run() throws Exception {
        // 初始化
      appMaster.init(conf);
        // 启动
      appMaster.start();
      ...
    }
});
}
```

（4）init()方法中的 serviceInit()方法为真正的初始化方法，按 Ctrl+Alt+B 快捷键，定位 MRAppMaster 实现类中的 serviceInit()实现方法，关键代码如下：

```
protected void serviceInit(final Configuration conf) throws Exception {
    ...

    // 创建调度者
    dispatcher = createDispatcher();

    // 创建提交路径
    clientService = createClientService(context);

    // 创建调度器
    clientService.init(conf);

    // 创建 Job 提交 RPC 客户端
    containerAllocator = createContainerAllocator(clientService, context);
    ...
}
```

（5）start()方法中的 serviceStart()方法为真正的启动方法，按 Ctrl+Alt+B 快捷键，定位 MRAppMaster 实现类中的 serviceStart()实现方法，关键代码如下。其中，startJobs()方法主要用于在初始化成功后，将 Job 提交到队列中。

```
protected void serviceStart() throws Exception {
    ...
    if (initFailed) {
        JobEvent initFailedEvent = new JobEvent(job.getID(), JobEventType.JOB_INIT_FAILED);
        jobEventDispatcher.handle(initFailedEvent);
    } else {
        // 在初始化成功后，将 Job 提交到队列中
        startJobs();
    }
}
```

（6）startJobs()方法中的代码"dispatcher.getEventHandler().handle(startJobEvent)"主要用于将 Job 存储于 YARN 调度队列中，getEventHandler()方法返回的是 GenericEventHandler 对象。按 Ctrl+Alt+B 快捷键，定位 GenericEventHandler 实现类中的 handle()实现方法，关键代码如下。可以看到，在 handle()方法中，可以将 Event 对象添加到 eventQueue 资源队列中。

```
class GenericEventHandler implements EventHandler<Event> {
    public void handle(Event event) {
        ...
        try {
```

```
      // 将 Job 存储于 YARN 调度队列中
      eventQueue.put(event);
    } catch (InterruptedException e) {
      ...
    }
  }
};
}
```

9.5.4 调度器任务执行

任务在被添加到资源队列中后，就会等待 YARN 的资源调度。根据资源调度策略，在轮到该任务时，YARN 会启动 YarnChild 进程，用于执行对应的 MapTask 和 ReduceTask，进而与 Mapper 与 Reducer 中的业务逻辑衔接上，从而执行对应的 Mapper 业务与 Reducer 业务，具体流程如图 9-26 所示。

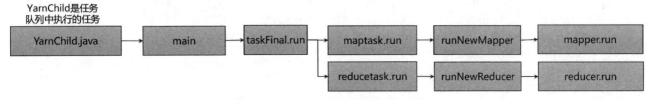

图 9-26 调度器任务执行的具体流程

（1）按 Ctrl+N 快捷键，查找 YarnChild 类，搜索 main()方法，找到 YarnChild 的主程序，关键代码如下：

```
public static void main(String[] args) throws Throwable {
  Thread.setDefaultUncaughtExceptionHandler(new YarnUncaughtExceptionHandler());
  LOG.debug("Child starting");

  ...

  task = myTask.getTask();
  YarnChild.taskid = task.getTaskID();
  ...

  // Create a final reference to the task for the doAs block
  final Task taskFinal = task;
  childUGI.doAs(new PrivilegedExceptionAction<Object>() {
    @Override
    public Object run() throws Exception {
      // use job-specified working directory
      setEncryptedSpillKeyIfRequired(taskFinal);
      FileSystem.get(job).setWorkingDirectory(job.getWorkingDirectory());
      // 执行 Task (MapTask 或 ReduceTask)
      taskFinal.run(job, umbilical); // run the task
      return null;
    }
  });
}
  ...
}
```

在 main()方法中，使用 taskFinal 对象调用 run()方法，启动对应的 MapTask 与 ReduceTask。taskFinal 对象是 Task 抽象类的对象，按 Ctrl+Alt+B 快捷键，可以发现 Task 抽象类有两个实现类，分别是 MapTask 类和 ReduceTask 类。其中，MapTask 类主要提供启动 MapTask 的源码，ReduceTask 类主要提供启动

ReduceTask 的源码。

（2）查看 MapTask 的启动过程。进入 MapTask 类内部，查看相应的 run()实现方法，关键代码如下。在 run()实现方法中，首先判断该任务是否为 MapTask，如果是 MapTask，那么判断 ReduceTask 的数量，如果 ReduceTask 的数量为 0 个，那么 MapTask 占用整个任务资源的 100%，否则 MapTask 占用整个任务资源的 66.7%，sort 阶段占用整个任务资源的 33.3%。在判断结束后，调用新的 API 的 runNewMapper()方法，用于执行 MapTask。

```java
public void run(final JobConf job, final TaskUmbilicalProtocol umbilical)
  throws IOException, ClassNotFoundException, InterruptedException {
  this.umbilical = umbilical;

  // 判断是否是 MapTask
  if (isMapTask()) {
    // If there are no reducers then there won't be any sort. Hence the map
    // phase will govern the entire attempt's progress.
    // 如果 ReduceTask 的数量为 0 个，那么 MapTask 占用整个任务资源的 100%
    if (conf.getNumReduceTasks() == 0) {
      mapPhase = getProgress().addPhase("map", 1.0f);
    } else {
      // If there are reducers then the entire attempt's progress will be
      // split between the map phase (67%) and the sort phase (33%).
      // 如果 ReduceTask 的数量不是 0 个，那么 MapTask 占用整个任务资源的 66.7%，sort 阶段占用整个任务资源的 33.3%
      mapPhase = getProgress().addPhase("map", 0.667f);
      sortPhase  = getProgress().addPhase("sort", 0.333f);
    }
  }
  ...
  if (useNewApi) {
    // 调用新的 API 的 runNewMapper()方法，用于执行 MapTask
    runNewMapper(job, splitMetaInfo, umbilical, reporter);
  } else {
    runOldMapper(job, splitMetaInfo, umbilical, reporter);
  }
  done(umbilical, reporter);
}
```

（3）进入 runNewMapper()方法内部，查看执行 MapTask 的具体过程，关键代码如下。在 runNewMapper()方法中，首先获取运行的相关对象。例如，获取任务上下文对象，利用任务上下文对象创建一个 Mapper 对象。Mapper 对象是启动 MapTask 的关键。为了使 MapTask 可以正确获取数据，需要创建输入对象，并且重建切片信息。此外，需要判断是否存在 reduce 阶段，如果不存在 reduce 阶段，则使用 NewDirectOutputCollector()方法创建一个输出对象；如果存在 reduce 阶段，则使用 NewOutputCollector()方法创建一个输出对象。不同的输出对象为后续的不同流程提供服务。

```java
private <INKEY,INVALUE,OUTKEY,OUTVALUE>
void runNewMapper(final JobConf job,
                  final TaskSplitIndex splitIndex,
                  final TaskUmbilicalProtocol umbilical,
                  TaskReporter reporter
                  ) throws IOException, ClassNotFoundException,
                           InterruptedException {
  // 获取任务上下文对象
```

```java
org.apache.hadoop.mapreduce.TaskAttemptContext taskContext =
  new org.apache.hadoop.mapreduce.task.TaskAttemptContextImpl(job,
                                                  getTaskID(),
                                                  reporter);
// 创建 Mapper 对象
org.apache.hadoop.mapreduce.Mapper<INKEY,INVALUE,OUTKEY,OUTVALUE> mapper =
  (org.apache.hadoop.mapreduce.Mapper<INKEY,INVALUE,OUTKEY,OUTVALUE>)
    ReflectionUtils.newInstance(taskContext.getMapperClass(), job);
// 获取输入对象
org.apache.hadoop.mapreduce.InputFormat<INKEY,INVALUE> inputFormat =
  (org.apache.hadoop.mapreduce.InputFormat<INKEY,INVALUE>)
    ReflectionUtils.newInstance(taskContext.getInputFormatClass(), job);
// 重建切片信息
org.apache.hadoop.mapreduce.InputSplit split = null;
split = getSplitDetails(new Path(splitIndex.getSplitLocation()),
    splitIndex.getStartOffset());
LOG.info("Processing split: " + split);

org.apache.hadoop.mapreduce.RecordReader<INKEY,INVALUE> input =
  new NewTrackingRecordReader<INKEY,INVALUE>
    (split, inputFormat, reporter, taskContext);

job.setBoolean(JobContext.SKIP_RECORDS, isSkipping());
org.apache.hadoop.mapreduce.RecordWriter output = null;

// 创建输出对象
if (job.getNumReduceTasks() == 0) {
  output =
    new NewDirectOutputCollector(taskContext, job, umbilical, reporter);
} else {
  output = new NewOutputCollector(taskContext, job, umbilical, reporter);
}

org.apache.hadoop.mapreduce.MapContext<INKEY, INVALUE, OUTKEY, OUTVALUE>
mapContext =
  new MapContextImpl<INKEY, INVALUE, OUTKEY, OUTVALUE>(job, getTaskID(),
      input, output,
      committer,
      reporter, split);

org.apache.hadoop.mapreduce.Mapper<INKEY,INVALUE,OUTKEY,OUTVALUE>.Context
    mapperContext =
      new WrappedMapper<INKEY, INVALUE, OUTKEY, OUTVALUE>().getMapContext(
        mapContext);

try {
  input.initialize(split, mapperContext);
  // 执行 MapTask
  mapper.run(mapperContext);
  ...
}
```

```
   ...
}
```

（4）查看 ReduceTask 的启动过程。进入 ReduceTask 类内部，查看相应的 run()实现方法，关键代码如下。在 run()实现方法中，最后调用新的 API 的 runNewReducer()方法，用于执行 ReduceTask。

```
public void run(JobConf job, final TaskUmbilicalProtocol umbilical)
  throws IOException, InterruptedException, ClassNotFoundException {
  job.setBoolean(JobContext.SKIP_RECORDS, isSkipping());

  ...

  if (useNewApi) {
  // 调用新的 API 的 runNewReducer()方法，用于执行 ReduceTask
    runNewReducer(job, umbilical, reporter, rIter, comparator,
              keyClass, valueClass);
  } else {
    runOldReducer(job, umbilical, reporter, rIter, comparator,
              keyClass, valueClass);
  }

  shuffleConsumerPlugin.close();
  done(umbilical, reporter);
}
```

（5）进入 runNewReducer()方法内部，查看执行 ReduceTask 的具体过程，关键代码如下。在 runNewReducer()方法中，先创建用户自定义的 Reducer 对象，再使用 Reducer 对象调用 run()方法。

```
void runNewReducer(JobConf job,
              final TaskUmbilicalProtocol umbilical,
              final TaskReporter reporter,
              RawKeyValueIterator rIter,
              RawComparator<INKEY> comparator,
              Class<INKEY> keyClass,
              Class<INVALUE> valueClass
              ) throws IOException,InterruptedException,
                   ClassNotFoundException {
...
try {
  // 调用 Reducer 对象的 run()方法
  reducer.run(reducerContext);
} finally {
  trackedRW.close(reducerContext);
}
}
```

9.6　Hadoop 的源码编译

Hadoop 是使用 Java 开发的，但是有一些需求和操作并不适合使用 Java 编写，因此引入了本地库（Native Library）的概念。

本地库是使用 C/C++编写的动态库.so 文件，可以通过 JNI（Java Native Interface）机制为 Java 层提供接口。在通常情况下，应用程序会出于对性能、安全等的考虑，使用 C/C++实现相关逻辑并将其编译为本地库，从而提供相应的接口，以供上层或其他模块调用。也就是说，Hadoop 的某些功能，必须通过 JNI 协调 Java 类文件和 Native 代码生成的库文件才能实现。要在 Linux 操作系统上运行 Native 代码，需要先将本地库代码编译成目标处理器架构的动态库.so 文件。要正确地执行不同的处理器架构，需要编译相应平台的动态库.so 文件。因此，用户在使用 Hadoop 前，最好重新编译一次 Hadoop 源码，让动态库.so 文件与处理器相对应。当然，在平时的学习或工作中，也可以根据需求修改 Hadoop 的源码，并且在修改后对源码重新进行编译。

9.6.1 前期准备工作

对于编译源码的过程，建议在 Linux 操作系统中采用 root 用户进行操作，可以避免发生权限问题，并且可以访问互联网。此外，需要提前准备好所需的包文件，具体如下。

- hadoop-3.1.3-src.tar.gz。
- jdk-8u212-linux-x64.tar.gz。
- apache-maven-3.6.3-bin.tar.gz。
- cmake-3.17.0.tar.gz。
- protobuf-2.5.0.tar.gz。

以下操作流程都是在本书前面用到的 CentOS 中进行的，在/opt/software 目录下创建 hadoop_source 目录，将以上安装包上传到该目录下。在/opt/module 目录下创建 hadoop_source 目录，将相关安装包解压缩到该目录下。

```
[root@hadoop104 hadoop_source]# pwd
/opt/software/hadoop_source
[root@hadoop104 hadoop_source]# ll
总用量 240428
-rw-r--r--. 1 root root   9506321 8月  23 2020 apache-maven-3.6.3-bin.tar.gz
-rw-r--r--. 1 root root   9466484 8月  23 2020 cmake-3.17.0.tar.gz
-rw-r--r--. 1 root root  29800905 8月  23 2020 hadoop-3.1.3-src.tar.gz
-rw-r--r--. 1 root root 195013152 7月  10 2020 jdk-8u212-linux-x64.tar.gz
-rw-r--r--. 1 root root   2401901 8月  23 2020 protobuf-2.5.0.tar.gz
```

9.6.2 安装工具包

在真正编译源码前，需要提前将工具包安装到 Linux 操作系统中，包括 JDK、Maven、相关依赖、cmake、protobuf 等。

1. 安装 JDK

（1）执行以下命令，将 JDK 安装包解压缩到/opt/module/hadoop_source 目录下。

```
[root@hadoop104 hadoop_source]# tar -zxvf /opt/software/hadoop_source/jdk-8u212-linux-x64.tar.gz -C /opt/module/hadoop_source/
```

（2）为了在全局使用 JDK 的相关命令，需要配置环境变量。打开/etc/profile.d/my_env.sh 文件，输入以下内容，保存文件并退出。

```
#JAVA_HOME
export JAVA_HOME=/opt/module/hadoop_source/jdk1.8.0_212
export PATH=$PATH:$JAVA_HOME/bin
```

（3）执行以下命令，使环境变量生效。

```
[root@hadoop104 ~]# source /etc/profile
```

（4）验证 JDK 是否安装成功。在任意位置执行 java -version 命令，如果得到以下结果，则说明 JDK 安装成功。

```
[root@hadoop104 ~]# java -version
java version "1.8.0_212"
```

```
Java(TM) SE Runtime Environment (build 1.8.0_212-b10)
Java HotSpot(TM) 64-Bit Server VM (build 25.212-b10, mixed mode)
```

2. 安装 Maven

（1）执行以下命令，将 Maven 安装包解压缩到/opt/module/hadoop_source 目录下。

```
[root@hadoop104 hadoop_source]# tar -zxvf /opt/software/hadoop_source/apache-maven-3.6.3-bin.tar.gz -C /opt/module/hadoop_source/
```

（2）为了在全局使用 Maven 的相关命令，需要配置环境变量。打开/etc/profile.d/my_env.sh 文件，输入以下内容，保存文件并退出。

```
#MAVEN_HOME
export MAVEN_HOME=/opt/module/hadoop_source/apache-maven-3.6.3
export PATH=$PATH:$MAVEN_HOME/bin
```

（3）执行以下命令，使环境变量生效。

```
[root@hadoop104 ~]# source /etc/profile
```

（4）将 Maven 默认镜像修改为阿里云镜像，便于以后使用 Maven 下载 jar 包。打开/opt/module/hadoop_source/apache-maven-3.6.3/conf/settings.xml 文件，在 mirrors 节点中添加阿里云镜像。

```
<mirrors>
    <mirror>
        <id>nexus-aliyun</id>
        <mirrorOf>central</mirrorOf>
        <name>Nexus aliyun</name>
<url>https://maven.aliyun.com/repository/public</url>
    </mirror>
</mirrors>
```

（5）验证 Maven 是否安装成功。在任意位置执行 mvn -version 命令，如果得到以下结果，则说明 Maven 安装成功。

```
[root@hadoop104 ~]# mvn -version
Apache Maven 3.6.3 (cecedd343002696d0abb50b32b541b8a6ba2883f)
Maven home: /opt/module/hadoop_source/apache-maven-3.6.3
Java version: 1.8.0_212, vendor: Oracle Corporation, runtime: /opt/module/jdk1.8.0_212/jre
Default locale: zh_CN, platform encoding: UTF-8
OS name: "linux", version: "3.10.0-862.el7.x86_64", arch: "amd64", family: "unix"
```

3. 使用 yum 命令安装相关依赖

系统需要访问互联网，以下安装顺序不可改变，否则可能会出现找不到依赖的情况。

（1）执行以下命令，安装 gcc make。

```
[root@hadoop104 ~]# yum install -y gcc* make
```

（2）执行以下命令，安装压缩工具。

```
[root@hadoop104 ~]# yum -y install snappy* bzip2* lzo* zlib* lz4* gzip*
```

（3）执行以下命令，安装一些基本工具。

```
[root@hadoop104 ~]# yum -y install openssl* svn ncurses* autoconf automake libtool
```

（4）执行以下命令，安装扩展源。

```
[root@hadoop104 ~]# yum -y install epel-release
```

（5）执行以下命令，安装 zstd。

```
[root@hadoop104 ~]# yum -y install *zstd*
```

4. 安装 cmake

（1）执行以下命令，将 cmake 安装包解压缩到/opt/module/hadoop_source 目录下。

```
[root@hadoop104 hadoop_source]# tar -zxvf /opt/software/hadoop_source/cmake-3.17.0.tar.gz
-C  /opt/module/hadoop_source/
```
（2）在解压缩好的 cmake 目录下，执行./bootstrap 命令进行编译，此过程需要 1 小时左右，请耐心等待。
```
[root@hadoop104 cmake-3.17.0]# pwd
/opt/module/hadoop_source/cmake-3.17.0
[root@hadoop104 cmake-3.17.0]# ./bootstrap
```
（3）在编译完成后，执行以下命令，进行安装。
```
[root@hadoop104 cmake-3.17.0]# pwd
/opt/module/hadoop_source/cmake-3.17.0
[root@hadoop104 cmake-3.17.0]# make && make install
```
（4）验证 cmake 是否安装成功。在任意位置执行 cmake -version 命令，如果得到以下结果，则说明 cmake 安装成功。
```
[root@hadoop104 ~]# cmake -version
cmake version 3.17.0
CMake suite maintained and supported by Kitware (kitware.com/cmake).
```

5. 安装 protobuf

（1）执行以下命令，将 protobuf 安装包解压缩到/opt/module/hadoop_source 目录下。
```
[root@hadoop104 hadoop_source]# tar -zxvf /opt/software/hadoop_source/protobuf-2.5.0.tar.gz
-C  /opt/module/hadoop_source/
```
（2）在解压缩好的 protobuf 目录下进行编译，执行--prefix 命令，指定安装目录。
```
[root@hadoop104 protobuf-2.5.0]# pwd
/opt/module/hadoop_source/protobuf-2.5.0
[root@hadoop104 protobuf-2.5.0]# ./configure --prefix=/opt/module/hadoop_source/protobuf-2.5.0
```
（3）在编译完成后，执行以下命令，进行安装。
```
 [root@hadoop104 protobuf-2.5.0]# pwd
/opt/module/hadoop_source/protobuf-2.5.0
[root@hadoop104 protobuf-2.5.0]# make && make install
```
（4）配置环境变量，打开/etc/profile.d/my_env.sh 文件，输入以下内容，保存文件并退出。
```
export PROTOC_HOME=/opt/module/hadoop_source/protobuf-2.5.0
export PATH=$PATH:$PROTOC_HOME/bin
```
（5）执行以下命令，使环境变量生效。
```
[root@hadoop104 ~]# source /etc/profile
```
（6）验证 protobuf 是否安装成功。在任意位置执行 protoc --version 命令，如果得到以下结果，则说明 protobuf 安装成功。
```
[root@hadoop104 ~]# protoc --version
libprotoc 2.5.0
```
至此，所有的准备工作都完成了，接下来到了最激动人心的时刻——编译源码。

9.6.3 编译源码

将 Hadoop 的源码包解压缩到/opt/module/ hadoop_source 目录下，并且执行以下命令，切换到该目录下进行操作。
```
[root@hadoop104 hadoop-3.1.3-src]# pwd
/opt/module/hadoop_source/hadoop-3.1.3-src
[root@hadoop104 hadoop-3.1.3-src]# mvn clean package -DskipTests
```
第一次编译需要下载很多依赖 jar 包，编译时间会很久，预计 1 小时左右，如果最后全部显示 SUCCESS，则表示编译成功，如图 9-27 所示。

```
[INFO] Apache Hadoop Client Packaging Integration Tests ... SUCCESS [  0.080 s]
[INFO] Apache Hadoop Distribution ......................... SUCCESS [ 27.180 s]
[INFO] Apache Hadoop Client Modules ....................... SUCCESS [  0.044 s]
[INFO] Apache Hadoop Cloud Storage ........................ SUCCESS [  0.886 s]
[INFO] Apache Hadoop Cloud Storage Project ................ SUCCESS [  0.023 s]
[INFO] ------------------------------------------------------------------------
[INFO] BUILD SUCCESS
[INFO] ------------------------------------------------------------------------
[INFO] Total time:  01:05 h
```

图 9-27　编译成功

编译成功后的 64 位 Hadoop 包存储于/opt/module/hadoop_source/hadoop-3.1.3-src/hadoop-dist/target 目录下，可以直接使用。

9.7　本章总结

在 Hadoop 的使用过程中，会遇到各种各样的问题，有的问题不能找到现成的解决方案，这时需要用户对源码进行深入解析，有时甚至需要修改 Hadoop 源码。本章主要带领读者对 Hadoop 的关键源码进行了一次梳理解析，通过对源码的解析，可以更直观、清楚地了解 Hadoop 的内部运行过程。本章讲解的源码只是一小部分关键流程源码，主要目的是教会读者如何主动阅读源码，并且对源码进行解析。